Behavioural Ecology of Siberian and European Roe Deer

Chapman & Hall Wildlife Ecology and Behaviour Series

Series editor: Dr R.J. Putman, Department of Biology, University of Southampton, UK

This series brings together a number of authoritative monographs on ecology and behaviour written by experts in their field. They are concisely presented, scholarly works at research level, but written in such a way as to be readable and affordable for the interested layperson.

Each book presents an up-to-date and authoritative review of the ecology and behaviour of a particular animal species or group, and in addition uses detailed case studies to explore a number of more general conceptual issues or principles in behavioural ecology. Developed from the original research of the author, each account offers a definitive review of the species or group 'under the spotlight' with that unmistakable 'immediacy' of style gained from being written by someone directly involved with ongoing research on the animals and issues concerned. In addition, through this detailed case study approach, they also offer unique insights into a number of more general principles of behavioural ecology and can often provide a critical test of accepted theory.

Published

Ecology and Behaviour of the African Buffalo: Social inequality and decision making
H.H.T. Prins

Behavioural Ecology of Siberian and European Roe Deer
A. Danilkin (in association with *A.J.M. Hewison*)

In preparation or being planned

Comparative Behavioural Ecology of Madagascan Lemurs
J. Ganzhorn, P. Kappeler and S. O'Connor

Behavioural Ecology of Black Bears: An analysis of factors determining home range behaviour
R.A. Powell

Fallow Deer: Behavioural plasticity in contrasting environments
J. Langbein and S. Thirgood

Behavioural Ecology of Muntjac Deer
N. Chapman and S. Harris

Competition and Resource Partitioning in Ungulate Assemblies
R.J. Putman

Behavioural Ecology of Siberian and European Roe Deer

A. Danilkin

Institute of Animal Evolutionary Morphology and Ecology, Russian Academy of Sciences, Russia.

in association with

A.J.M. Hewison

Department of Biology, University of Southampton, UK.

CHAPMAN & HALL

London · Glasgow · Weinheim · New York · Tokyo · Melbourne · Madras

Published by Chapman & Hall, 2–6 Boundary Row, London SE1 8HN, UK

Chapman & Hall, 2–6 Boundary Row, London SE1 8HN, UK

Blackie Academic & Professional, Wester Cleddens Road, Bishopbriggs, Glasgow G64 2NZ, UK

Chapman & Hall GmbH, Pappelallee 3, 69469 Weinheim, Germany

Chapman & Hall USA, 115 Fifth Avenue, New York, NY 10003, USA

Chapman & Hall Japan, ITP-Japan, Kyowa Building, 3F, 2-2-1 Hirakawacho, Chiyoda-ku, Tokyo 102, Japan

Chapman & Hall Australia, 102 Dodds Street, South Melbourne, Victoria 3205, Australia

Chapman & Hall India, R. Seshadri, 32 Second Main Road, CIT East, Madras 600 035, India

First edition 1996

Typeset in 10/12pt Palatino by Saxon Graphics Ltd

Printed in Great Britain by St Edmundsbury Press, Bury St Edmunds, Suffolk

ISBN 0 412 63880 0

A catalogue record for this book is available from the British Library

Library of Congress Catalog Card Number: 95–74660

∞ Printed on permanent acid-free text paper, manufactured in accordance with ANSI/NISO Z39.48-1992 and ANSI/NISO Z39.48-1984 (Permanence of Paper).

Contents

Series editor's foreword

Over the past 30 years or so, research effort in behaviour and ecology has progressed from simple documentation of the habits or habitats of different species to asking more searching questions about the adaptiveness of the patterns of behaviour observed; it has moved from documenting simply *what* occurs, to trying to understand *why*. Increasingly studies of behaviour or ecology explore the function of particular responses or patterns of behaviour in individuals or populations – looking for the adaptiveness that has led to the adoption of such patterns either at a proximate level (what environmental circumstances have favoured the adoption of some particular strategy or response from within the animal's repertoire at that specific time) or at an evolutionary level (speculating upon what pressures have led to the inclusion of a particular pattern of behaviour within the repertoire in the first place).

Many common principles have been established – common to a wide diversity of animal groups, yet showing some precise relationship between a given aspect of behaviour or population dynamics and some particular ecological factor. In particular, tremendous advances have been made in understanding the foraging behaviour of animals – and the 'decision-rules' by which they seek and select from the various resources on offer – and patterns of social organization and behaviour: the adaptiveness of different social structures, group-sizes or reproductive tactics. Equal progress has been made in understanding the relationship of many population processes – rates of increase or mortality, relative stability of population size and structure – to environmental character.

Such principles of *explanations* of behaviour or ecology are now widely recognized and find treatment in almost any undergraduate text. Yet, important though it is is to recognize such generalities of pattern, in the very derivation of such general principles, often much subtlety and elegance apparent in an individual case is lost and it may be that very subtlety which is of critical importance in fine-tuning adaptation. Indeed, it has become increasingly apparent over more recent years that there is a tremendous flexibility of behaviour and ecology even within a species. While earlier research had concentrated on exploring more general patterns and had sought to explain interspecific differences in behaviour, in individual or population ecology, in relation to observed environmental differences, in practice almost equal variation is apparent between different populations of one and the same species.

This new Series tries to develop a focus on that subtlety of adaptation by exploring in detail the behaviour and ecology of particular species or closely related groups and trying to show how particular aspects of the behaviour or ecology described may be adaptive. In focusing on the subtlety of adaptation shown within a single species or even a single population, however, the series may in addition add new perspective and new depth to our understanding of even those more general principles. Many of the earlier breakthroughs in developing the broader conceptual framework in behavioural ecology derived from comparative studies of closely related species – or by investigation of the wealth of increasingly recognized intraspecific variation in behaviour and ecology – and such comparative studies often reveal important new insights into the shaping of relationships between behaviour, ecology and environment.

This series, then, brings together a sequence of authoritative monographs on behaviour and ecology of particular species or species groups. They are concisely presented, scholarly works at research level, written by experts in their field – but written in such a way as also to be readable by and accessible to the interested layperson. Each book in the series presents an up-to-date and comprehensive review of the behaviour and ecology of some particular animal species or group but in addition uses these detailed case-studies to explore a number of more general conceptual issues or principles in behavioural ecology or population ecology.

In each case developed upon the original research of the author, each account thus offers a definitive review of the species of group 'under the spotlight', with that unmistakable immediacy of style and understanding from being written by someone directly involved with ongoing research on the animal or issues concerned. In addition, through this detailed case-study approach, each offers unique insight into a number of more general principles of behaviour or ecology and can often provide a critical test of accepted theory.

In this volume, Aleksey Danilkin and Mark Hewison focus upon the behaviour and ecology of European and Siberian roe deer. The book offers an authoritative review of the taxonomy of this confused genus and a definitive review of current knowledge of the biology of these two distinct species. Supported by Danilkin's many years of personal research experience in Russia and the Far East, it presents a great deal of new information on Siberian roe – and eastern populations of the more familiar European roe. A collaboration with Mark Hewison, who has himself contributed a great deal in recent years to our knowledge of the ecology, population genetics and reproductive biology of European roe in western Europe, ensures that Danilkin's studies of this latter species are brought into perspective within the context of a more established western literature. Together they offer a balanced and integrated overview of the biology of the two species based upon both western and eastern experience.

This book is important at a number of levels. First, it distils the knowledge and understanding gained in over 20 years of research by Aleksey Danilkin on eastern European and Siberian roe deer. It makes available for the first time in the English language the results of this often exciting and ground-breaking work; in placing this work in context it also offers a review in English of a large body of other Russian and eastern European literature hitherto inaccessible to a western readership. Mark Hewison's familiarity with the western literature, together with his own extensive experience of working with European roe in Great Britain and France, brings to the work an overall balance that assures a definitive and integrated analysis of the current state of our knowledge on roe deer worldwide.

But however important this 'window' on the eastern European literature may be, this is not simply an updated review of the biology of roe deer. Perhaps the most important – and unusual – difference between eastern and western experience of roe is that while the majority of roe populations in western Europe are sedentary, or undertake only small seasonal movements between adjacent summer and winter ranges, many eastern populations (of both Siberian and European roe) are strongly migratory, travelling very substantial distances in loose 'herds' in clear seasonal movements between summer and winter ranges. Almost all populations of Siberian roe undertake these synchronized seasonal migrations; but it is not a difference between the two species of Siberian and European roe – for eastern populations of European roe also undertake these long-distance migrations between discrete summer and winter ranges. In analysing these movements and considering what environmental factors trigger migratory behaviour in some populations and not in others, this book also addresses some fundamental questions about how environmental factors may influence home-ranging and dispersal patterns in animals.

Summary

This monograph offers a comparative analysis of the biology of the European and the Siberian roe deer. It considers phylogeny and evolution, ecology and behaviour, geographical distribution, population numbers and the factors affecting them.

The European roe has been the subject of a great number of scientific publications but these have primarily focused on western populations. By contrast, little account has been taken of the very different ecology of this species towards the east of its range, and still less has been published outside the former USSR on the Siberian roe deer. The principal focus of this work is therefore on roe deer populations inhabiting eastern Europe and Asia and presents results of research that has not previously been published in English. In particular, a body of data is presented from comparative studies in the former USSR on Siberian and European roe deer resulting from over 20 years of study by A. Danilkin.

Developing on these new data, the book also offers a re-examination of the taxonomy of roe deer, showing that Siberian roe carry additional B-chromosomes and demonstrate the phenomenon of chromosomal mosaicism in some individuals and populations. Updated information on the roe deer's range and distribution is provided, together with details of population structure and ranging behaviour established through extensive radiotracking studies. The migratory behaviour of populations in the east is discussed. The dynamics of roe deer populations are examined and the response to human-imposed environmental change is discussed. This book is of interest to biologists, ecologists, ethologists, taxonomists and naturalists.

Dr Aleksey Danilkin is a chief research associate at the Institute of Animal Evolutionary Morphology and Ecology, the Russian Academy of

Sciences and author of more than 100 publications on the ecology, behaviour and systematics of roe deer.

Dr Mark Hewison, formerly a lecturer in ecology at Southampton University, with research activity concentrating on the ecology of roe deer, is currently a full-time researcher at the Centre d'Études Biologiques de Chizé (CNRS), in France.

Preface

Roe deer are the most numerous wild ungulate species in Eurasia. Their amazing plasticity and adaptability to the human-altered landscape have allowed them not only to withstand the rapid changes in their habitat, but also to colonize new regions and to reach extremely high population densities in optimal conditions. The widespread distribution and high levels of abundance of this species, together with its importance as a game animal, have combined to make the roe deer a most popular subject of scientific study. Thousands of scientific papers and dozens of monographs dealing with this species have already been published and it might seem that further studies could add nothing significantly new. However, studies published so far have principally concerned the European roe deer only, concentrating on populations in western and central Europe. Far less attention has been paid to the roe deer of eastern Europe, Siberia and the Far East which together constitute a far greater proportion of the geographical range of this species. Indeed, very little has been published on the Siberian roe deer outside the former USSR and such a limitation of perspective hampers the development of a comprehensive understanding of the evolution, systematics and ecology of the genus *Capreolus*.

Clearly, western scientists have great problems accessing the considerable body of data that has been published in Russian and they are therefore not familiar with this work, although they are aware of its existence. It is the aim of this book to rectify this situation. It offers a new analysis of the taxonomy, behaviour and ecology of roe deer, particularly focusing on more than 20 years of comparative studies by A. Danilkin on the Siberian and European roe deer in the former USSR. While the major part of this book deals with a functional analysis of the results of this and other research, it may be appropriate here to summarize briefly some of the methods employed in these long-term studies.

DATA SOURCES AND METHODOLOGY

Six forest-steppe populations were selected for field studies in Russia. Over the period of study, we were able to recognize individually up to 30 or 40 animals each year in each population. We also gradually managed to habituate some animals to human presence, which enabled us to follow them more closely, recording all their behaviours directly in a natural environment.

In addition to those animals recognizable by individual characteristics, a number were caught in nets or from a motor boat at river crossings (Figure A) and were individually marked (when necessary, the deer were anaesthetized). The animals were marked with coloured ear tags and numbered collars, and were fitted with radio transmitters (transmitter weight 300 g). Pulse signals were transmitted at 167 MHz for up to 2 years, with a range of between 6 and 15 km when using hand-held antennae or automatic systems, but as high as 55 km when deer were tracked from the air (Figure B).

A total of 108 roe were marked and tracked for a substantial period and 89 of these were fitted with radio transmitters. Radiotracking enabled us to accumulate a total of 20 000 animal locations, including 108 sets of 24-hour observation periods (fixes taken at 15-minute intervals). Where possible, a transmitter was replaced once it had run out, enabling us to track some individuals continuously for as long as 5 years.

Data on the distribution and number of roe deer in the former USSR are derived from statistics collected by the Game Management Departments and State Committees for Nature Conservation, also original data from the Game Animals Census Service, other published work and A. Danilkin's own data.

Information on genetics concerning chromosomal analyses is based on the procedure of Orlov *et al.* (1976). For biochemical (zone electrophoresis, isoelectric focusing) and immunological (immunodiffusion, immunoelectrophoresis and other techniques) analyses, samples of muscle tissue and blood serum were forwarded frozen to specialist laboratories.

The statistical processing of morphometric data (dimensions of more than 1500 skulls and antlers of roe deer) was carried out using various methods (e.g. correlation, stepwise discriminant analysis, group and cluster analysis, analysis of variance, Mayr's CD-criterion, Fisher's variance ratio). Morphometric measurements are illustrated in Figure 2.3. Samples for morphological analysis were obtained from our own material and collections of the zoological museums of Moscow University, the Russian Academy of Sciences' Zoological Institute (Leningrad), the Kazakhstan Academy of Sciences' Institute of Zoology (Alma-Ata), the Georgian

(a)

(b)

Figure A Capture of roe deer: (a) with nets; (b) at a river crossing.

(a)

(b)

(c)

Figure B Roe deer radiotracking equipment: (a) hand-held aerial; (b) mobile and (c) stationary automatic systems.

Figure B Roe deer radiotracking equipment: (d) tracking from aircraft.

Academy of Sciences' Institute of Zoology (Tbilisi), the Armenian Institute of Zoology (Erevan), the Azerbaijanian Academy of Sciences' Institute of Zoology (Baku), the Irkutsk Agricultural Institute, the Institute of Biology of the Siberian Branch of the USSR Academy of Sciences (Novosibirsk), the Lithuanian Academy of Sciences' Institute of Zoology and Parasitology, the Ukrainian Academy of Sciences' Schmalhauzen Institute of Zoology (Kiev), the Mongolian Academy of Sciences' Institute of General and Experimental Biology (Ulan-Bator), the Alekhin Central Chernozem Preserve (Kursk), the Bulgarian Academy of Sciences' Forest Institute (Sofia), Le Museum National de Sciences Naturales (Paris) and the British Museum of Natural History (London). Some further data concerning roe deer morphology were kindly provided by G. Markov, C. Stubbe, W. Bresinski, M. Bek, Yu.A. Darman, V.G. Shvets, K.A. Morozov, A.A. Kiselev, P.P. Bluzma, A.B. Arabuli, V.E. Loktionov, V.B. Pole, G.G. Sobanski and many hunters.

Heart-felt gratitude is expressed to all those colleagues who shared their materials and data. Special acknowledgement is due to Danilkin's co-authors and field work companions (G.G. Markov, C. Stubbe, A. Sempere, R. Prior, Yu.A. Darman, A.N. Minayev, A.V. Shurkhal, M.I. Baskevich, T.Yu. Chistova, O.V. Chekalina, V.M. Gudkov, Yu.D. Vulis) and to A. Nadyozhin for translations. At all stages of this work, invaluable aid was provided by V.E. Sokolov. Sincere thanks are also due to Dr R.J. Putman for valuable notes and comments.

1

Phylogeny, evolution and systematics

1.1 PHYLOGENY AND EVOLUTION

1.1.1 Classification and relationships within the Cervidae

The classification of the family Cervidae Gray, 1821 (like that of many other artiodactyls) is somewhat unclear, suffering from many inconsistencies. Evolutionary and taxonomic relationships are often far from being resolved definitively. This seems to be particularly the case for roe deer, the modern *Capreolus*, Gray, 1821 and the ancient *Procapreolus* Schlosser, 1924. Modern roe deer are included by some authors within the subfamily Cervinae, Baird, 1857 (Flerov, 1952; Sokolov, 1959; Vislobokova, 1981), but have more generally been classified within the subfamily Odocoileinae, Pocock, 1923 (Simpson, 1945; Heptner *et al.*, 1961; Haltenorth, 1963; Sokolov, 1979; Barishnikov *et al.*, 1981; Korotkevich, 1984; Groves and Grubb, 1987; Putman, 1988; Cronin, 1991a; Grubb, 1993). Within the latter family, Simpson (1945) first distinguished the tribe Capreolini, comprising not only *Capreolus* but also the ancient genus *Procapreolus*.

Recently, new studies on the phylogenetic relationships of the Cervidae family have been published. Based on a detailed analysis of the evolution and distribution of cervids, Harrington (1985a) came to the conclusion that the telemetacarpal and plesiometacarpal groups of deer are of independent, polyphyletic origin, with roe deer clearly belonging to the former group (Table 1.1). Helminthological data suggest that roe deer branched off from the common trunk of Cervidae before the initial form of protostrongylid nematodes evolved into elaphostrongylines (found in most cervid species but to which roe deer are naturally immune) (Pryadko, 1972), i.e. roe deer differentiated before the separation of the Cervidae family into Cervinae and Odocoileinae occurred.

Other characteristics such as the lack of any brow-tine processes, the unusual cycle of antler growth, the absence of a well developed preorbital gland and the occurrence of a period of embryonic diapause during pregnancy (Aitken, 1974) all distinguish roe deer from other cervids (Table 1.1). This may indicate that roe deer should be classified as an independent taxonomic group, perhaps even distinguished at the subfamily level.

However, through the analysis of morphological (Groves and Grubb, 1987) and, more recently, mitochondrial DNA (Cronin, 1991a) data, roe deer were assigned to one of four separate lineages within the Odocoileinae which diverged about 2.5–5.0 $\times$ 10^6 years ago. Furthermore, on the basis of morphological characteristics (Flerov, 1952), the structure of the skull, metacarpal bones and teeth (Korotkevich, 1970, 1974) and genetic distances (Baccus *et al.*, 1983 – see Table 1.2; Cronin, 1991a), *Capreolus* are more closely related to the Alcini and the Odocoileini than they are to the Cervini (Groves and Grubb, 1987), supporting Simpson's (1945) classification of cervids. Thus, the classification of the ancient *Procapreolus* and the modern *Capreolus* within the tribe Capreolini of the subfamily Odocoileinae seems to be the most coherent option. Nevertheless we are of the opinion that final conclusions should be based on more detailed comparative studies of cervids using modern techniques.

1.1.2 Evolution of the genus *Capreolus*

It is probable that the genus *Capreolus* stems from a Pliocene species of the genus *Procapreolus* (Korotkevich, 1970, 1974), as is indicated by skull and antler morphology (Figure 1.1). The similarity is most apparent in the Siberian roe deer, *Capreolus pygargus* Pall., where limb bones such as the metapodia, tibia, hoof phalangae, etc. are morphologically similar to those of *Procapreolus* and the structure of the teeth of these two forms is virtually identical.

The skull and tooth morphology of *Procapreolus* indicates that they were better adapted to living on soft juicy parts of trees and shrubs rather than grazing. Most probably, these animals lived in thickets fringing lakes or other bodies of water which were common in the subtropical climate characteristic of the *Procapreolus* period. The subsequent climatic changes, becoming drier and milder in the second half of the Pliocene, and the consequent alteration of the landscape may possibly have contributed to evolutionary transformations of *Procapreolus*.

The earliest examples of the European *Capreolus* have been found in the Kotlovina Faunistic Group which dates back approximately to the Lower Villafrank, and their presence is apparent throughout the period

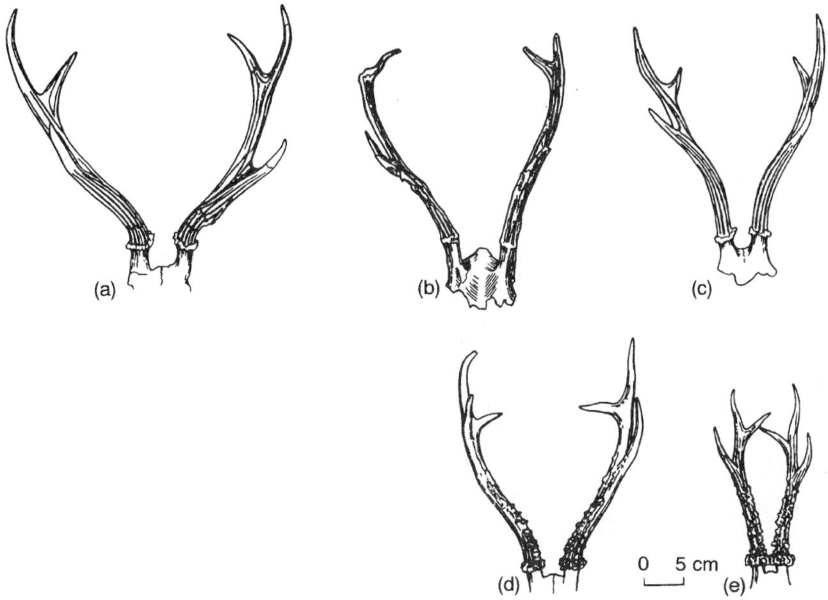

Figure 1.1 Antler shapes: (a)–(c) *Procapreolus* (from Korotkevich, 1970, 1974); (d)–(e) *Capreolus*.

of the Taman Faunistic Complex. No evidence of their presence has been found in the Tiraspol and Singil Complexes, but they were clearly a part of the Khazar Complex (Wangenheim and Flerov, 1982). According to the latest reports, the ossiferous deposits of Kotlovina were formed at different times; roe deer remains probably originate from deposits of the Khaper Faunistic Complex dating from the Middle Villafrank or Late Pleistocene. Roe deer remains from Lower Anthropogenic deposits are reported from the south of Ukraine, Moldavia, Anterior Asia and Transcaucasia (Gromova, 1948; Pidoplichko, 1951a, b; Tatarinov, 1970; Alekseyeva, 1977).

In western Europe, *C. suessenbornensis* Kahlke, 1956, is the form closest to the European roe deer (*C. capreolus* L.) judged on morphometric characters (Kahlke, 1956, 1958a,b, 1969). The Sussenborn fauna dates back to the lower sections of the Middle Pleistocene and is of comparable age to that of the Tiraspol Complex fauna dating back to the Lower Pleistocene, according to the stratigraphic scheme adopted in the USSR (Kahlke, 1971).

In eastern Europe, *Capreolus* forms which are closely related to modern roe deer date back to the Late Pliocene, and in Central Europe, back as far as the Early Anthropogene. The most ancient Asian *Capreolus* species,

Table 1.1 Important distinctive features of some cervid species

Species	Group	Chromosome set (2*n*)	Shed antlers	Antler growth	Rut	Eye canthus glands
Capreolus pygargus	T	70+ (1–14)	Nov	Dec	Jul–Sep	–
Capreolus capreolus	T	70	Nov	Nov	Jun–Aug	–
Odocoileus hemionus	T	70	Jan–Feb	Apr–May	Sep–Nov	+
Odocoileus virginianus	T	70	Jan–Feb	Apr–May	Oct–Nov	+
Rangifer tarandus	T	70	Nov–Dec	Aug–Apr	Sep–Oct	+
Alces alces	T	68–70	Nov–Dec	Apr–May	Sep–Oct	+
Cervus elaphus	P	68	Mar–Apr	Apr	Sep–Oct	+
Cervus nippon	P	64–68	Apr–May	Apr–May	Sep–Nov	+
Dama dama	P	68	Apr	Apr–May	Sep–Oct	+
Axis axis	P	66	Apr	Apr–May	Sep–Oct	+

T = metacarpal deer
P = plesiometacarpal deer
+ = character present
– = character not present

Table 1.2 Genetic similarity (I, above diagonal) and difference (D, below diagonal) of some cervids (from Baccus *et al.*, 1983)

Species	*Antil. amer.*	*Bos bison*	*Bos taurus*	*Alces alces*	*Capr. capr.*	*Cervus elaphus*	*C. e. canad.*	*Odoc. hemion.*	*Odoc. virg.*	*Rangif. taran.*	*R. t. caribou*
Antilocapra americana	–	0.241	0.188	0.267	0.312	0.396	0.322	0.324	0.283	0.274	0.264
Bos bison	1.422	–	0.918	0.295	0.243	0.519	0.406	0.245	0.225	0.357	0.347
Bos taurus	1.670	0.085	–	0.319	0.325	0.517	0.430	0.269	0.249	0.377	0.371
Alces alces	1.321	1.221	1.143	–	0.539	0.445	0.440	0.460	0.468	0.626	0.522
Capreolus capreolus	1.164	1.215	1.123	0.619	–	0.439	0.487	0.445	0.488	0.599	0.435
Cervus elaphus	0.925	0.656	0.661	0.809	0.824	–	0.290	0.411	0.392	0.454	0.431
C. e. canadensis	1.133	0.902	0.843	0.822	0.719	0.749	–	0.372	0.408	0.494	0.427
Odocoileus hemionus	1.127	1.406	1.313	0.771	0.811	0.889	0.989	–	0.862	0.514	0.457
Odocoileus virginianus	1.164	1.491	1.392	0.759	0.718	0.936	0.896	0.149	–	0.578	0.465
Rangifer tarandus	1.294	1.030	0.975	0.469	0.512	0.791	0.706	0.665	0.548	–	0.174
R. t. caribou	1.330	1.059	0.993	0.650	0.832	0.841	0.850	0.783	0.766	0.840	–

whose age is comparable with that of the Taman Faunistic Complex, was found in the Pavlodar Region, near the Irtish (Vislobokova, 1973). Wangenheim (1977) places this *Capreolus* sp. in the Kizikh Complex which is characteristic of the first half of the Late Eopleistocene of Western Siberia which he notes is close to the Odessa Complex and Upper Villafrank faunas of central and western Europe. However, most Asian roe deer finds date back to the Upper Pleistocene and more recent periods (Figure 1.2). It should be noted that the scarcity of fossil remains and their inaccurate dating do not allow us to state definitively that roe deer appeared earlier or later in a particular region. The body of available data on the Pleistocene mammals of Europe drew Kurten (1968) to conclude that the roe deer, as well as some other mammals, immigrated into Europe from Asia.

The antler remains of Eopleistocene *Capreolus* sp. from the Irtish Zone of the Pavlodar Region stored in the Russian Academy of Sciences' Paleontological Institute indicate that these were very large animals. The relatively large size of Asian fossil roe deer and smaller size of European specimens has been recorded in many instances. Indeed, fossilized remains of a 'big' *Capreolus* found in Europe (Boessneck, 1956) in fact fall within the accepted size range of characters of the European roe deer (Paaver, 1965) but are significantly smaller than fossil Siberian roe deer species (Kosintsev, 1981). Based on known group genesis history and following on from the fact that natural (glacial and water) barriers existed along the frontier between Europe and Asia, it may be concluded that European and Siberian roe deer have, for some considerable time, developed along different lines, which has resulted in great genetic differentiation and partial reproductive isolation (see below).

1.2 *CAPREOLUS* TAXONOMY IN THE LIGHT OF MODERN RESEARCH

1.2.1 European and Siberian roe deer

It is clear that there exists an enormous geographic variability within *Capreolus*, which has been the basis for the identification of one or several species, as well as a number of races and subspecific forms, over the range of this group (Pallas, 1809; Satunin, 1906; Dinnik, 1910; Brauner, 1915; Flerov, 1928, 1929, 1952; Lavrov, 1929; Allen, 1930; Dahl, 1930; Jacobi, 1932; Barclay, 1933; Ellerman and Morrison-Scott, 1951; Pasternak, 1952, 1955; Sokolov, 1959; Heptner *et al.*, 1961; Corbet, 1978; Sokolov, 1979; Barishnikov *et al.*, 1981; Honacki *et al.*, 1982; Sokolov and Gromov, 1985a; Danilkin, 1986a, 1989; Lehmann and Sagesser, 1986). However, most scientists have considered *Capreolus* to represent a single species of the genus, while nevertheless distinguishing the European and Siberian roe deer as separate forms, groups or subspecies with distinct character-

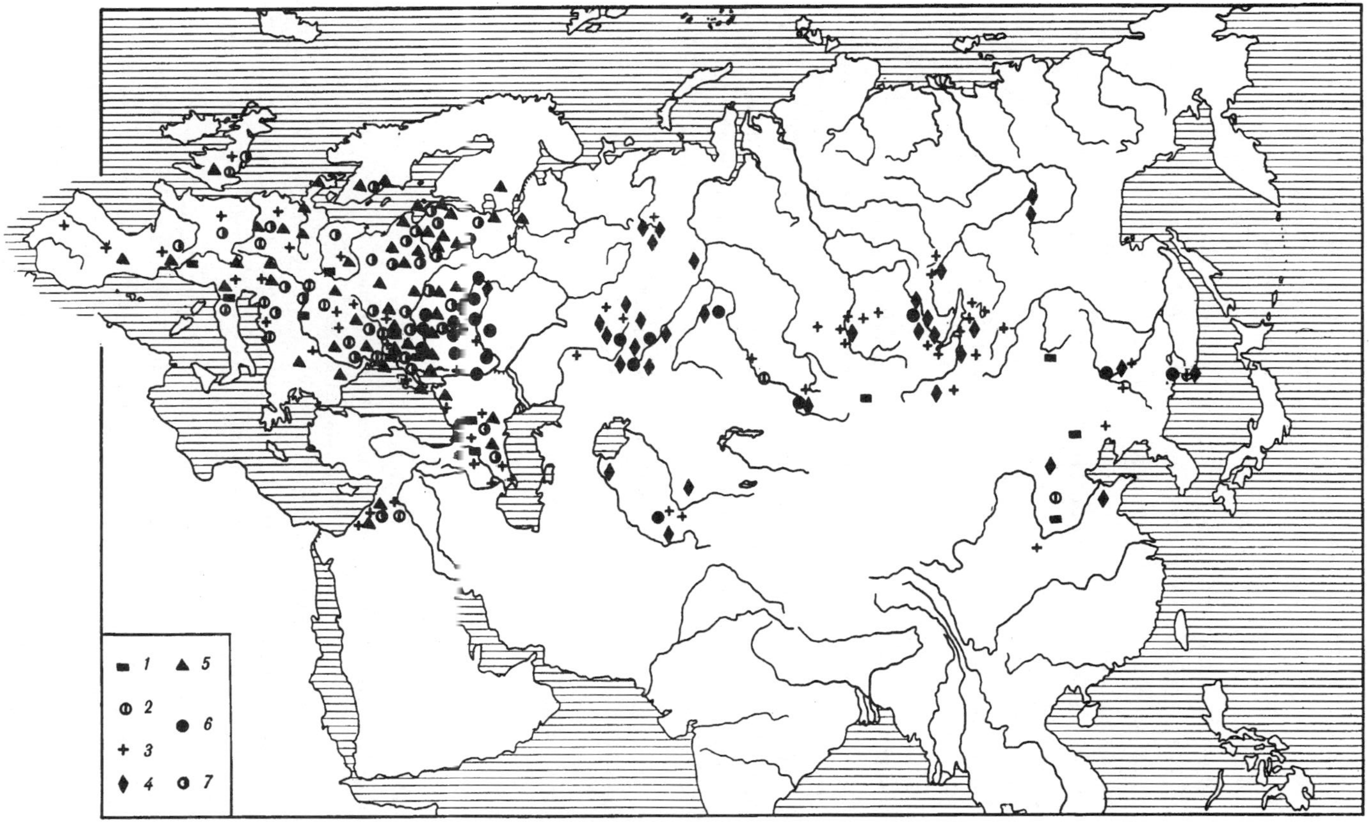

Figure 1.2 Sites of remains of: (1) *Procapreolus* (Ye.L. Korotkevich, personal communication); (2)–(7) *Capreolus*. 1 = Pliocene; 2 = Early Pleistocene; 3 = Pleistocene; 4, 5 = Recent; 6, 7 = 1000 AD to present (Siberian and European roe deer, respectively).

istics. Doubts concerning the classification of these two forms as a single species have re-emerged only in recent years following the discovery of differences between European and Siberian roe deer in chromosome number (Sokolov *et al.*, 1975, 1978; Sokolov and Danilkin, 1980; Zernahle, 1980), general metabolic activity (Grayevskaya *et al.*, 1980) and the recognition of reproductive barriers (Stubbe and Bruchholz, 1979, 1980).

Thus, both genetic and morphometric dissimilarity (Chapter 2), as well as reproductive barriers and geographical isolation (Chapters 3 and 8), indicate that European and Siberian roe deer are distinct species. However, it has also become evident that it is not a valid strategy to classify roe deer of a particular geographical area into a given group. This is illustrated by the discovery that on the basis of karyotype, Northern Caucasian deer are classified as European and not Siberian roe (Sokolov *et al.*, 1980, 1986a; Sokolov and Danilkin, 1981; Danilkin *et al.*, 1983; Danilkin, 1986a,c, 1989).

The definition of what constitutes a species, particularly from the standpoint of evolutionary ecology, is unclear as no absolute criteria exist and therefore classification often proceeds on a somewhat ad hoc basis (Mayr, 1968, 1974; Timofeyev-Resovski *et al.*, 1977; Haffer, 1986). From classical systematics a species should be distinguished, from biological independence and morphophysiological uniqueness, under the following three criteria:

- genetic isolation (lack of hybridization with other groups);
- hiatus (discontinuity in character variation) between the closest groups;
- independence of geographical distribution.

Schwarz (1980) considered a fourth criterion to be the ecological specificity or degree of adaptation of a species to a particular environment. From this point of view, a new species would be distinguished from the ancestral species by its superior adaptation to local conditions, not only in functional efficiency but also in energy dynamics. Thus, profound biochemical changes play the leading role in species' adaptation and fundamental differences between species are biochemical distinctions or tissue modifications, which determine the specific nature of a species' relationship with the environment. As Schwarz (1980, p.72) observes:

> There is not a single species (at least a higher vertebrate one) in existence, whose morphophysiological features cannot be found at the tissue level, too. All forms qualifying under the requirements the present-day systematics imposes on species, reproductive isolation being the most important of them, are specific at such a low level as the tissue. In this respect, the difference between the species and well-marked subspecies is principal... The species is a

subspecies which is significantly distinct at the tissue level; tissue incompatibility being a measure of that significance.

The level of reproductive isolation, often viewed as the most important criterion of species integrity, to some extent measures the phylogenetic propinquity of a species. Hybrids produced by crossing representatives of different genera are usually sterile, if any are born at all (Gelder, 1977). Crossed forms which produce sterile or non-prolific hybrids are generally classified as a 'subgenus', while when hybrids are fertile but their reproductive success is low and mortality is high, parental forms can be considered to be 'good species' (Khaveson, 1960). Judged by this criterion it seems that the European and Siberian roe deer are 'good species' as hybrids are inviable (Chapter 8). In comparison, hybridization is not infrequent among other deer species. Hybrids have been described, for example, between fallow and red deer, sika and red deer (Bartos *et al.*, 1981; Chegorka, 1986; Steklenev, 1986), white-tailed, black-tailed and mule deer (Wishart, 1980; Cronin 1991b). Furthermore, hybrids are often easy to produce and are relatively fertile, forming hybrid populations in the wild (Steklenev, 1975; Harrington, 1985b).

However, as Schwarz states (1980, p. 66): 'Species are not species merely owing to the fact that they do not cross, but they do not cross owing to the fact that they are species.' On what basis can the European and Siberian roe deer be differentiated as separate species? Undoubtedly, they have geographically independent ranges (Chapter 3: Figures 3.2, 3.4) and are morphophysiologically unique (Chapter 2) in terms of body size, craniometry (Sokolov *et al.*, 1985a) and non-metric character variations (Zima, 1989). Furthermore, this discontinuity in character variation is well marked in adjacent populations of the two species. In addition, there are significant differences at the cytogenetic, immunochemical and biochemical level (Sokolov *et al.*, 1986a,b) and a certain degree of histo-incompatibility may exist. The Siberian roe is evidently better adapted to more severe living conditions, especially to deep snow and long periods of low temperature, than is the European form.

Siberian and European roe deer also differ in terms of their energy metabolism, with different basal metabolic rates (Grayevskaya *et al.*, 1980, 1988). Basal metabolic functions produce high-energy compounds which are necessary for supplying energy. A group of hormones called catecholamines are responsible for adapting such metabolic processes to prevailing environmental conditions. The occurrence of these regulators in the adrenals of European roe is three times greater than in Siberian roe, but in the blood this ratio is the inverse. Blood glucose concentration also differs between the two forms. The level of liver glycogen in European roe deer is half that of Siberian roe, while the activities of glucose-6-phosphatase, a key enzyme in carbohydrate metabolism, and of alkaline phosphatase in European roe are also substantially lower (one seventh and one third, respectively). In contrast, indices of carbohydrate metabolism regulation and of energy reserves are substantially higher in European roe deer. To maintain homeostasis, the European roe probably expends less energy; thus, metabolically, European roe seem more adaptable than Siberian roe.

Despite the marked differences noted above, in certain morphometric and morphological characters, the European roe deer seems to be merely a smaller copy of the Siberian roe (Figure 1.3), with similar behaviour and ecology (see below) indicating a close phylogenetic relationship. Thus, by practically all the criteria of classical systematics, as well as the additional criterion of ecological specificity, the European and Siberian roe deer are different species, *Capreolus capreolus* L. and *Capreolus pygargus* Pallas, which are very closely related.

1.2.2 Subspecific classification

It is clear that the earlier divisions of the *Capreolus* genus into subspecific groups was also inconsistent, and substantial revision is required. Unfortunately, there is still no accepted absolute criterion of the definition of such taxonomic groups. According to Mayr *et al.* (1956), subspecies are geographically isolated groups of local populations which are taxonomically different from similar divisions of the species. More recently, the subspecies has generally been defined as a group of individuals occupying some ecological or geographical area and differing from other individuals of the same species by one or several diagnostic characters peculiar to all or most members of the group. Generally discerned as subspecies are 'large' and 'clear' groups occupying extensive areas (Timofeyev-Resovski *et al.*, 1977).

The European roe deer cannot easily be broken down into such categories but the Siberian roe deer readily falls into two large and clearly different taxonomic groups, occupying extensive areas which are separated by the geographical barriers imposed by the mountain ridges of the Altai, western and eastern Sayan, Baikal Lake, the Stanovoy Upland and Stanovoy Ridge (Chapter 3). One subspecies group (Siberian roe) can be defined as including the largely identical populations of the Ante-Urals, Trans-Urals, northern Kazakhstan, western and eastern Siberia (but not farther than Baikal) and most probably Yakutia, and a second subspecies (Tien Shan) as including the populations of southern Kazakhstan, Central Asia (Uzbekistan, Kirghizia), Mongolia, the Trans-Baikal region, the Far East and China. Populations of the Altai, eastern Kazakhstan and probably the southern part of the Irkutsk region inhabit an area where the ranges of these two groups meet and hybridization of

(a)

(b)

Figure 1.3 (a) European roe deer (*Capreolus capreolus* L.), (photo by C. Stubbe); (b) Siberian roe deer (*Capreolus pygargus* Pall.).

the Siberian and Tien Shan subspecies is quite possible in this region (see Figure 3.4).

In summary, we suggest the systematic classification of *Capreolus* as follows:

Order **Artiodactyla**
Suborder **Ruminantia**
Family **Cervidae** Gray, 1821
Subfamily **Odocoileinae** Pocock, 1923
Tribe **Capreolini** Simpson, 1945
Genus **Capreolus** Gray, 1821 – the roe deer

- Capreolus capreolus Linnaeus, 1758 – European roe deer
- Capreolus pygargus Pallas, 1771 – Siberian roe deer

1.3 CHARACTERIZATION OF THE GENUS *CAPREOLUS*

The roe deer is a small telemetacarpalian deer with a mean body length and mass which vary among populations from 100 to 145 cm and from 18 to 49 kg, respectively. The front legs are rather shorter than the hind ones, no mane is present, the ears are fairly large (12–16 cm), the tail is rudimentary (2–3 cm), and there are no preorbital glands. Coloration in winter is light grey to dark brown, with a large white caudal patch. In summer, the coat becomes reddish to red-brown and the white caudal patch is less conspicuous or is absent. Roe deer moult twice a year, once in spring and once in autumn. Kids are spotted. Males are slightly larger than females and have three-tined antlers, which are tuberculate. Basal rosettes or burrs are well defined. Antlers are shed in autumn or early winter and begin to regrow immediately afterwards. There is no preorbital tine.

The skull is small (population average ranging from 180 to 245 mm) but relatively elongated, with a maximum width (75–106 mm) less than half its length. Lacrimal bones are shorter than the orbital cavity diameter. The adorbital fossa on the lacrimal bone is wide and normally shallow, slightly pronounced and situated close to the orbit margin. The ethmoidal aperture is variable in size. Intermaxillary bones evenly taper forward without forming blade-shaped projections on their exterior border. Anterior ends of nasal bones are forked and touch admaxillary bones. The dental section of the facial region is relatively short; the anterior orbital margin is level with M2 or M3. Orbits are medium-sized; their vertical diameter is equal to or slightly greater than the length of the molar row. The maxillary bone is comparatively high and its length is equal to that of the molar row. The dental formula is: i 0/3, c 0/1, p 3/3, m 3/3, total 32.

The karyotype ($2n$ = 70–84) comprises 70 main chromosomes plus, in Siberian roe deer only, 1–14 accessory B-chromosomes.

1.3.1 ***Capreolus capreolus*** Linnaeus, 1758 – the European roe deer

Capreolus capreolus is distinguished from *C. pygargus* by its shorter body length (population average between 100 and 126 cm), smaller cranium (condylobasal length averages 180–200 mm) and shorter antlers (length 17–26 cm, span 7–14 cm). The average body weight is 18–32 kg. The length of the tooth row of the lower mandible is between 58 and 66 mm. Auditory bullae of European roe deer skulls are small and do not protrude from the bullar fossa (Figures 1.3a, 1.4). When in summer coat, the hair of the head and the metatarsal gland is grey or grey-brown (usually much darker than that of the body). Unlike the Siberian roe deer, B-chromosomes are non-existent, but two additional blood serum antigens are present.

This species consists of a single taxonomic group which is widely distributed in Europe (not farther than the Volga) and is also found in Asia Minor (Figure 3.4). However, a variant form, *Capreolus capreolus garganta* Meunier, 1983, from the south of Spain differs from typical European roe deer in coloration as well as some body proportions. The taxonomic status of this form requires further detailed investigation.

1.3.2 ***Capreolus pygargus*** Pallas, 1771 – the Siberian roe deer

Capreolus pygargus is larger, with a total average population body length of between 127 and 145 cm, an average body mass of 32–49 kg, a bigger skull (condylobasal length 201–231 mm), a lower tooth row of average length 71–76 mm, and antlers generally longer than 27 cm with a span averaging in excess of 17–20 cm. Auditory bullae of Siberian roe deer are larger and protrude noticeably from the bullar fossa (Figures 1.3b, 1.5). In summer coat, the hair of the head and metatarsal gland, like that of the rest of the body, is generally reddish. The chromosome set includes between 1 and 14 accessory B-chromosomes. This species is distributed through Eastern Europe and Asia (Figure 3.4) but comprises two separate subspecies.

(a) Capreolus pygargus pygargus *Pallas, 1771 – the Siberian roe deer*

Capreolus pygargus pygargus is the larger of the two subspecies (body length 140–145 cm, body mass 41–49 kg, condylobasal length 223–231 mm; all figures are population averages). The chromosome set generally includes no more than 4 B-chromosomes: $2n = 70 + (1–4)$. This group ranges across the Volga area, the Urals, northern Kazakhstan, western and eastern Siberia (not farther than Baikal), the Altai and Yakutia (Figure 3.4).

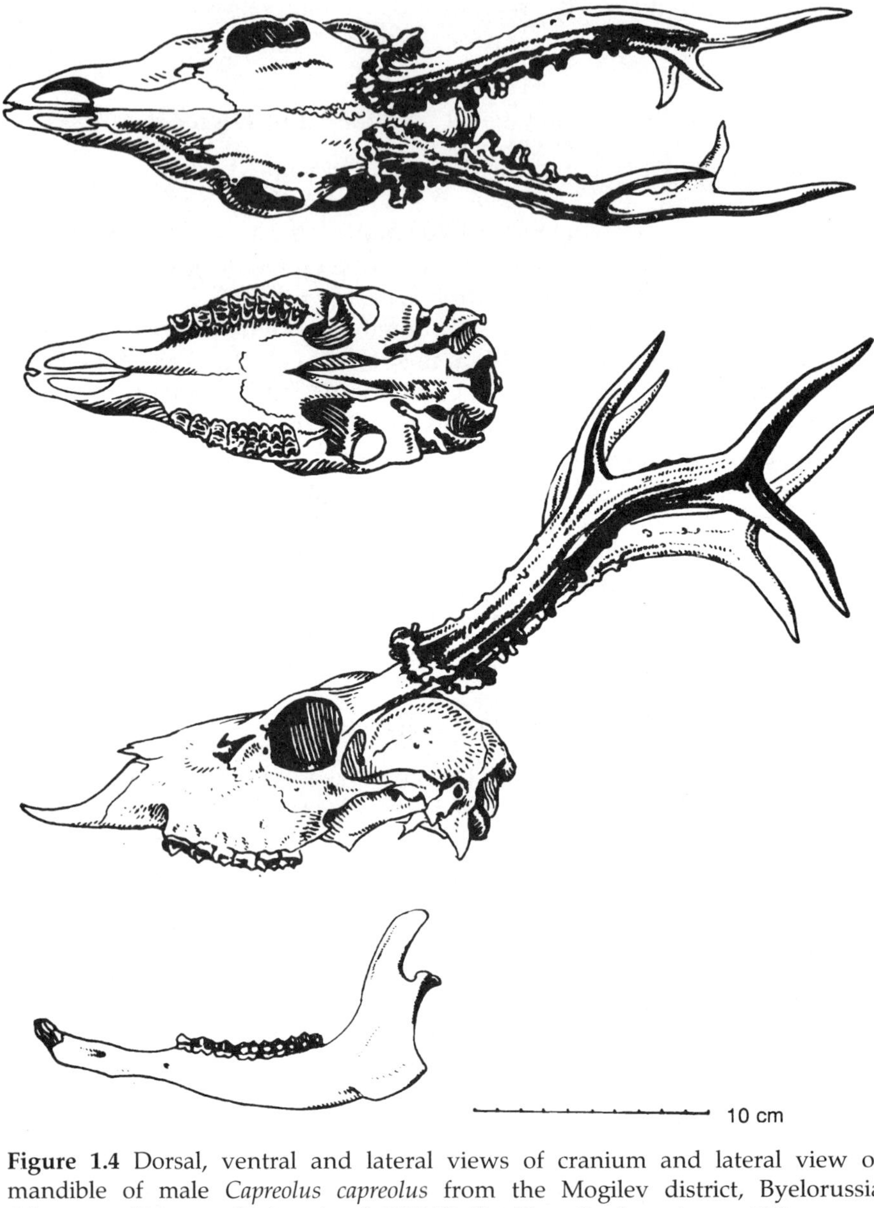

Figure 1.4 Dorsal, ventral and lateral views of cranium and lateral view of mandible of male *Capreolus capreolus* from the Mogilev district, Byelorussia (Museum of Moscow University, S-150060). Total length of cranium = 208 mm.

(b) Capreolus pygargus tianschanicus *Satunin, 1906 – the Tien Shan roe deer*

Capreolus pygargus tianschanicus (sometimes referred to as *C. p. bedfordi*) is smaller than *C. p. pygargus* (body length 126–137 cm, body mass 32–40

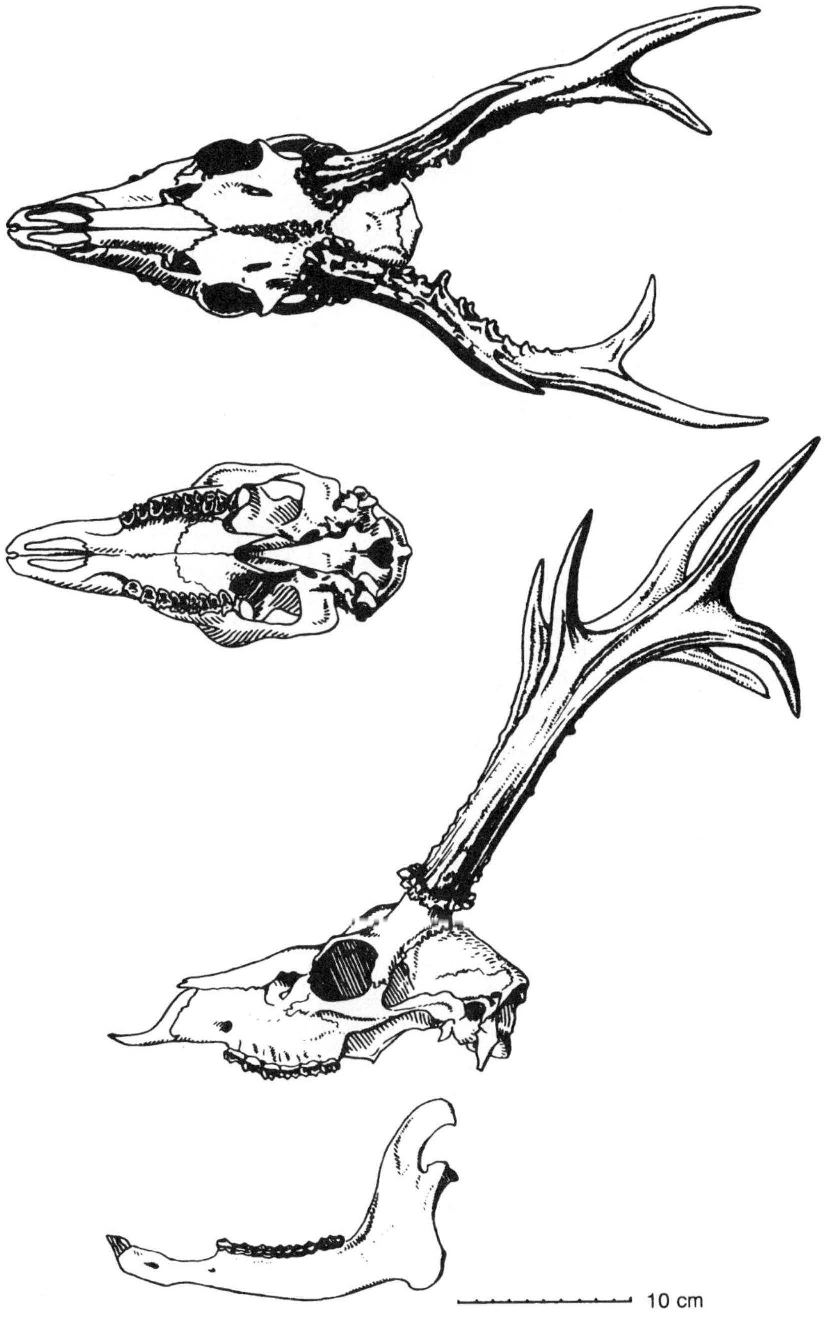

Figure 1.5 Dorsal, ventral and lateral views of cranium and lateral view of mandible of male *Capreolus pygargus* from the Samara district, Trans-Volga region, Russia (personal collection). Total length of cranium = 247.5 mm.

kg, condylobasal length 201–218 mm; all figures are population averages). The chromosome set generally includes more than 4 B-chromosomes: $2n = 70 + (5–14)$. This subspecies is found in southern Kazakhstan, the Tien Shan, southern Altai, Mongolia, the Trans-Baikal area, the Far East region of Russia, Korea, and northern and central China (Figure 3.4). At the extremes of its range, it can reach to Yakutia and the Ante-Baikal area.

2

Variation

2.1 MORPHOMETRIC VARIABILITY

2.1.1 Variation in body dimensions and weight

Unfortunately, morphometric characteristics of roe deer reported in the literature are not always comparable owing to non-standard definitions (e.g. 'dressed' carcass weight may be used to indicate complete evisceration, only stomach and intestine removed, head and feet also removed, etc.), as well as actual errors of measurement. In a simple experiment, the measurement of body length of the same individual roe by six experienced specialists produced an error of 2.5 cm. Body dimensions and weight vary with age (Angibault *et al.*, 1993; Hewison *et al.*, in press a) and depend strongly on the living conditions an animal experiences, particularly availability of food resources (Sagesser, 1966a; Wandeler and Huber, 1969; Ellenberg, 1978; Stubbe and Passarge, 1979; Hanus and Fiser, 1981; Gaillard *et al.*, in press; Hewison *et al.*, in press b). Furthermore, roe body weight varies between seasons, particularly among males (Holand, 1992; Hewison *et al.*, in press a), and even during a single day may change by between 1 and 3 kg, depending on the amount of food and water consumed. Thus phenotypic variation, which does not have a genetic basis, may be expected due to the influence of environmental conditions.

In spite of this, certain patterns in the morphological variability of roe deer do exist. Firstly, it is clear that the European roe deer is generally smaller than the Siberian form (Table 2.1), with pronounced differences in average body dimensions but some overlap in size of the largest European deer with the smallest Siberian specimens (Figures 2.1, 2.2). Geographically, size variation of roe deer generally follows Bergman's rule, with larger animals found in the north and east, where the climate is colder, and smaller animals inhabiting the south and west, where the climate is milder. Indeed, the largest European roe deer are found in Sweden, in the northern extreme of their range (Essen, 1966), and also to the east of the Dnieper and in the mountains of northern Caucasus. This generality, however, does not always hold, as a comparison among populations of British roe deer showed a decrease in size with increasing latitude, and the lowest population average body weight recorded for this species (carcass weight for adult females: 12.1 kg ± 0.35) was at Queens, Scotland, at a latitude equivalent to that of southern Sweden (Hewison, 1993). However, even the largest animals, at the eastern limit of the total range of European roe, are significantly smaller than deer from the adjacent Siberian roe populations of the Volga Area and Urals, highlighting the discontinuity of morphological variation between these two groups. Furthermore, morphometrically large populations of Siberian roe inhabiting a vast region stretching from the Volga to Baikal, to the north of the Altai–Sayani mountain system, are significantly larger than Siberian roe populations of southern Kazakhstan, Mongolia, the Trans-Baikal Territory and the Far East (Pole, 1974; Kiselev, 1975; Bromlei and Kucherenko, 1983; Darman, 1986; Danilkin *et al.*, 1992a).

Growth, development and metabolic rates differ even in closely related species due to some degree of genetic influence. If the European and Siberian roe deer are to be considered as different forms, it should be possible to distinguish between them on the basis of these parameters,

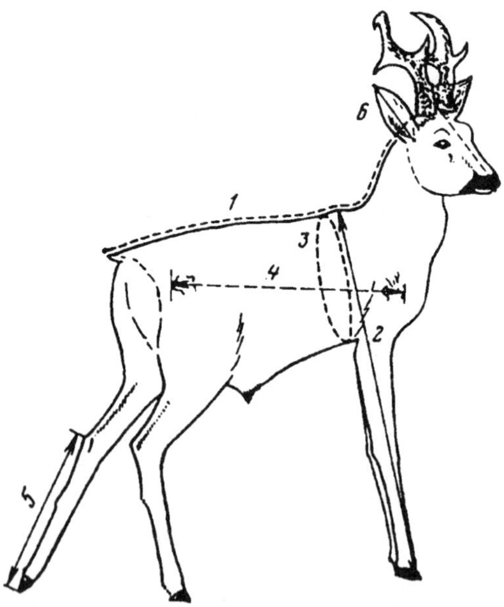

Figure 2.1 Roe deer body dimensions. (1) Total length; (2) height at shoulder; (3) body girth; (4) body length; (5) hind foot length; (6) ear length.

Table 2.1 Some morphometric characteristics of European and Siberian roe deer (adapted from Danilkin *et al.*, 1992a)

Character[1]	Sex	Range of means for different populations (with standard error)	
		European roe deer	Siberian roe deer
Body length (cm)	Male	108.2–126.5 (±2.4)	128.1 (±1.1)–143.8 (±0.8)
	Female	107.1–125.7 (±1.9)	126.7 (±1.1)–144.4 (±1.2)
Shoulder height (cm)	Male	66.0 (±0.2)–80.7 (±0.9)	83.1 (±0.4)–94.1 (±1.7)
	Female	66.0 (±0.2)–83.3 (±0.8)	81.7 (±0.6)–91.0 (±0.8)
Body mass (kg)	Male	23.7–32.0 (±0.8)	34.9 (±1.2)–48.6 (±1.6)
	Female	18.0 (±0.2)–30.0 (±0.7)	32.0 (±0.7)–46.9 (±2.1)
Condylobasal skull length (mm)	Male	182.9 (±0.7)–200.1 (±1.6)	201.0 (±2.4)–231.4 (±0.8)
	Female	179.4 (±0.5)–199.5 (±1.6)	207.0 (±3.1)–226.5 (±0.7)
Maximum skull width (mm)	Male	87.2 (±1.9)–95.7 (±0.8)	94.8 (±1.0)–106.1 (±0.8)
	Female	84.3 (±0.4)–91.5 (±0.9)	92.2 (±1.1)–99.5 (±0.4)
Maximum nasal bone length (mm)	Male	50.7 (±3.5)–65.6 (±2.1)	68.9 (±1.6)–80.3 (±1.5)
	Female	53.4 (±0.4)–64.2 (±1.0)	68.8 (±0.8)–80.0 (±1.8)
Mandible length (mm)	Male	153.0 (±1.5)–167.2 (±0.9)	168.7 (±1.6)–193.8 (±0.8)
	Female	147.4 (±0.2)–168.3 (±1.1)	172.7 (±3.2)–190.4 (±1.7)
Lower tooth row length (mm)	Male	61.9 (±0.5)–66.0 (±1.0)	70.9 (±0.9)–76.3 (±0.7)
	Female	60.9 (±0.7)–67.2 (±0.8)	71.0 (±1.3)–76.0 (±2.3)
Maximum antler length (mm)	Male	184.1 (±8.0)–258.6 (±3.3)	275.7 (±5.6)–333.5 (±8.5)
	Female		
Antler–antler distance (mm)	Male	75.8 (±7.0)–138.7 (±15.0)	168.4 (±3.5)–257.2 (±8.4)
	Female		

[1] See Figures 2.1, 2.3

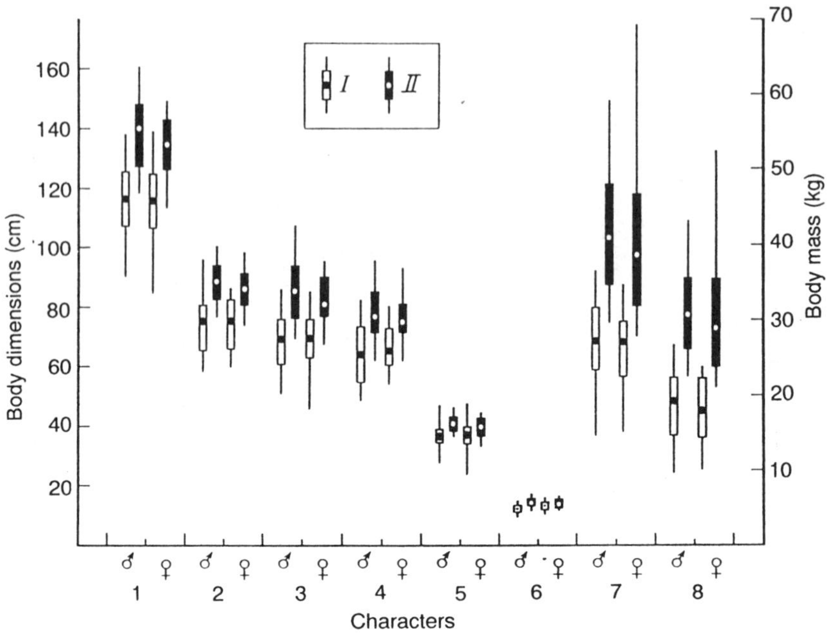

Figure 2.2 Variation of body dimensions and weight of (I) European and (II) Siberian roe deer. Central dot denotes average values for each species; rectangle defines limits of population averages; vertical line illustrates range of extreme values for single individuals. (1) Total length; (2) height at shoulder; (3) body girth; (4) body length; (5) hind foot length; (6) ear length; (7) body mass; (8) carcass mass. Dimensions taken as in Figure 2.1.

even when environmental conditions are identical. In an experiment on captive deer kept near Moscow, the body weight of newborn kids of European and Siberian roe deer averaged 1.65 kg and 1.85 kg, respectively. During the initial growth period, weight gain averaged 4 kg per month for European roe and 6 kg per month for Siberian roe. By the age of 4 or 5 months, the maximum body weight among European roe was 22 kg, while Siberian roe weighed as much as 28 kg, and subsequently, as adults, body weight reached maximums of 33 kg and 42 kg, respectively. In all seasons, this difference was of approximately the same magnitude, averaging 8–9 kg, or about 20% of Siberian roe deer weight (Gromov, 1988b). These figures are consistent with other published data on birth weight and growth rate of roe deer in western Europe (Gaillard *et al.*, 1993a,b). Thus, even under identical living conditions, Siberian roe deer are larger with a faster rate of development, indicating differences at the genetic level.

2.1.2 Variation in cranial morphology

On the basis of numerous craniometric characteristics, European and Siberian roe deer are distinct groups in terms of absolute size (Table 2.1, Figures 2.3 to 2.5), although the maximum values of some European roe deer dimensions overlap with the minimum values of Siberian animals (Figure 2.6a). For example, differences between the two groups in the length of the tooth row are particularly marked. Furthermore, significant differences in the growth rates of skulls of the two groups have been recorded (Sokolov *et al.*, 1985a,b) (Figure 2.7).

Among European roe, skull size is largest in eastern Europe, including populations in the northern Caucasus, while deer of western Europe and the Transcaucasia have the smallest skulls. Again, morphometric discontinuity, indicative of the presence of two separate species of roe deer, is evident, with significant differences in skull size between the populations of eastern Europe (which include the largest examples of European roe) and the adjacent Urals population of Siberian roe. Skull dimensions also discriminate between the two groups or subspecies of Siberian roe deer. Populations either side of the Altai–Sayani mountain system differ markedly in skull size: skulls from populations of the Urals–Ante-Baikal

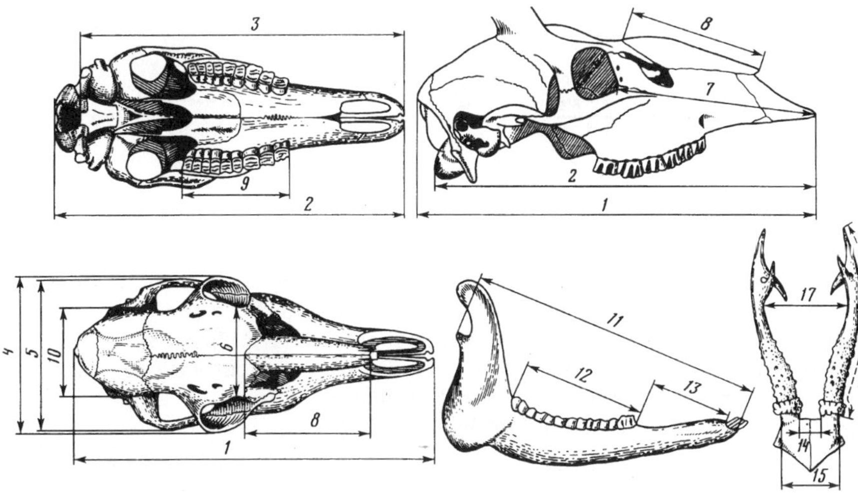

Figure 2.3 Roe deer skull and antler dimensions. (1) Maximum skull length; (2) condylo-basal length; (3) basilar length; (4) maximum skull width; (5) cheek-bone width; (6) interorbital width; (7) rostrum length; (8) maximum nasal bone length; (9) length of upper tooth row; (10) maximum width of brain case; (11) length of mandible; (12) length of lower tooth row; (13) diastema length; (14) distance between inner sides of antler beams; (15) distance between outer sides of antler beams; (16) maximum antler length; (17) maximum antler-to-antler distance.

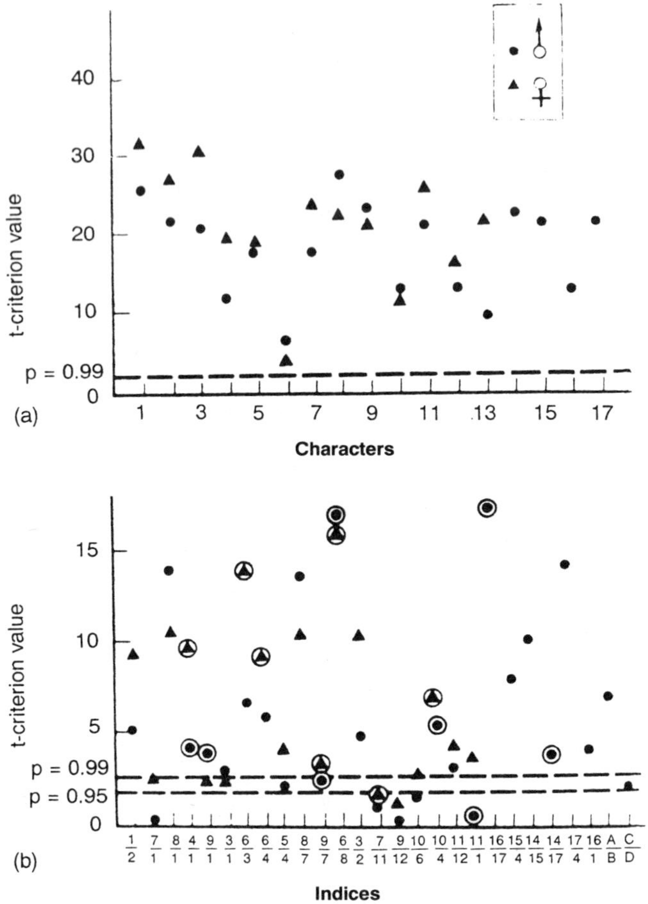

Figure 2.4 Values of *t* for Student's *t*-test comparing Siberian and European roe deer skull and antler dimensions in terms of (a) absolute size and (b) indices, for both males and females. For designation of characters (1)–(17), refer to Figure 2.3. Circled symbol indicates a negative *t*-value. Points above dashed lines indicate a significant difference at that particular threshold of probability. Complex indices: A = (1 x 17); B = (4 x 16); C = (0.5 x 16); D = (15 - 14).

are much larger than those from populations of the Tien Shan and the Far East. The smallest Siberian roe skulls are those of central China.

In order to examine the patterns of variation not only in absolute size but also in skull shape, the multivariate technique of cluster analysis can be employed. The shape of each skull is described by measurement of 13 different distances, and skulls are compared with all others according to these measurements. In such analyses, skulls of 417 bucks and 488 does

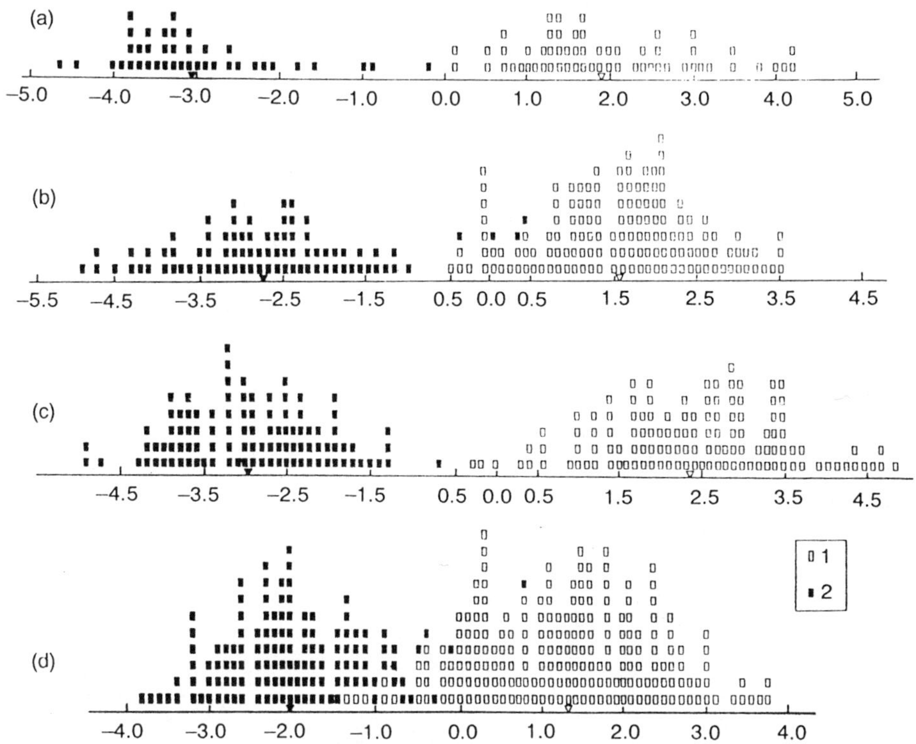

Figure 2.5 Relative distribution and overlap of (1) Siberian and (2) European roe deer according to Discriminant Keys of: (a) male skulls and antlers; (b) male skulls; (c) female skulls; (d) male antlers. Values of discriminant function Y are given on horizontal axis. Triangle indicates centre of a group distribution. (From Sokolov *et al.*, 1985a.)

of various geographical origins were compared and two clear groups were identified from the results (Abramov *et al.*, 1989; Danilkin *et al.*, 1989, 1992a). The first comprised all European specimens, several skulls of some regions of Asia and half the individuals of a sample from central China. The second group included Siberian roe deer skulls only (Table 2.2). When the analysis was restricted to include only the five characters most commonly used in taxonomy (skull length, skull width, nasal bone length, and the length of the upper and lower tooth rows), the differentiation of the European and Siberian groups is even clearer. In this case, the only anomalies are the classification of one sample from a 'hybrid population' of the Dnepropetrovsk Region as Siberian roe and two specimens, one from Tien Shan and the other from China, as European roe (Table 2.2).

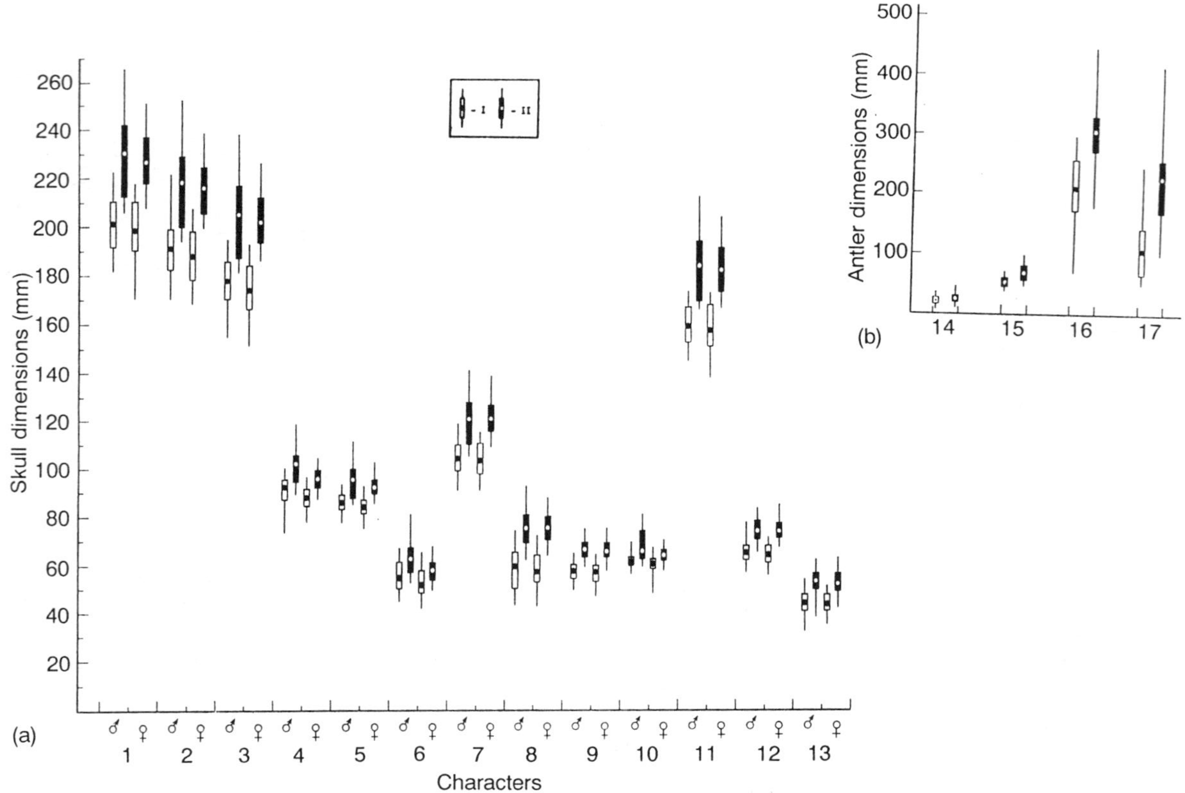

Figure 2.6 Variation of (a) skull and (b) antler dimensions for (I) European and (II) Siberian adult roe deer. Central dot denotes average values for each species; rectangle defines limits of population averages; vertical line illustrates range of extreme values for single individuals. For designation of characters (1)–(17), refer to Figure 2.3.

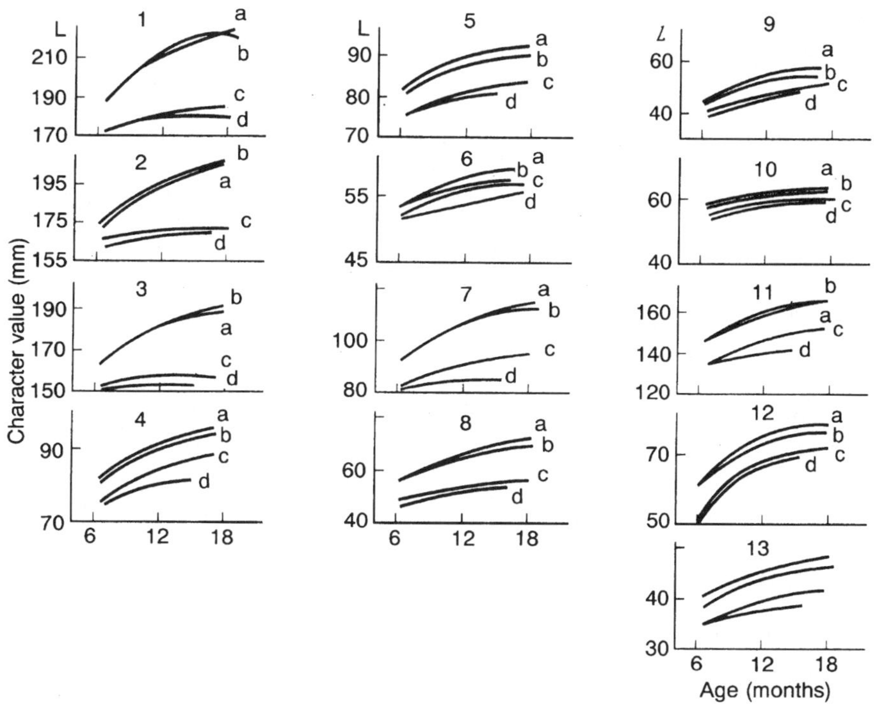

Figure 2.7 Absolute growth of skull in roe deer: (a) Siberian male; (b) Siberian female; (c) European male; (d) European female. (From Sokolov *et al.*, 1985b). Each graph plots growth of a single craniometric character. For designation of characters (1)–(13), refer to Figure 2.3.

These results provide strong evidence of a pronounced morphological differentiation between European and Siberian roe deer, supporting conclusions made in Chapter 1 that they should be recognized as separate species (Sokolov *et al.*, 1985a). At the same time, the relative similarity between the roe deer of central China and European roe is, to a certain extent, indicative of the close phylogenetic relationship and monophyletic origin of these different forms.

Within each of the two main groups (the European roe and the Siberian roe) some degree of heterogeneity in skull morphology exists and each group can be divided into two separate 'morphs'. Among European roe deer, eastern European populations are generally different from western European populations, although it is not possible to identify a clear geographical boundary separating these two groups, or to discern taxonomically different subspecific forms. However, a separate comparative study of roe deer skull morphology in Britain has provided some support for the identification of a taxonomically discrete form, native to Britain, currently restricted in range to central Scotland

Table 2.2 Classification of skulls as 'European type' or 'Siberian type' from different regions of the roe deer's geographic range (percentage of skulls correctly classified by cluster analysis)

Region	Sample size (n)	Percentage of skulls classified			
		13 characters[1]		5 characters[1] (1, 4, 8, 9, 12)	
		'European'	'Siberian'	'European'	'Siberian'
Great Britain	4	100	–	100	–
France	9	100	–	100	–
Germany	220	100	–	100	–
Bulgaria	66	100	–	100	–
Poland	46	100	–	100	–
Sweden	9	100	–	100	–
Baltic States	75	100	–	100	–
Ukraine-Byelorussia	41	100	–	98	2
Kursk-Belgorod	19	100	–	100	–
N. Caucasus	12	100	–	100	–
Transcaucasia	13	100	–	100	–
Urals[2]	113	–	100	–	100
W. Siberia	17	–	100	–	100
Ante-Baikal[3]	16	–	100	–	100
Altai	80	–	100	–	100
Tien Shan	38	5	95	3	97
Mongolia	32	3	97	–	100
Trans-Baikal	15	7	93	–	100
Far East	67	–	100	–	100
China	10	50	50	10	90

[1] For descriptions of characters see Figure 2.3.
[2] Includes the Ante-Urals, Trans-Urals and northern Kazakhstan.
[3] Includes the Krasnoyarsk and Irkutsk regions.

(Hewison, 1993, in press b). In contrast, the Siberian roe deer falls into two large, fairly well defined subgroups which are relatively isolated and taxonomically different: the Urals–Ante-Baikal group (*Capreolus pygargus pygargus*) and the Tien Shan–Far East group (*C. p. tianschanicus*, formerly *C. p. bedfordi*), the latter including populations of central China (Danilkin *et al.*, 1985; Danilkin, 1989). On the basis of certain characteristics, Altai roe deer and, to a lesser extent, roe from the Irkutsk Region appear to be of an 'intermediate' type, indicating that populations from these regions contain a mixture of Siberian roe deer forms (Danilkin and Markov, 1985, 1987; Markov, 1985; Markov *et al.*, 1985a; Sokolov *et al.*, 1986c; Danilkin *et al.*, 1992a).

2.1.3 Variation in antler characteristics

The degree of variability of antler form according to age, animal condition and environmental factors is so pronounced that they are an unreliable taxonomic tool (Figure 2.6b). Nevertheless, practically all roe deer texts present the antler form of European and Siberian roe as a primary distinguishing feature. Indeed, on the basis of population averages and indices, antler form differs significantly between the two groups (Sokolov *et al.*, 1985a), but the degree of overlap is so great that only the largest antlers of Siberian roe, measuring more than 30 cm in length and 25 cm span, can be reliably identified (Table 2.1). Siberian roe with the largest antlers are found in the Altai, Tien Shan and the Ante-Baikal area, while European roe of the Baltic republics, the Ukraine and Caucasus also possess relatively large antlers.

A comparative cluster analysis of a sample of antlers ($n = 460$) revealed that, from the antler characteristics measured, roe deer cannot be divided into European and Siberian forms with the same precision as that provided by a classification based on cranial morphometrics. The antler array separated into three groups with no obvious geographical association (Danilkin *et al.*, 1992a). Groves and Grubb (1987) note that Siberian roe possess antlers which are strongly tuberculated, while European roe antlers are only weakly so, but this distinction does not provide a particularly powerful tool for taxonomic classification.

2.2 VARIATION IN BODY COLORATION

In many mammals, body colour is one of the most important taxonomic characteristics; however, researchers are divided on the question of whether roe deer populations can be characterized in this way. Many putative races and subspecies were first distinguished on the basis of particular colorations, which were subsequently found in other populations too. Despite this, colour has figured as a distinguishing characteristic in numerous studies.

The range of roe deer colouring, which may vary according to both age and season (Flerov, 1952), is so broad that it almost defies accurate description, particularly since the definition of different hues and shades, which also depend on the degree of illumination, is subjective. Pasternak (1955) studied the geographical variability of roe deer colour in great detail and concluded that this character was unsuitable for discerning taxonomic groups. Generally, from west to east and from south to north the colour of the coat becomes lighter and subtle changes of hue occur. At the same time, at the extremes of the species range, the coat becomes darker and brighter, a feature also characteristic of other deer species, due to an increase in the number of phenomelanins with climate humidity. Lehmann (1971) supported this idea that differences in hair structure and coloration are dependent on the climate. Towards the east and north the summer coat of European roe deer becomes less bright, the deep red colour gradually turns yellowish or a shade of ochre, the hair becomes longer and thicker and the colour at the hair base also alters to some extent.

Having noted this clinal variation of coat colour across the total range of roe deer, Flerov (1952) nevertheless made some distinctions between European and Siberian roe which seem accurate. In the European form, the winter coat is grey-brown (sometimes darker with grey sides). When in summer coat, the superior aspect of the head is grey or brown, in sharp contrast with the red colour of the back and sides. The base of the hairs are grey-brown or dark-brown, the epidermis is light and not pigmented. In Siberian roe, the winter coat is grey or ochre-grey, while the abdomen and interior aspects of the limbs are creamy. In summer, the head, back and sides are uniformly red, the base of the hairs are light, the epidermis is brown-grey and pigmented. In addition, European roe have a noticeably light nose patch and brown or dark brown metatarsal glands (Meyer, 1968; Raesfeld, 1956; Stubbe and Passarge, 1979). These distinctive features have been shown to develop early, and are present even in kids (Sokolov, 1959; Heptner *et al.*, 1961; Gromov, 1986a,b).

From studying the skins of roe deer stored in museums of the former USSR and some other countries, using Bondartsev's scale to compare the coloration of more than 600 culled animals of both sexes and various ages, it seems that Pasternak's description of clinal variation of coat colour across the range of roe is generally born out. Despite this, European and Siberian roe deer are markedly different in coloration. This may be of some taxonomic importance but it should be noted that, since coat characteristics vary by individual, age and season, they are far from reliable, even for differentiating between these two very different forms. Furthermore, unlike Gromov (1986a,b), we have not found any significant colour differences between Siberian roe subspecies. The difficulty of distinguishing between groups on the basis of coat coloration is illustrated by the example of variability in the colour of the head. In some European roe in summer, the head is the same colour as the upper side of the body, a characteristic typical of Siberian roe deer; conversely, some young Siberian roe have a grey-brown nasal bridge and forehead which is more usual among European roe. The light nose patch, characteristic of European roe, is also sometimes observed on some older Siberian animals. Similarly, numerous Siberian roe deer possess a well marked and sometimes large 'caudal disc' or 'target' when in summer coat, a typical feature of European roe deer (e.g. Sokolov and Danilkin, 1981 – Figures 4, 15, 39, 40). In contrast, some European roe deer often lack this feature (e.g. Honak, 1970; Stubbe and Passarge, 1979). In the course of long-term observations of the Siberian roe deer of the southern Urals, it has been established that the summer-coat target is a hereditary feature but, contrary to Gromov (1986a,b), it seems an unreliable way to differentiate between European and Siberian roe.

The colour of the metatarsal glands (which is the same as that of the leg hair in Siberian roe but darker in European roe) may offer the most reliable distinction between the two forms. This feature is not totally dependable, as six of 34 European roe examined from the Belgorod Region had 'Siberian' coloration of the glands, while Siberian animals of the southern Urals and Far East have included individuals with the darker coloration typical of the European form. Equally, the colour and position of lip markings or 'moustaches' and the colour of the lower area of the head and neck do not provide definitive criteria for classification, as animals from both European and Siberian populations are similar to specimens from the Trans-Baikal area described by Smirnov (1978).

Abnormally coloured individuals (albino, melanic and maculate forms) are sometimes observed among both Siberian and European roe deer, with most reports concerning European populations, probably because they are generally more intensively monitored. Of particular interest is a population of 'black' roe deer inhabiting the low marshes of Niedersachsen, West Germany, and the surrounding locality. These animals were first documented in 1591. The typical form of this area is red, but black roe deer constitute between 2% and 80% of certain populations. Following the spring moult, adult animals are a shiny black colour during summer. The winter coat is also black, but dull, and becomes lead-grey towards the abdomen, with frequent flecks of silver. The target is smaller than that of the red form and is reddish, brown or black. Fawns are born black or, more rarely, skewbald. As many as 33% of the black form but only 7% of the normal population have maxillary canine teeth. The body dimensions and weight of the two forms are not different, although the black animals have rather longer tails. Their behaviour is similar and they will form mixed pairs.

Unlike the red form, black animals exhibit a preference for the moist woodland habitats of the plain rather than the mountains. This black coloration may be a recessive characteristic resulting from inhabiting the cold and humid climate of the lowlands but their canine teeth, longer tail and colouring may indicate that this population contains genes of an ancestral type (Stubbe and Passarge, 1979; Meyer-Brenken, 1986).

The roe deer of southern Spain also exhibit distinctive coloration, differing from typical European roe in that their summer coat is grey and they carry a large white patch. Overall, it is clear that colour variation between individuals within a single population or group may be greater than reported differences in coloration between the European and the Siberian roe deer.

2.3 CYTOGENETIC VARIABILITY

The karyotype is one of the most important characters in taxonomic studies of animals, especially when the groups concerned are closely related. The study of chromosomal differences enables inferences to be made on microevolutionary processes, speciation and reproductive isolation, and allows us to explore genetic relationships between populations and phenotypically similar groups.

It has been established that the karyotypes of the Siberian and European roe deer are distinctly different (Figure 2.8). All populations of the European roe deer have the same chromosome set: $2n = 70$ (Gustavsson, 1965; Amrud and Nes, 1966; Wurster and Benirschke, 1967; Gustavsson and Sundt, 1968; Zivkovic and Isakovic, 1972; Sokolov *et al.*, 1978; Danilkin *et al.*, 1983; Sysa and Kaluzinski, 1984; Markov and Dobrianov, 1985; Zima and Havrankova, 1987), whereas all populations of the Siberian roe deer are characterized by the presence of one or more microchromosomes belonging to a group of accessory chromosomes known as B-chromosomes: $2n = 70 + (1–14)$ (Sokolov *et al.*, 1975, 1978; Danilkin, 1985b; Sokolov and Danilkin, 1980, 1981; Zernahle, 1980; Danilkin *et al.*, 1983), probably formed as the result of fissions of autosomes (Groves and Grubb, 1987).

There are some cytogenetic similarities between the European roe deer and Siberian roe deer: autosomes are acrocentric, female sex chromosomes are XX, males are XY, and the X-chromosome is the largest of the set, two-limbed, and submetacentric. Despite their outwardly morphologically similar karyotypes, morphometric differences between European and Siberian roe at the chromosomal level are also evident: 11 of the 34 chromosome pairs constituting the main chromosome set, as

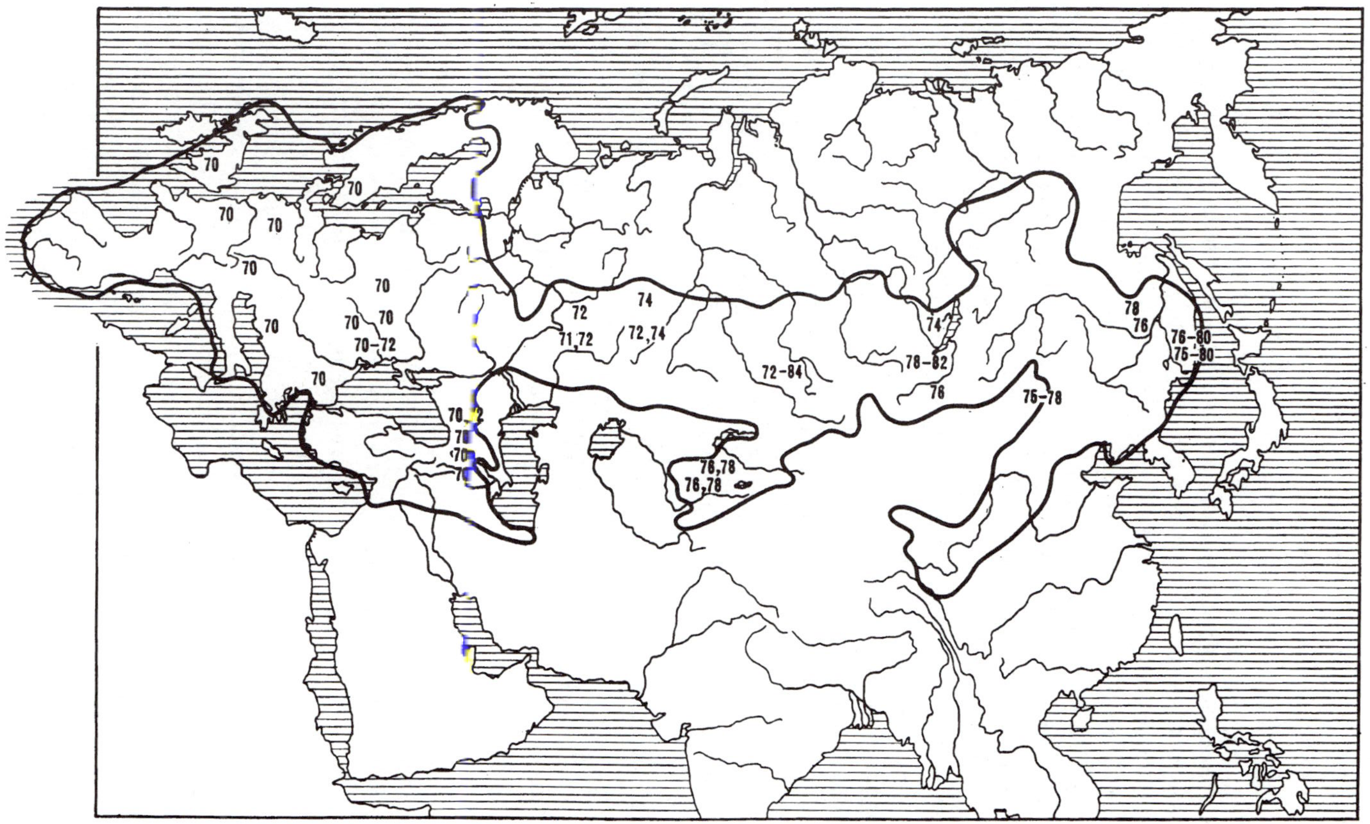

Figure 2.8 Variation of chromosome number in roe deer. Values shown are diploid number across its geographical range (black outline).

well as the Y-chromosome, all differ significantly in size between the two groups (Sokolov *et al.*, 1986a). Moreover, karyotypes of the European roe deer are stable (i.e. they have the same number of chromosomes on each of the metaphase plates), whereas those of the Siberian roe deer are either stable or are mosaic (where different numbers of B-chromosomes may occur in a single animal, or in different animals of a single population).

Mosaicism is especially evident in roe of the Far East, where 4- or 5-clone mosaics have been found. For example, in one particular buck the number of accessory chromosomes varied from 5 to 10 ($2n$ = 75–80), clones with 6 or 8 B-chromosomes prevailing (75%). In does, their number varied from 6 to 10 (the respective diploid number varying from 75 to 80) and clones comprising 8 (40%) and 10 (30%) accessory chromosomes were predominant. In a further single Siberian buck, up to 12 B-chromosomes were found within a 3-clone mosaic ($2n$ = 78, 80, 82), the clone with 12 accessory chromosomes prevailing (60%). Unfortunately, the origin of this animal is not known for certain, but it presumably came from the Trans-Baikal area. Karyotypes of deer from the Ante-Urals and Trans-Urals are more stable, with only a few (1–4) microchromosomes generally present (Danilkin and Baskevich, 1987). The maximum variability of karyotype has been found in the Altai roe deer population, where individuals had a total of between 72 and 84 chromosomes (Grafodatski *et al.*, 1990).

Hybridization experiments conducted in East Germany involving European and Siberian roe have shown that B-chromosomes can be inherited by hybrids (Zernahle, 1980). When Siberian roebucks, with 10 B-chromosomes, were crossed with European does, F1 hybrids were produced which carried only 3 accessory chromosomes (i.e. $2n = 70 + 3$), less than the expected number ($2n = 70 + 5$). Back-crosses between such hybrid females and the Siberian buck increased the number of accessory chromosomes present up to 6 or 7, but this was still less than expected. This result indicates unequal segregation of these 'microchromosomes' in meiosis, and the potential for 'enrichment' of the Siberian genotype by back-crossing. Hybrids produced in these experimental crosses inherit the phenotype of the Siberian parent which could suggest that roe deer phenotype is controlled by the presence and number of accessory chromosomes. Indeed, there is a marked morphometric similarity between populations of Siberian roe of the Trans-Volga area, the Trans-Urals and Ante-Baikal area, which all carry 1 to 4 B-chromosomes, but a great difference between these deer when compared with populations characterized by the presence of 5 to 12 B-chromosomes in southern Kazakhstan, Kirghizia, Mongolia and the Far East, which are also morphometrically similar to each other.

However, a single population may include individuals with different numbers of B-chromosomes, as well as mosaics. Furthermore, there were no significant differences in body dimensions or coloration between animals carrying accessory chromosomes ($2n = 71, 72$) and those without them ($2n = 70$) in populations of the left bank of the Ukraine and northern Caucasus, whereas differences between these deer and animals of the Ante-Urals with the same karyotype ($2n = 71, 72$) were observed. It should also be noted that these large and relatively isolated geographical populations are genetically similar across vast stretches of their range, and the change in the number of accessory chromosomes often occurs at geographical barriers.

Overall, it is evident that the number of accessory chromosomes diminishes from east to west across the roe deer's range, being entirely absent in central and western Europe, which is clearly clinal variation (Figure 2.8). This change in the number of accessory chromosomes implies that there should exist a geographical zone of 'intermediate' genotype, where the Siberian and European roe deer merge, and where the karyotype would include a minimum number of such chromosomes, if any at all. Such a zone should theoretically be situated to the west of the Volga. This karyotype has indeed been found in the taxonomically ill-defined roe populations of the Samara Forest, the Dnepropetrovsk Region of the Ukraine, and in the northern Caucasus. Of nine animals of the Samara Forest population examined, two did not have accessory chromosomes, three had one such chromosome each, and four individuals had two each. In the northern Caucasus, accessory chromosomes have been found in only one roe killed near the town of Apsheronsk, in the Krasnodar Territory. In all the other regions stretching from the Volga to the Dnieper, roe deer were exterminated at the beginning of the twentieth century and the 'intermediate' genotype has virtually disappeared.

The genetic role of accessory chromosomes, their nature and the basis of their polymorphism in roe deer and other mammals are to a large extent unclear. As a taxonomic tool, this characteristic of the karyotype appears to be useful at the interspecific level but less so at the intraspecific level. The wide variability of chromosome number among mosaics and members of a single population means it is not possible to differentiate definitively between subspecies and populations of Siberian roe. Accessory chromosomes in the diploid set of the Siberian roe deer, with the support of the palaeontological data presented in the last chapter, may suggest that this species is of more ancient origin than the European form. Taking into account the discontinuity of chromosomal differences, we may perhaps speculate about possible centres of origin for *Capreolus*, tracing the evolutionary relationships between the two forms. Roe deer probably originated in central Asia, mainly to the south of the Altai–Sayani mountain system, and, from this point, moved northward, then eastward to Baikal and westward to Europe. Judging from the exceptional variability of B-chromosome number in the Altai, this may well have been such a centre of origin (Figure 2.8).

An alternative hypothesis should be considered which seems to be supported by the latest research. This is that the formation of the modern European roe deer karyotype was greatly influenced by climatic factors such as the glaciation of the Russian plains, which physically separated populations. Through such isolation, genetic differences accumulated, such as the possession of accessory chromosomes, and gradually European and Siberian populations acquired characters which ensured partial reproductive isolation from each other. Thus, one parental gene complex gave rise to two well-integrated gene complexes (Danilkin, 1985b). In the postglacial period the Siberian roe deer moved westward, to colonize the areas stretching up to the Dnieper and the northern Caucasus, which may well have been inhabited by the European roe deer. This resulted in the occurrence of 'mixed' genotype populations. This hypothesis of allopatric speciation of roe deer agrees with established facts of genetic and phenotypic variability, and may explain the origin of some taxonomically obscure populations.

2.4 BIOCHEMICAL AND IMMUNOLOGICAL ANALYSIS

In a comparative study of protein polymorphism among roe deer populations using zone electrophoresis, it has been established that of 11 enzyme systems coding for 14 enzyme loci, three (EST-D, IDG and G6FDG) were polymorphic among European roe, while only two (EST-D and G6FDG) were polymorphic among Siberian animals (Sokolov *et al.*, 1986b). Genotype frequency differences have been discovered for IDG only. In the enzyme systems so far investigated, no differences in allelic distribution have been found between the Siberian roe of the Urals and those of the Far East.

Isoelectric focusing of blood plasma proteins has also revealed marked differences in some fractions in the pre-albumin zone (PAZ) of the spectrum, at a pH of around 5.2, between the European and Siberian roe deer (Figure 2.9). The differences in these fractions are fixed, with each European roe different from each Siberian animal. Furthermore, although European and Siberian roe deer are similar in the IEF spectra of membrane proteins of the muscle tissue, differences have also been established in the IEF spectra of soluble proteins of muscle tissue at a pH of around 6.6 and 7.8 (Figure 2.10). The analysis of data on the optical density and distribution of pH values of these fractions allows us to hypothesize that, in each case, these different protein fractions are the products of alternative alleles at particular loci. In addition, based on the

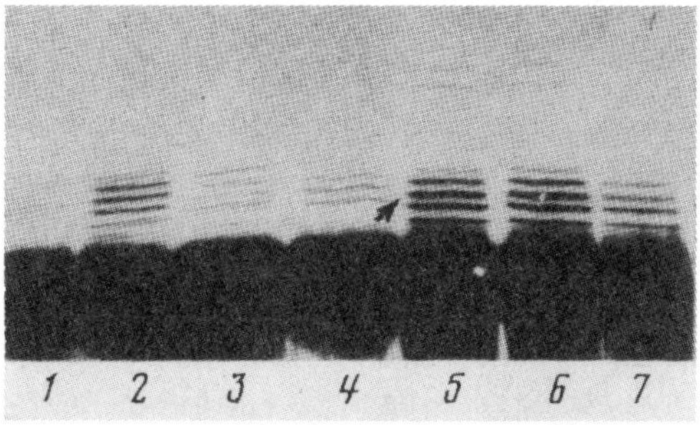

Figure 2.9 Spectrum from isoelectric focusing of blood proteins of roe deer. (1)–(4) Siberian; (5)–(7) European.

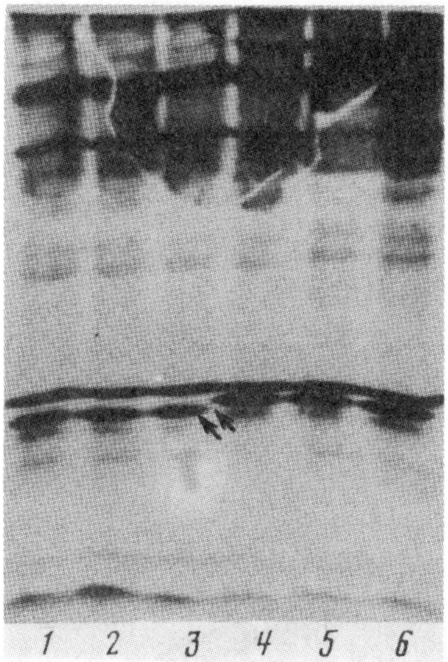

Figure 2.10 Spectrum from isoelectric focusing of muscle tissue soluble proteins of roe deer. (1)–(3) Siberian; (4)–(6) European. (From Sokolov *et al.*, 1986b.)

IEF spectrum of plasma proteins, populations of Siberian roe from the right bank and the left bank of the Amur have been found to differ in the distribution of PRA locus genotypes: the left bank population has three alleles at this locus, whereas the right bank population has only one.

Thus, of 14 enzyme-coding loci, plus eight loci coding for the synthesis of proteins of the IEF spectrum, three are polymorphic in each of the European and Siberian roe. Among the former group of deer, all three polymorphisms concern enzyme loci, while among the latter group one of the three concerns a locus in the IEF spectrum of plasma proteins. The percentage of polymorphic loci is estimated as 13.5% both for European and Siberian roe deer, and the average heterozygosity as 1.2% and 1.1%, respectively. These relatively low values of polymorphism are in good agreement with data reported by Hewison (1995) for British roe deer (13.3% and 3.2% respectively), and by other researchers who have also noted that the level of heterozygosity is generally low among ungulates (e.g. Baccus *et al.*, 1983). However, this contrasts strongly with the results of Hartl *et al.* (1991), who calculated the proportion of polymorphic loci as 30% and the expected average heterozygosity as 4.9% for roe of central Europe, and hypothesized that *Capreolus capreolus* is one of the genetically most variable deer species yet studied.

How great is the degree of genetic differentiation between European and Siberian roe deer? Traditional techniques of studying protein variation by zone electrophoresis do not reveal any fixed differences between these two forms. Allele frequency differences have been shown for one locus only, and the generalized estimate of genetic differences, expressed in terms of Nei's genetic distance, is $D = 0.0044$. Such D-values are more characteristic of the level of differentiation between poorly isolated populations, than indicative of true species. However, other techniques such as isoelectric focusing have revealed species-specific characters, with alternative fractions of isoelectric spectra of blood plasma proteins and soluble muscle proteins exclusively present in one or other form of the roe deer (Shurkhal *et al.*, 1985; Sokolov *et al.*, 1986b). It is highly probable that these fractions are the protein products of alternative alleles of three different loci, indicating a fairly high degree of reproductive isolation between European and Siberian roe deer. If all these data from enzyme studies and IEF studies are pooled, Nei's genetic distance is $D = 0.7183$, a value which usually characterizes fairly reliable species (Table 1.2).

Similar data have been obtained through comparative immunochemical analysis (Markov *et al.*, 1985a). The testing of native antiserums and those absorbed (through saturation with a serum from the alternative form of roe deer) by Ouchterllony's double immunodiffusion method and Scheideger's micro-immunoelectrophoresis method has shown that the antiserum for Siberian roe serum proteins (anti-*C.p.*) reacts with the homologous serum, forming numerous precipitation bands in agar. Similar band numbers and banding patterns have been produced by reaction of the antiserum with European roe serum (anti-*C.c.*), indicating that Siberian roe blood serum contains all those antigens present in European roe. This is also confirmed by the fact that upon absorption of the anti-*C.p.* by European roe blood serum, the antiserum ceases to react with the serums of both European and Siberian roe deer. The complete similarity of the antigens is also established by immunoelectrophoretic study of the anti-*C.p.* In contrast, investigation of the anti-*C.c.* has revealed antigens which are characteristic of European roe blood serum only. Upon absorption of the anti-*C.c.* by Siberian roe blood serum, this antiserum ceases to react with Siberian roe blood serum, but forms two precipitation bands in the immunodiffusional or immunoelectrophoretic reaction with European roe serum.

Additional data on differences in blood serum antigen composition between the European and Siberian roe deer have been obtained by comparative polyacrylamide electrophoresis of the sera on a gradient gel. The European roe blood serum protein spectrum differs in one medium- and one fast-migrating fraction, which may correspond to two accessory antigens. Gel column chromatography has confirmed that these antigens have very different molecular weights. Immunochemical analysis has revealed specific differences in the blood serum of European and Siberian roe deer but the general similarity between the two sets of blood serum antigens confirms that they are closely related phylogenetically.

The nucleotide sequences of mitochondrial DNA cytochrome *b* gene of Siberian roe deer samples are very different from European roe deer, as expected (Ettore Randi, personal communication).

From the data presented so far it is evident that, despite an extremely pronounced geographical variability, on the basis of morphological, cytogenetical, biochemical and immunological criteria, the European and the Siberian roe deer are distinct: it is possible to classify them as different but phylogenetically neighbouring species. The distribution and ecological features of these two roe deer forms are outlined in the following chapters.

3

Distribution and numbers

3.1 DISTRIBUTION OF THE GENUS *CAPREOLUS*

3.1.1 Historical distribution

The fossil remains of the Lower Pleistocene indicate that roe deer closely resembling modern forms (Figure 1.2) inhabited central and western Europe, including Great Britain (but they failed to colonize the islands of Ireland, Corsica and Sardinia which had by then separated from the mainland), as well as Asia Minor, southern Siberia and the Far East. Unfortunately, the scarcity of *Capreolus* remains from that period mean that we can only approximate the extent of their distribution (Figure 3.1).

From analysis of the literature containing data on roe fossil records, it seems that at least two forms, the smaller *Capreolus capreolus* of Europe and the bigger *Capreolus pygargus* of Asia, have definitely existed since the Pleistocene period. Suggestions that the Siberian roe deer may have occurred in central Europe during the Pleistocene and up until historical times (Kahlke, 1955; Raesfeld, 1956; Lehmann, 1958, 1960a) are probably not true. Paaver (1965) has demonstrated that even very large bones found in central Europe are still smaller than the maximum dimensions of European roe deer from the Late Recent period in the eastern Baltic region. At the same time, comparison of the latter samples with remains of Siberian roe from the Recent period in the Trans-Urals reveals significant differences (Kosintsev, 1981).

Pleistocenic glaciers on the Russian Plain and Caspian Sea floods extending far northward along the Volga (Bader, 1950; Vereshchagin, 1959; Markov *et al.*, 1965) were probably serious obstacles to the westward movement of Siberian roe for a considerable time. Only after these barriers had receded did the Siberian roe colonize the Russian Plain, moving from the east, up to the Dnieper (Flerov, 1952) and, in the late Middle Ages, possibly reaching the northern Caucasus (Vereshchagin,

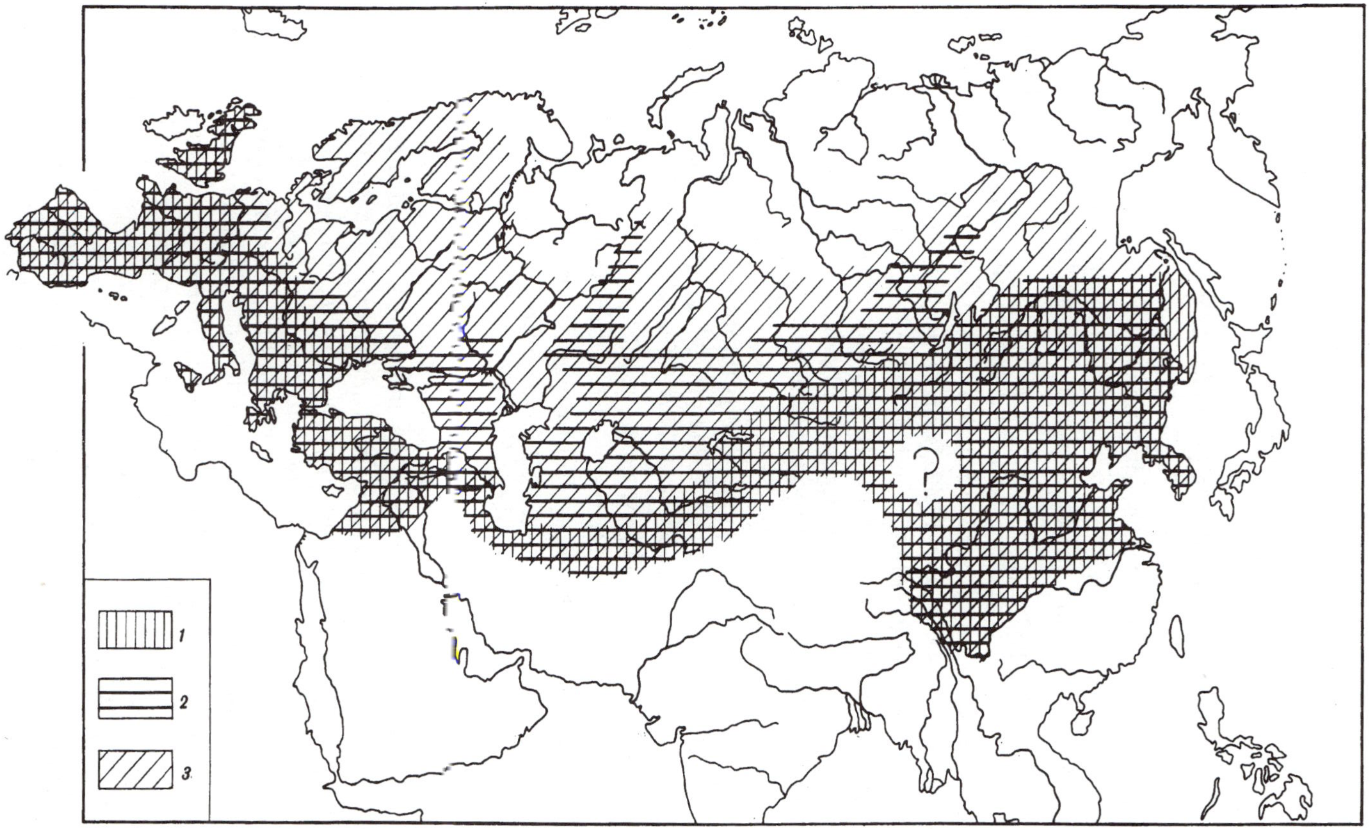

Figure 3.1 *Capreolus* distribution during the Quaternary Period: hypothetical reconstruction based on present distribution and occurrence of fossil remains (see Figure 1.2). (1) Late Pliocene and Early Pleistocene; (2) Pleistocene; (3) Recent Period.

1959). By the Middle and Upper Pleistocene periods, the distribution of *Capreolus* had greatly expanded (Figure 3.1). Modern roe deer colonized the Crimea, Azov area, Lower Don region and the northern Caucasus (Gromov, 1948; Vereshchagin, 1959). In the Urals and Siberia, they moved a great distance northward, to reach the upper Pechora in the Ante-Urals and Yakutia in Siberia (Yermolova, 1963; Ravski *et al.*, 1964; Kuzmina, 1971; Wangenheim, 1977; Alekseyeva, 1980; see also Chapter 1). In Europe, they were widely distributed, but restricted to the more southerly latitudes during the colder phases of this period (Lister, 1984).

More recently, as the climate became warmer and glaciers shrank, forested areas began to increase, resulting in an explosion in numbers of forest-dwelling ungulates. This was not accompanied by an expansion in geographical range until somewhat later, in the nineteenth and early twentieth centuries (Kirikov, 1959). From records and our knowledge of the current roe deer distribution, it is possible to estimate the maximum range of the genus *Capreolus* relatively accurately (Figure 3.2).

In the northern Caucasus, ranges of European and Siberian roe deer overlapped until the early twentieth century (Dinnik, 1910), and, until recently, local populations of Siberian roe deer also inhabited the Dnepropetrovsk and Kirovograd regions (Brauner, 1915, 1923; Migulin, 1927, 1929; Charlemagne, 1937; Korneyev, 1952; Sokolov and Danilkin, 1981). Siberian roe remains have also been found in the Orel region of Russia (Gorbachev, 1915; Flerov, 1952), in the Sarkel Site on the Lower Don, dating from the eighth to thirteenth centuries (Vereshchagin, 1959), in Slav sites of the ninth and tenth centuries, near Voronezh (Gromova, 1948), in peat layers near the Zavorich Station of the Kiev region (Korneyev, 1952), and on the Middle Dnieper (Timchenko, 1972).

Based on comparative studies, specimens taken from the Orel, Kursk and Voronezh regions dating from the late first and early second millennium AD, as well as specimens from the Troitsk site in the Mozhaisk District of the Moscow region, were attributed by Tsalkin (1963) to the Siberian form. However, contemporary bone remains from the right bank of the Dnieper belong to the European roe deer. A hybrid zone probably existed on the left-bank Ukraine in the Recent period, complicating the identification of bone remains. In this period some individuals or groups probably left the Dnieper and migrated westward as far as central Europe and northward to the Baltic region, where mixed Siberian and European roe deer populations may well have formed, resulting in hybridization. Individuals with 'Siberian' form antlers are occasionally found there even now (Cotta, 1969; Lehmann, 1976); however, reproductive barriers and the abundance of the European form were such that these hybrids did not persist.

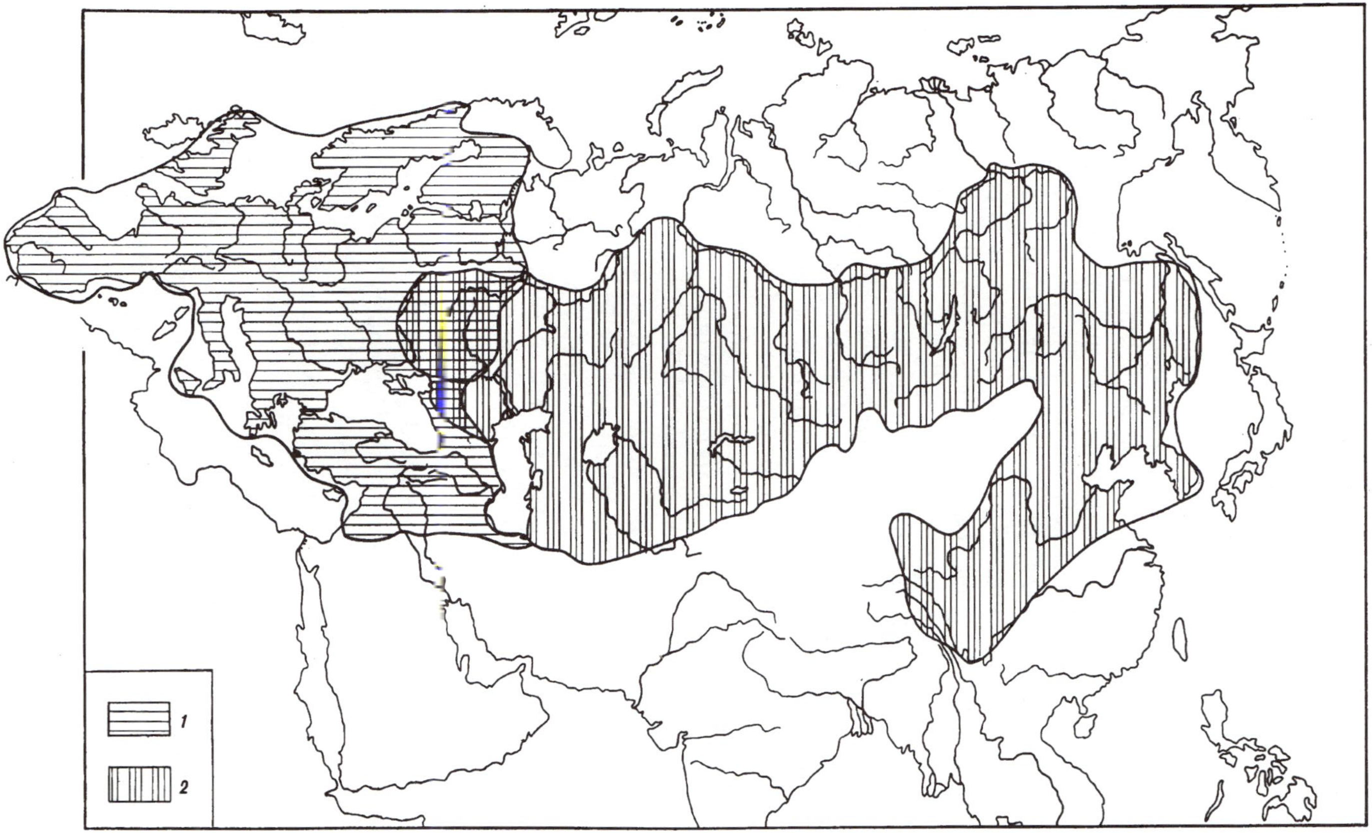

Figure 3.2 Maximum range of *Capreolus*, Gray. Distribution of (1) *Capreolus capreolus* L. and (2) *Capreolus pygargus* Pall.

Prior to the nineteenth century, the roe deer populations of western Europe had experienced a decline due to deforestation, cattle breeding and intensive hunting. The range of roe deer in Britain was reduced to a remnant population in the Scottish highlands (Whitehead, 1964), about 100 roe remained in the southernmost area of Sweden, while the species had been entirely eradicated in Switzerland. In eastern Europe and Asia the reduction in range occurred somewhat later, with the Siberian roe being virtually exterminated by humans within eastern Europe (Figure 3.3), except for some local populations on the Dnieper which, in recent decades, have mixed with European roe, being naturally absorbed by the latter. In the middle of the nineteenth century, the roe deer was exterminated from the central Chermozem regions of Russia (Barabash-Nikiforov, 1957), the Volga region and the southern Ukraine, resulting in discontinuities in its geographical range. By the end of that century, roe had also disappeared from numerous parts of the European territories of the former USSR (Soloviev, 1922; Lavrov, 1929; Heptner *et al.*, 1961) and excessive hunting and an abundance of predators had eliminated it completely from most parts of Kazakhstan and western Siberia (Shukhov, 1929). At this time the European roe experienced opposite fortunes, with the restoration of its former distribution and recovery of numbers in western Europe, initially in Scotland, following over-hunting during the Middle Ages (Whitehead, 1964; Meikle, 1975).

The twentieth century saw great changes in the distribution of *Capreolus*, particularly in the territories of the former USSR (Figure 3.3). At the very beginning of the century, intensive hunting and an increasing level of human interference resulted in such a wide reduction in available habitat that roe deer range became fragmented into areas which were completely isolated from each other (the west, the Crimea, Caucasia, the Urals and western Siberia, eastern Siberia and the Far East), with only a few small occupied pockets remaining between them (Lavrov, 1927, 1929; Troitski, 1930; Dementiev, 1933; Zalesski, 1934; Belishev, 1934; Flerov, 1935; Podarevski, 1936; Dmitriev, 1938; Demidov, 1940; Kuznetsov, 1948; Nasimovich, 1955; Barabash-Nikiforov, 1957; Heptner *et al.*, 1961). The roe deer was virtually exterminated from vast expanses stretching from the Dnieper to the Urals, western Siberia and northern and central Kazakhstan, and the northern limit of its distribution in eastern Siberia retreated southward. This reduction in range reached its peak in the 1920s and may have been largely due to the wars of 1914–1920, when arms were often available.

However, as early as the 1930s an increase in range of roe deer again became evident, perhaps due to more moderate levels of hunting, a warmer climate with only limited snow depths, and intensive exploitation of the closed-canopy coniferous forests which were subsequently replaced

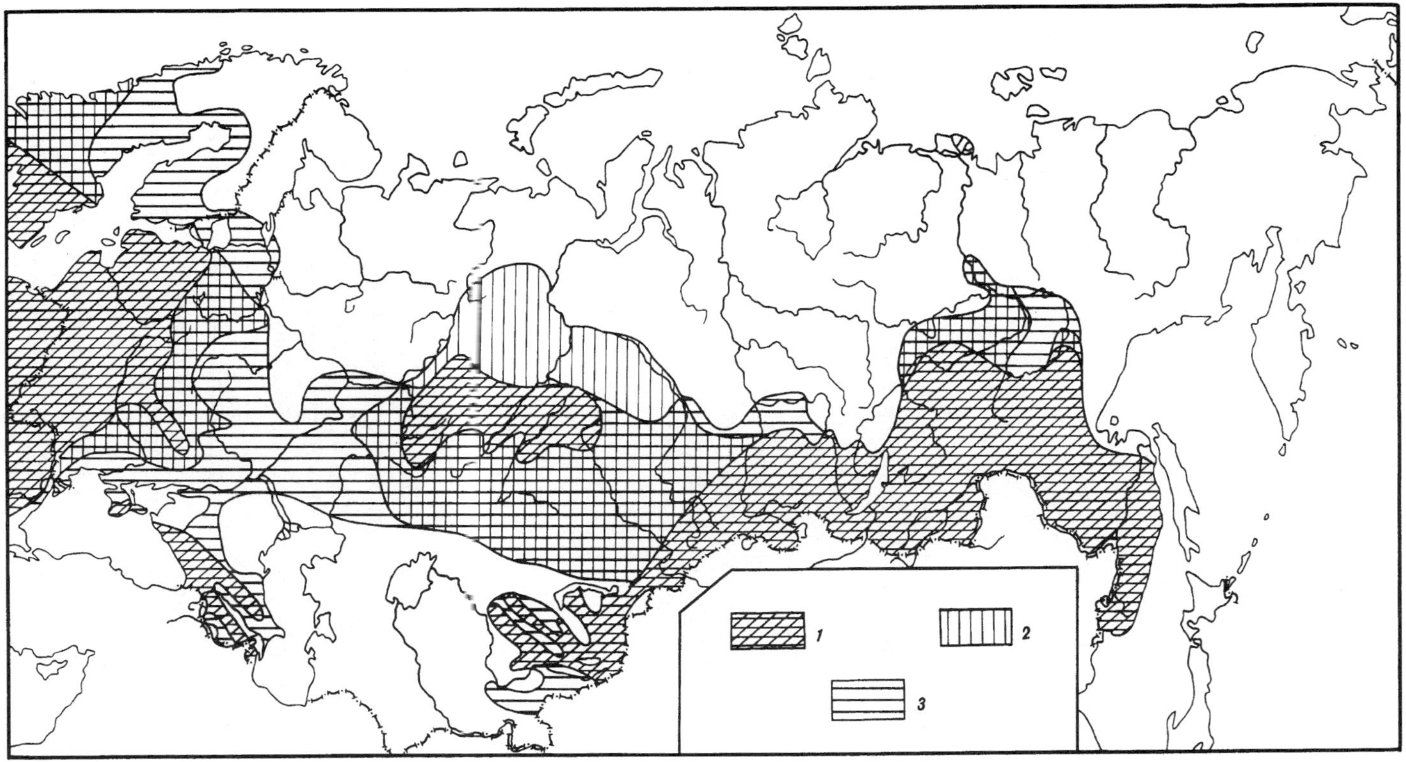

Figure 3.3 Twentieth century changes in distribution of roe deer in the former USSR and northern Europe. (1) Beginning of century (numbers at a minimum); (2) mid century (from Heptner *et al.*, 1961, with revisions and additions); (3) second half of 1980s.

by deciduous trees. Roe deer spread quickly and in all directions and, as early as the middle of the 1950s, previously isolated groups within the overall range merged in Siberia (Zverev, 1937; Antipin, 1941; Korsh, 1941; Skalon, 1946; Yanushevich and Blagoveshchenski, 1952; Belik, 1953; Laptev, 1955, 1958; Yegorov, 1965; Shargayev, 1966; Tavrovski *et al.*, 1971; Kiselev, 1979). However, groups occupying western regions, Crimea and Caucasia remained separated from each other and from the Siberian roe (Figure 3.3).

3.1.2 Recent changes in the distribution of Siberian roe

The Siberian roe deer has been steadily advancing westward from the Urals since the middle of this century. By the end of the 1960s, Siberian roe had reached the Volga (Belyanin, 1975, 1980a; Zlobin, 1975), and by the end of the 1970s it had moved even farther west, reaching the Khoper and Don rivers. Here, along the Khoper and Don, in the Volgograd region, the geographic ranges of European roe and Siberian roe now overlap and thus the European and Siberian roe deer have come into contact again (Figure 3.4).

In the Urals, Kazakhstan and Siberia, range expansion reached its peak at the beginning of the 1960s but poaching, the ploughing up of land and overgrazing caused a rapid decline once more (Tsarev, 1969; Lobachev, 1972; Pavlov *et al.*, 1974; Cheltsov-Bebutov, 1976; Belov *et al.*, 1980; Kapitonov, 1980; Nikolayev, 1982; Yelski, 1983; Sobanski, 1986). Only very recently has the population recovered and begun to increase its range again.

In central Asia, the changes recorded during the twentieth century have not been so pronounced as they were in Europe, the Urals, Kazakhstan and Siberia. The present pattern of distribution is approximately the same as that of the late nineteenth century, with the exception of an area on the Amu Darya river where roe deer were probably present until the 1930s (Rasewig, 1909; Zarudni, 1915; Lavrov, 1929; Kashkarov, 1931; Flerov, 1935, 1952; Gladkov and Nikolski, 1935; Antipin, 1941; Kuznetsov, 1948; Tolstov, 1948; Korelov, 1956; Ishunin, 1961; Aizin, 1969; Yanushevich *et al.*, 1972; Zakirov, 1974; Sludski and Baidavletov, 1984). Compared with the distribution pattern in the middle of this century, roe deer have increased their range somewhat over the past few decades in the south of the region (Figure 3.3).

In the Trans-Baikal area, the Far East and Mongolia, changes in the distribution of Siberian roe deer in historical times do not appear to have been significant. Its status in China unfortunately remains unknown.

3.1.3 Recent changes in the distribution of European roe

During the 1960s and 1970s, due to certain protective measures, the absence of large predators and the availability of suitable habitat, the

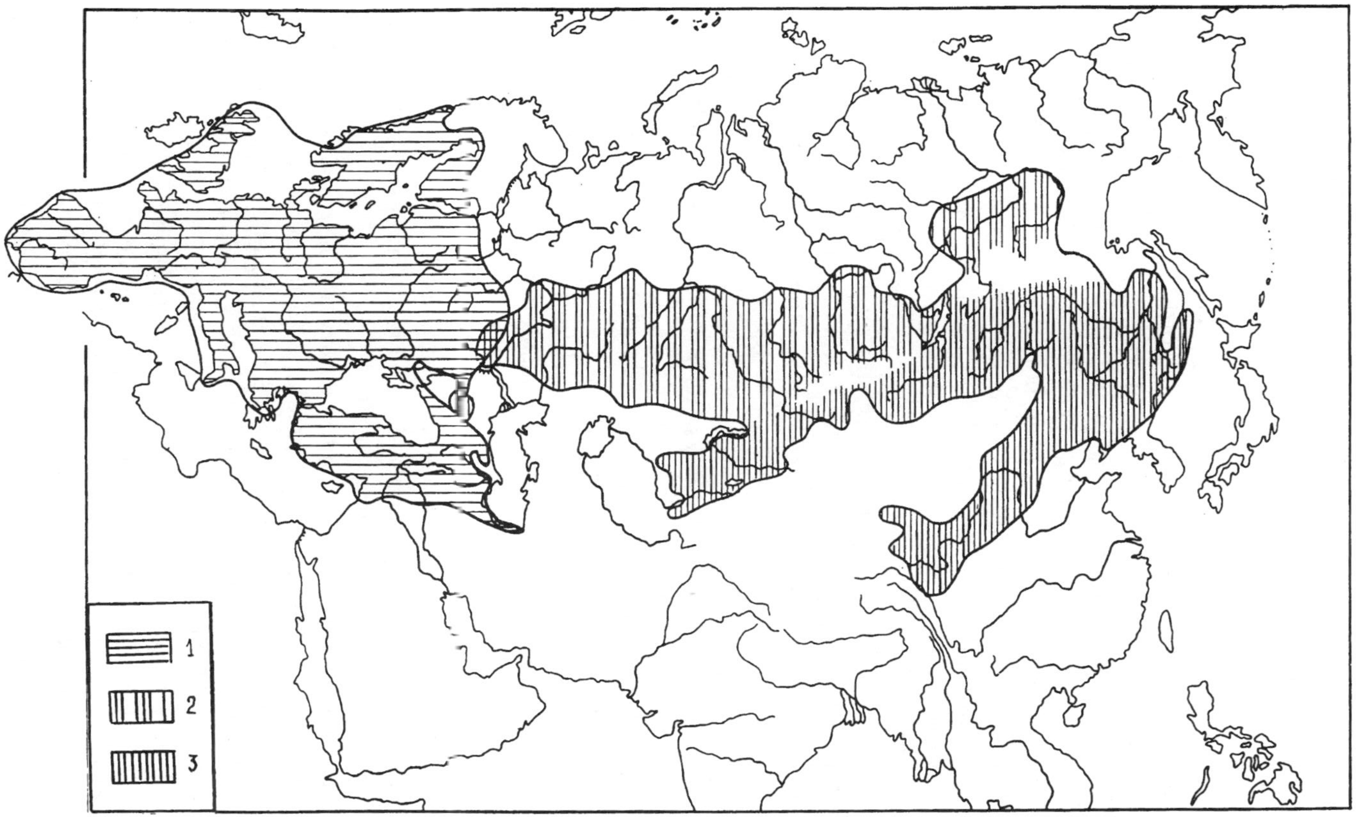

Figure 3.4 Present ranges of roe deer. (1) European (*Capreolus capreolus* L.); (2) Siberian (*C. pygargus pygargus* Pallas); (3) Siberian (*C. pygargus tianschanicus* Satunin).

number of European roe deer greatly increased in western and central Europe. Populations colonized all suitable biotopes, including open fields, and expanded substantially towards the north, in particular, on the Scandinavian peninsula. At the beginning of this century, roe were found only in the southern part of Sweden, to approximately 60° N (Scharff, 1918), but by the 1950s they had advanced as far as 65° (Heptner *et al.*, 1961) and are now recorded as far north as 70° or 71° N, at the northernmost tip of the peninsula and far beyond their historical range. In 1970 roe deer were recorded in Finland for the first time and subsequently have colonized this country from three directions: from the former USSR into southern Finland, from northern Norway into northern Finland, and from Sweden along the Gulf of Bothnia coast (Pulliainen, 1980). Over the same period a similar expansion in the distribution of European roe deer has been observed in the north-west of the former USSR (Danilov, 1974; Sapozhenkov, 1975; Vereshchagin and Rusakov, 1979). In the southern Ukraine, an equally rapid range expansion has resulted in the merging of the western deer range with the Crimean populations (Gurski and Nazarenko, 1969; Gurski, 1975; Dulitski *et al.*, 1975). However, it was the expansion of range to the east that was most marked in European roe, with deer occupying all suitable biotopes, although some re-introduction programmes in this region were not particularly successful (Fadeyev, 1981).

In the 1980s, the European roe deer of the Lower Don reached the Stavropol Upland of the Ante-Caucasus, inhabited by an introduced local population of Siberian roe deer (Yemelianov and Ribalov, 1968) but neighbouring a northern Caucasian population of European roe deer. Thus, the Caucasian section of the overall distribution again became continuous with the main body of roe deer range (Figures 3.3 and 3.4). However, this link is still unstable and is maintained largely by occasional immigrants. A similar convergence of the main European range with that of the Ante-Caucasus is yet to occur.

In summary, during the twentieth century we have witnessed the most severe restriction in the geographical distribution of the genus *Capreolus*, its fragmentation into several sections, and its subsequent recovery to practically its former limits, with an additional expansion into northern Europe (Figure 3.3). The present distribution of the European and Siberian roe deer is illustrated in Figure 3.4. Obviously, the ranges of the European and Siberian roe deer overlap in the Volgograd region, on the west bank of the Volga (Danilkin, 1985c), and a 'hybrid zone' has formed naturally to the east of the Khoper. An isolated Siberian roe deer population, formed by artificial introduction, exists in the Ante-Caucasus, in the Stavropol Upland (Figure 3.4).

For more detailed descriptions of specific areas of the distribution, refer to (Ishunin, 1961; Harrison, 1968; Aizin, 1969; Nikiforov, 1969; Yanushevich *et al.*, 1972; Zakirov, 1974; Kumerloeve, 1975; Smirnov, 1975a, 1978, 1980, 1985; Kucherenko, 1976; Kapitonov, 1980; Shen *et al.*, 1982; Sokolov *et al.*, 1982a; Zhuge, 1982; Bromlei and Kucherenko, 1983; Feng *et al.*, 1983; Sludski and Baidavletov, 1984; Zhang, 1984; Danilkin, 1985c; Prior, 1987; Gill, 1990).

3.1.4 Factors limiting the geographical distribution of *Capreolus*

As a whole, the geographical range of the genus *Capreolus* falls within the moderate zone of Eurasia. The majority of its surface area occurs in the forest steppe zone, but it also extends into the south of the pure forest biotope and the north of the steppe zone. The range boundaries are not fixed, even during a single year, due to seasonal fluctuations. This is particularly so in more northern or alpine regions in Asia, where animals migrate between summer and winter ranges. In summers when snow depth is reduced, the limits of distribution are also further to the north.

The actual structure of the ranges of the European and the Siberian roe deer are somewhat different. The European roe is distributed over a practically continuous area, while the Siberian roe has a rather more discontinuous distribution, particularly in Kazakhstan, central Asia and western Siberia (Figure 3.5). This discontinuity of Siberian roe deer range can be explained by the habitat heterogeneity of these regions and, probably, by a stronger tendency to migrate. It is clear that the main routes of such migrations leading to colonization are river valleys.

Figure 3.5 shows clearly that the range of Siberian roe deer is interrupted along the mountain ridges of the Altai, the western and eastern Sayans, Lake Baikal, the Stanovoye Highland and Stanovoi Ridge, dividing this area into a northern and a southern range. The existence of such a barrier is born out by differences in the deer themselves. Populations to one or other side of this dividing line are morphometrically very similar to each other, but deer from populations to the north and west of the line differ greatly from those to the south (see Chapter 2). Indeed, it was this very line which was taken to delineate the boundary separating the Siberian (*C. p. pygargus* Pall.) from the Tien Shan (*C. p. tianschanicus* Satunin) subspecies of Siberian roe (Danilkin, 1986c) prior to the more detailed studies described here. However, mountain ridges do not necessarily form a complete barrier to the movement of roe deer, as in the Caucasus and the Tien Shan there are populations at 3000 m above sea level. Therefore, other factors must be responsible for this break in Siberian roe range, and these may also determine roe distribution in general.

Snow depth is probably one of the most important factors limiting roe deer distribution. Deep snow may restrict mobility and increase the ener-

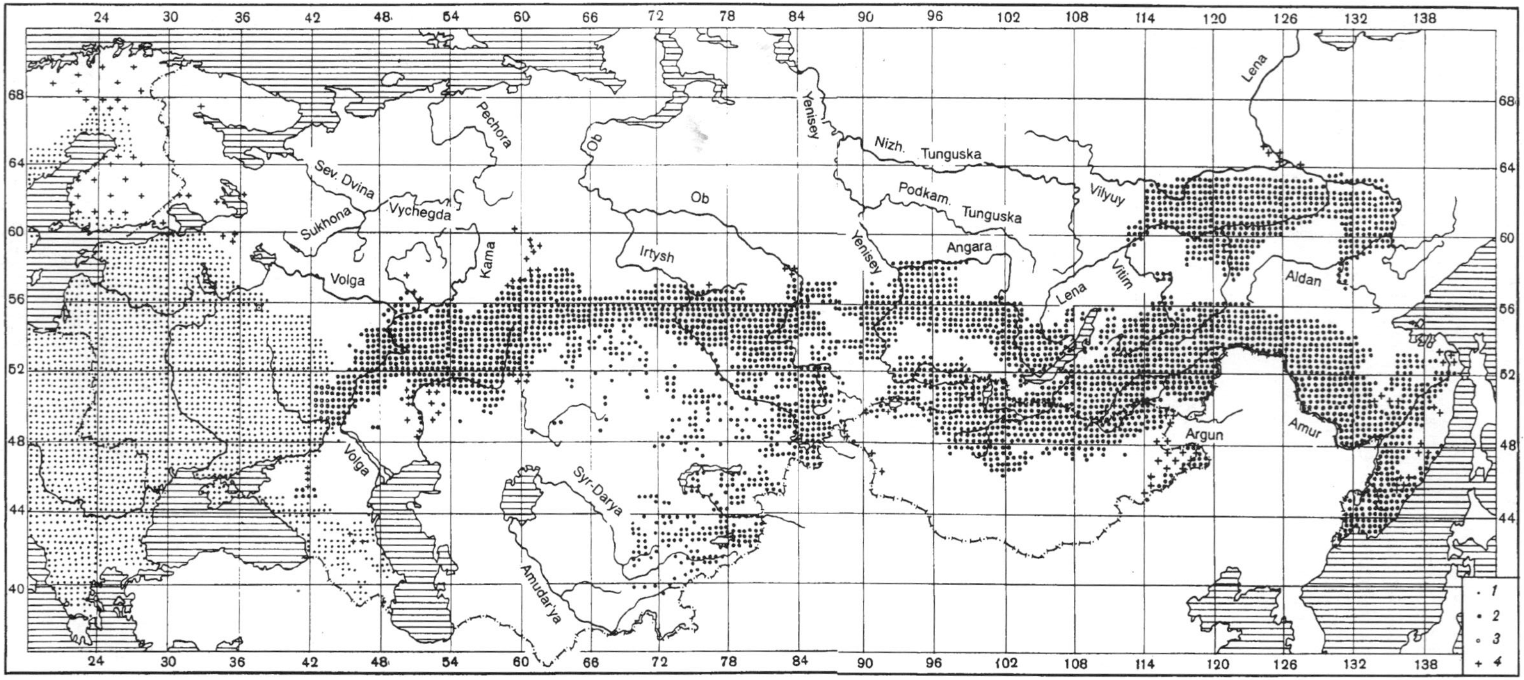

Figure 3.5 Structure of current distribution of roe deer in the former USSR and adjoining Republics. (1) European; (2) Siberian; (3) range overlap zone; (4) recent immigrations.

getic demands of movement (Borg, 1970) and also increase susceptibility to predation (Cederlund and Lindstrom, 1983). The weight load on each individual hoof is between 300 and 360 g/cm^2 of the limb-bearing surface, which is greater than that of musk deer, reindeer, chamois, lynx and wolf, but less than that of elk, red and sika deer. At snow density typical for forest and forest steppe, critical depth is 40-50 cm for adult Siberian roe deer and 20–40 cm for the smaller European roe deer (Nasimovich, 1955). However, snow depth and density are not uniform, enabling animals to live in areas where the depth exceeds the above figures.

Generally, the northern limit of roe deer distribution is taken to be dependent on snow depth, with an average depth of 50 cm thought to prevent the further spread of both European and Siberian forms (Formozov, 1946), and which has rarely been exceeded (Nasimovich, 1955; Heptner *et al.*, 1961). The northern boundary of the European roe distribution does in fact follow such a line defined by a snow depth of 50 cm but this depth of snow does not limit the range of the larger Siberian deer. Only snow depths of 60 cm or more restrict to some extent the distribution of this species (Figure 3.6). At the same time, snow depths are substantially lower than this figure in some areas of eastern Siberia where no roe deer are present, implying that other factors besides snow depth must be important. Roe deer distribution is also strongly dependent on the duration of the snow period, with a range limited to areas with less than 230–240 days snow lie per year for Siberian roe, and 140–150 days per year for European roe.

A third factor limiting roe deer distribution is the predominant vegetation type. Roe tend to avoid mountain tundras, closed coniferous taigas and mountain taigas. Their colonization of certain areas of closed coniferous taiga (Figure 3.6) was facilitated by logging operations over vast areas of its surface. To the south, roe deer distribution is limited by desert conditions; in Kazakhstan and central Asia, the range boundary follows the line of the desert zone (Figure 3.6).

Under the most extreme circumstances, each of the above factors may be the sole limiting factor. More usually, distribution limits will be set by the interaction of the first three of the above factors, particularly in the mountains of Asia. It is this interaction that limits the spread of roe deer in Siberia and in the Far East, and creates an ecological barrier of the Asian mountain ridges. The first snows, which fall in the mountains earlier that in the foothills or on the plains and which are especially deep in the taiga zone, stimulate the annual migration of deer towards milder zones where the snow is not so deep. As a result, the northern distribution limit of the roe deer populations to the south of this barrier recedes farther southward. Similarly, roe deer on the northern slopes of the mountain ridges retreat northward and the gap between the two separate ranges increases up to several hundred kilometres. In spring this sit-

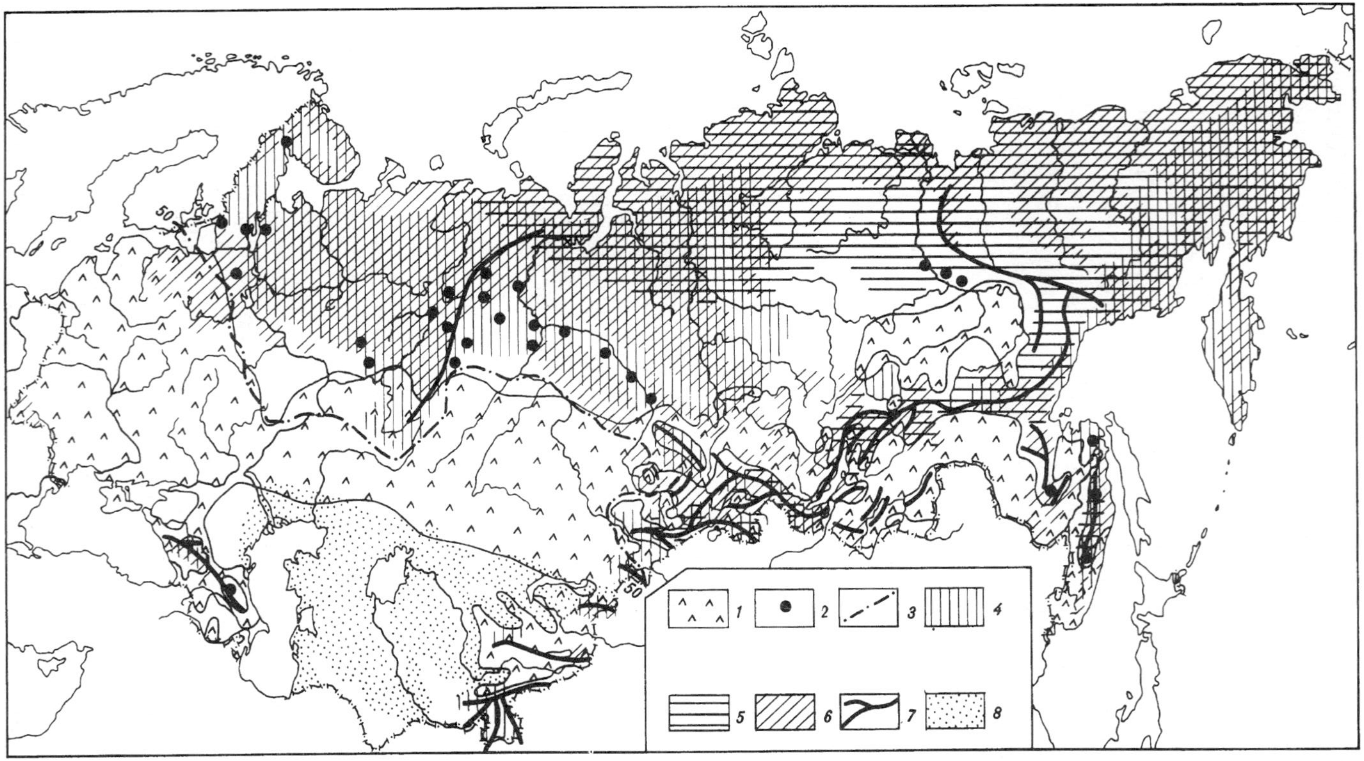

Figure 3.6 Factors limiting distribution of roe deer in the former USSR and adjoining Republics. (1) Core range of *Capreolus*; (2) migrations beyond core range; (3) limit of 50 cm snow depth; (4) regions where snow is deeper than 60 cm; (5) regions where period of snow lie lasts more than 230–240 days; (6) dark coniferous taiga, mountain taiga forest or arctic tundra; (7) main mountain ranges (axes); (8) desert.

uation is reversed but roe cannot pass the barrier of mountain ridges because snow melt in the mountains does not occur until after bucks have already begun establishing territories and does have given birth.

The Asian system of mountain ridges is extremely ancient and has undoubtedly had an effect on the evolution of the two forms of roe deer living on either side of this barrier, which extends from the Far East to the Pamir. From any source population, deer could probably have moved only along this barrier; however, a single breach in the mountain chain (the Djungar Gate) allowed a form closely resembling modern roe to spread both to the north and to the south of the mountains. The morphology of deer from populations of the Far East, Trans-Baikal, Mongolia and central Asia corroborates this hypothesis (Danilkin and Markov, 1987). Thus, ecological and geographical isolation has led to the morpho-physiological differentiation of the northern and the southern populations of Siberian roe deer, which may be regarded as the Siberian and Tien Shan subspecies. The main Caucasian Ridge may form a similar barrier separating two heterogeneous populations of European roe.

These patterns of roe deer distribution allow us to predict future range expansion and to identify areas with suitable available habitats and resources. It is clear (Figures 3.5, 3.6) that, having colonized all suitable biotopes except those of Ante-Caucasia, the European roe deer has reached the limits of its potential expansion. Favourable conditions for successful breeding exist only to the south of a line defining maximum snow depths of less than 40 cm. The optimum conditions for the European roe deer occur in western and central Europe, where the snow depth is less than 20 cm, the average monthly temperature is at least +10°C for no fewer than 150–160 days a year and the annual precipitation is between 450 and 800 mm. In terms of elevation, a zone between 300 and 600 m above sea level is the most favourable (Hell and Herz, 1970; Novakova and Hanzl, 1974; Hell, 1979; Pedroli *et al.*, 1981; Hromas, 1983).

It is evident that the Siberian roe deer is better adapted to severe weather conditions, inhabiting certain regions where the winter temperature falls as low as –60°C (e.g. Yakutia) and others where summer temperature may exceed +40°C (e.g. central Asia). If numbers increase further, it may well colonize the vast surface of the Middle Siberian Plateau and expand its range westward (through immigration into northern areas where it could replace the European roe deer of the Russian plain), also possibly reaching the Caucasus. Conditions favourable for the Siberian roe deer exist in the mountainous areas and foothills of central Asia, in the Altai, the southern Urals, the forest–steppe both to the north and south of the Sayans, the Trans-Baikal area, the southern region of the Far East and in the region surrounding the town of Yakutsk.

3.1.5 The effects of introductions on the current pattern of roe distribution

Conclusions concerning the effects of translocations and re-introductions on the distribution of roe deer should be examined in the light of the classification of the genus *Capreolus* into two species – the European and Siberian forms. Such operations were generally carried out for two main reasons: the restoration and expansion of the existing roe deer range, and the introduction of the larger Siberian roe to areas inhabited by the smaller European form. These introductions were justified in terms of introducing 'new blood' to increase body weights and improve trophy quality in European roe populations. From the last century, a number of releases of Siberian roe deer have been carried out around Europe, including in Great Britain, Germany, Belgium, Slovakia and the western regions of the former USSR. Beyond its natural range of the nineteenth century, the European roe deer was introduced into Ireland, where it flourished until changes in forestry practice led to eradication of the entire population. In addition, in the twentieth century some individuals were released in Texas, USA (Severtsov, 1941; Turcek, 1951; Bump, 1970; Stubbe and Passarge, 1979).

Within the natural range of these species, animals have been released on a number of occasions and in a variety of countries. For instance, an estimated 4780 roe deer were released in France alone between 1952 and 1979 (Boisaubert, 1980). Introductions within Britain are poorly documented but include the release of animals of both Austrian and German origin (Whitehead, 1964; Hewison, 1995). In the former USSR such introductions have been performed on a large scale (Figure 3.7), with more than 3000 animals (more than half of which were Siberian roe) released since 1925, mainly in the European territories of Russia. About 1500 head were released in the Moscow and Tver regions (Pavlov *et al.*, 1974; Vershinin, 1969; Fadeyev, 1969a, 1969b; Nasimovich and Shubnikova, 1976; Denisov, 1978; Shurupov, 1982), and more than 500 animals have been released in the Ukraine, 60 of which were Siberian roe. Overall, estimates are that no fewer than 10 000 animals were translocated, but this is almost certainly a conservative figure.

With little exception, this long-term and costly work has generally proved to be almost entirely unsuccessful due to poor organization, lack of scientific research and errors made during releases. Animals were often introduced in small groups whose sex ratios varied from 1 : 9 to 6 : 1. More than half these introductions took place in regions prone to deep snow cover where supplementary feed and protection against predators and poachers were not provided, so that the majority of these introduced deer perished within the first few years. Moreover, as noted above, introductions often placed roe deer belonging to different species

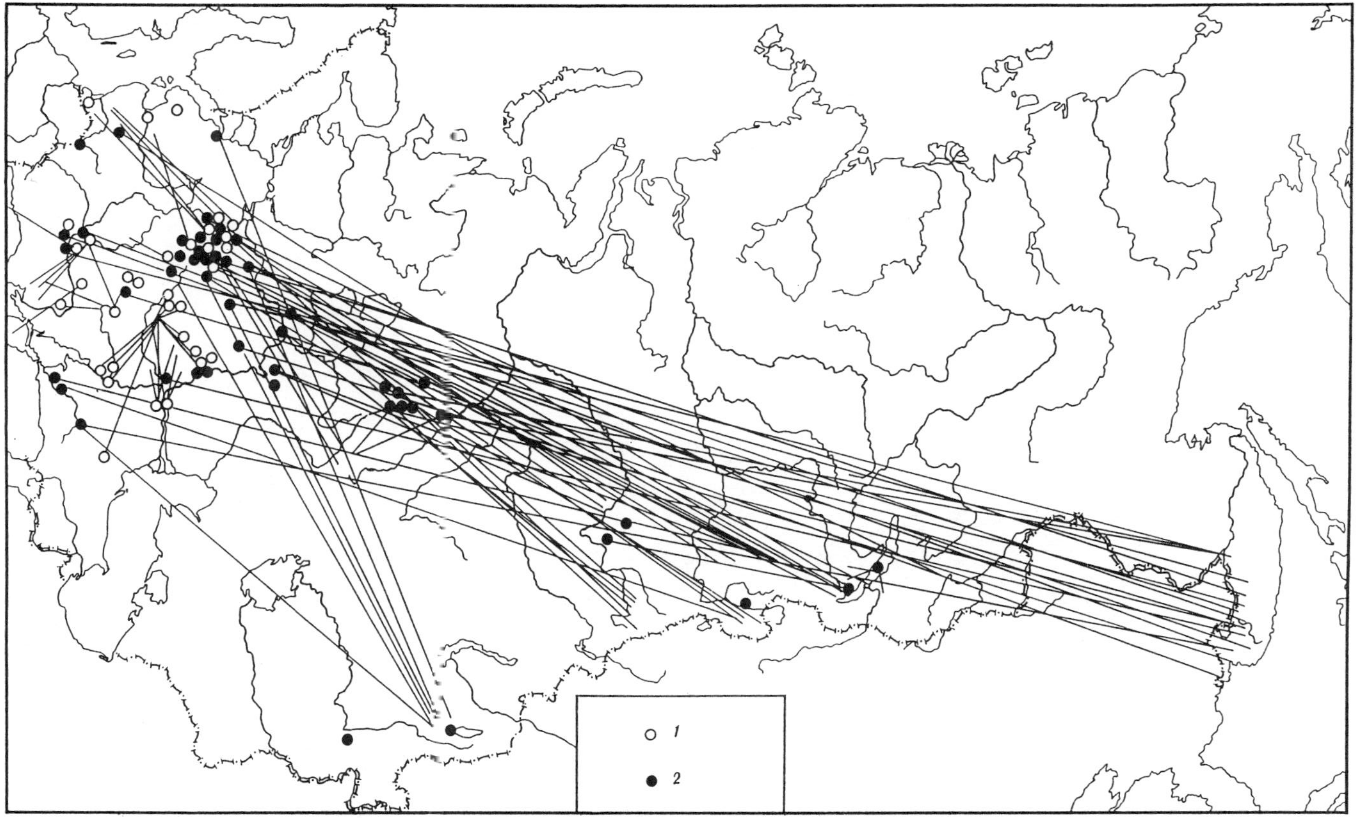

Figure 3.7 Translocations of roe deer in the former USSR. Introductions of (1) European and (2) Siberian roe deer. Lines originate at points of capture.

or subspecies in the same area. The introduction of the Siberian roe deer into the European roe deer range inevitably resulted in hybridization between these two forms.

Evolution has led to certain reproductive barriers and probably a degree of tissue incompatibility between the European and the Siberian roe deer. Studies have shown that many hybrid bucks are sterile, and not infrequently European does are unable to deliver the larger hybrid kids and die with their offspring in the process of giving birth. Such consequences, in addition to natural mortality, result in a rapid elimination of small groups of introduced animals and the loss of the 'Siberian' genotype. Hybrids of European and Siberian roe deer which have survived are heavier and have larger antlers than the pure European form, indicating that one of the main aims of such introduction programmes was achievable. However, this painstaking and ambitious work was, in practice, unsuccessful due to the low level of mixed-pair reproduction.

Only two releases of Siberian roe deer in Europe were successful, both involving sites where European roe were either entirely absent or present in only limited numbers. In 1886, 59 Siberian roe deer were released in the Tsar's Chase near Gatchina, in the Sanet-Petersburg region. These animals flourished at first and by 1909 the population totalled 1068 (Severtsov, 1941), but later the entire population died out. In 1956, 15 Siberian roe deer from Kirghizia and the Primorski Territory were released in the Temni Forest, situated in the Shpakovo District of the Stavropol region (Yemelianov and Ribalov, 1968). Within 10 years their number had increased to 200 head; they had colonized large areas and occupied practically the whole of the Stavropol Upland. This localized population of Siberian roe is now surrounded by European roe deer (Figures 3.4, 3.5) and hybridization is inevitable.

It is clear that any act which promotes the possibility of hybridization of the two different roe deer species is inadmissible, and even the mixing of individuals of different subspecies is undesirable. Speciation is preceded by a long period of population development under particular environmental conditions, and subspecies are forms which are specifically adapted to 'local' conditions. Therefore, the transfer of animals of one subspecies into the range of another, and the consequent mixing of the gene pools, may disrupt unique gene combinations among the indigenous population which has been adapting to local conditions over thousands of years.

3.1.6 Range overlap of European and Siberian roe deer

Since it has been shown experimentally that partial hybridization between European and Siberian roe deer is possible, the problem of their hybridization in nature must be addressed. The hybrid zone, where the European and Siberian roe deer ranges overlap each other, is currently situated on the right bank of the Volga (Figure 3.4). We might expect to observe hybrids in the northern Caucasus and on the Russian plain; in addition, hybrids may exist in the Moscow region and other areas where Siberian roe deer have been released within the European roe's range. What will be the fate of European and Siberian roe deer populations in this hybrid zone? Based on the results of controlled experiments, higher mortality among European does due to problems at parturition should be expected, along with a generally reduced level of fertility (particularly among hybrid males), and a great number of stillbirths, so that overall productivity in a mixed population will be very low. Where the Siberian roe is the more numerous species, the population phenotype will appear mainly 'Siberian'; where European roe are predominant, the phenotype will be mainly 'European' but the 'Siberian' karyotype will be present.

Modelling the dynamics of mixed populations of European and Siberian roe deer suggests that even the introduction of several dozen Siberian animals into a European roe deer population results in the gradual but complete disappearance of the Siberian form with time. These results are reflected in reality by the generally unsuccessful conclusion to artificial introduction experiments. Such a scenario probably occurred in the Samara Forest of the Dnepropetrovsk Region of the Ukraine. Between 30 and 40 years after European roe deer had immigrated into this area, the local Siberian roe deer possessed virtually none of the typical features of this species and were morphometrically and biochemically similar to European roe. Despite this, the population's karyotype contained B-chromosomes, a feature characteristic of Siberian deer (Sokolov and Danilkin, 1981; Danilkin, 1985b; Sokolov *et al.*, 1986b). In the hybrid zone where the natural ranges of European and Siberian roe deer overlap, the model predicts the decline of mixed populations (even where environmental conditions would not necessarily depress a monospecific population), with the European roe representation declining at a faster rate. Where survival rates are higher, it may be predicted that the boundary between the ranges of the two species will move westward and the Siberian form will supplant the European form. This process will probably be extremely slow due to the regular integration of migrants into the hybrid zone population, and the fact that high rates of survival are unlikely in reality. Therefore, the hybrid zone is perhaps more likely to be relatively static, a phenomenon observed in other animal species (Mayr, 1974).

3.2 NUMBERS AND DENSITY

Changes in roe deer population number and density have broadly followed the pattern of decline and expansion in geographical range which was described above. Most probably, in post-glacial times roe were most numerous in the second half of the second millennium when, according to eye-witness reports, European forests 'swarmed with roe deer' and they were taken 'in incredible quantities'. In the Urals, Siberia and the Far East, roe were also abundant, with migrating herds numbering hundreds (Przewalski, 1870; Sabaneev, 1872, 1875; Turkin and Satunin, 1902; Kirikov, 1959, 1966, 1983; Averin, 1960a,b). Even in the second half of the nineteenth century, their number in the Amur-Ussuri territory probably ran into millions (Bromlei and Kucherenko, 1983).

Habitat destruction and alteration resulted in a decline of both numbers and distribution in western Europe from the eighteenth century, in eastern Europe from the nineteenth century, in Siberia and Kazakhstan from the early twentieth century, but not until the middle of this century in the Far East. Stocks decreased with the intensification of human economic activity in these regions, but the decline was further aggravated by the advent of high-powered weapons and consequent over-hunting. Subsequent recovery in numbers occurred as early as the nineteenth century with the introduction of some protective measures and the destruction of large predators over much of the roe deer's range. During the war years, European population numbers underwent a second sudden decrease but soon recovered to former levels, growing from 10 000 animals in Hungary in 1946 to a current total of 220 000, from 24 000 head in 1925 to 120 000 in Bulgaria, and from 16 000 head in 1950 to 290 000 in Romania. The number of roe deer in other European countries also increased by about an order of magnitude over the same period (Table 3.1).

In eastern Europe, the Urals and Siberia, similar increases in the number of roe deer began in the 1930s. However, although the 1970s in particular saw a substantial expansion in geographical range up to its previous limits, actual numbers of roe had ceased to increase in most parts of the former USSR by the 1960s, and Siberian roe deer numbers actually decreased by about 100 000 head during this decade (Morozov, 1975; Metelski, 1976; Danilkin, 1982, 1985a; Bromlei and Kucherenko, 1983). By the beginning of the 1980s, this catastrophic decline was arrested through the prohibition of hunting, but not until 1985 was there any noticeable recovery of roe stocks.

Considering the situation during the 1980s, the total range of *Capreolus* can be estimated as measuring about 13 million km^2, with the European and the Siberian roe occupying 5.5 million km^2 and 7.4 million km^2 of this total, respectively. The total population numbered approximately 7.6 million animals, the vast majority of which (almost 7 million) were European roe, with around 700 000 Siberian roe (Table 3.1). A large proportion (approximately 85%) of the total roe deer population occurred in

Table 3.1 Approximate overall densities of roe deer in various countries during the mid 1980s[1]

Country	Population[2] size (1000)	Range area[3] (1000km²)	Forest area[3] (1000km²)	Deer/1000 ha Total range	Deer/1000 ha Forest
		European roe deer			
USSR	380	1916	478	2.0	8.0
Poland	500	313	84	16.0	59.2
Romania	290	237	64	12.2	45.2
Czechoslovakia	290	128	45	22.7	64.7
Yugoslavia	350	256	87	13.7	40.2
Hungary	220	93	14	23.7	157.1
Bulgaria	120	111	33	10.8	36.0
West Germany	1700	248	74	68.5	228.5
East Germany	310	108	30	28.7	102.6
Austria	1000	84	37	119.0	270.3
France	300	551	32	5.4	22.7
Denmark	140	43	5	32.6	269.2
Switzerland	120	41	10	29.3	116.5
Netherlands	25	41	3	6.1	75.8
Belgium	25	30	6	8.2	41.7
Luxembourg	13	3	–	50.0	–
Great Britain	(400)	173	20	(23.1)	(200.0)
Italy	100	58	14	17.3	71.4
Spain	20	89	40	2.3	5.0
Norway	150	302	72	5.0	20.8
Finland	5	53	32	1.0	1.5
Sweden	400	450	256	8.9	15.6
Albania	(10)	(14)	(7)	(7.1)	(14.3)
Greece	(5)	(7)	(3)	(7.1)	(15.7)
Turkey	(20)	(150)	(100)	(1.3)	(2.0)
Syria	(5)	(14)	–	(3.6)	–
Iran	(5)	(14)	–	(3.6)	–
Iraq	(5)	(14)	–	(3.6)	–
TOTAL	6918	5541	1560	12.5	44.3
		Siberian roe deer			
USSR	315	4460	2012	0.7	1.6
Mongolia	100	484	44	2.1	22.7
China	(250)	(2187)	(175)	(1.1)	(14.3)
Korea	(10)	(56)	(39)	(1.8)	(2.6)
TOTAL	675	7387	2270	0.9	3.0

1. Parentheses indicate very approximate estimations; hyphens (–) indicate no data are available.
2. Data taken from the literature.
3. Area estimates calculated from the *Geographic Encyclopedia Dictionary* (1986), the *Great Soviet Encyclopedia* (1970–1978) and other literature.

western Europe, particularly concentrated within Germany and Austria. While the former USSR included as much as 50% (6.5 million km^2) of the total geographical range, only 10% (700 000 animals) of the population occurred there. Similarly, Russia accounted for 37% (4.8 million km^2) of the entire roe deer range but only 4.7% (350 000 animals) of the estimated total stock (Tables 3.1 and 3.2).

In the mid 1980s, average population densities of European and Siberian roe were, respectively, about 12 deer and 1 deer per 1000 ha of the total range area. The highest population densities occurred in Austria (119 deer/1000 ha) and West Germany (69 deer/1000 ha), the lowest in the former USSR (2 deer/1000 ha) and some Asian countries (Table 3.1). The

Table 3.2 Approximate overall densities of roe deer in the former USSR at the end of the 1980s

Country	Population[1] size (1000)	Range area[2] (1000km^2)	Forest area[2] (1000km^2)	Deer/1000 ha Total range	Deer/1000 ha Forest
			European roe deer		
Russia	63.6	816	228	0.8	2.8
Ukraine	144.2	600	86	2.4	16.8
Byelorussia	22.8	207	72	1.1	3.2
Moldavia	9.8	34	3	2.9	32.7
Lithuania	40.0	65	16	6.1	25.0
Latvia	44.3	64	22	7.0	20.0
Estonia	38.4	45	16	8.5	24.0
Georgia	11.8	44	22	2.7	5.4
Armenia	4.1	13	3	3.2	13.7
Azerbaijan	3.1	28	10	1.1	3.1
TOTAL	382.1	1916	478	2.0	8.0
			Siberian roe deer		
Russia	291.2	3993	1950	0.7	1.5
Kazakhstan	14.8	576	47	0.3	3.1
Uzbekistan	0.5	8	8	0.6	0.5
Kirghizstan	7.3	83	7	0.9	10.4
TOTAL	313.8	4660	2012	0.7	1.6

1. Data taken from the state Game Management Departments and Committees for Nature Conservation.

2. Data taken from the *Forest Encyclopedia*, vols 1 and 2 (1985–1986) and the *Atlas of the USSR* (1984).

average roe deer population density in western Europe was about 20 animals/1000 ha, while in eastern Europe (the former USSR) it was only 2 animals/1000 ha. In the territory of the former USSR, the highest population density of European roe deer (6–9 animals/1000 ha) was recorded in the Baltic republics and the lowest (0.8 animals /1000 ha) in Russia. The average density of Siberian roe deer was much lower – generally less than 1 animal/1000 ha of total range area.

These density figures are calculated on the basis of deer numbers in relation to the total geographical range and therefore probably underestimate real densities, as distribution is patchy with deer occupying only those habitats which are suitable. Unfortunately, figures based on the actual extent of the area occupied by roe deer over large areas are impossible to calculate since such specific data on habitat structure are not available for most regions. Where overall density is low, a figure based on the area of available forest only will be closest to the true value, but where density is high, the figure is more likely to lie between that calculated on the basis of available forest area and that which considers the whole surface of the geographical range, as in this case deer may occupy open agricultural areas as well.

Considering such 'forest densities', on average each 1000 ha of woodland within the geographical range of the particular species was inhabited by about 44 European roe and 3 Siberian roe. The highest densities occurred in western and central Europe (average of 72 deer/1000 ha), with more than 100 roe/1000 ha in Hungary and Switzerland and more than 200 animals/1000 ha in West Germany, Austria, Denmark and Great Britain. The lowest densities occurred in the former USSR, Finland, Spain and countries of Asia Minor, with fewer than 10 deer/1000 ha. Roe density in the European territory of the former USSR was higher than in the Asian territory, with the highest densities in the Baltic Republics and Moldavia (20–33 deer/1000 ha) and the lowest in Uzbekistan (0.6 deer/1000 ha) and Russia (2.8 European roe and 1.5 Siberian roe/1000 ha).

Density is an important indicator in the development of any wildlife management programme of the overall state of the population. Clearly the forests of Austria and West Germany were overstocked, which may result in problems of damage and eventual habitat degradation. In contrast, roe deer density in eastern Europe, Siberia and the Far East was probably lower than the potential carrying capacity by about an order of magnitude at this time. However, since 1985 the number of Siberian roe deer in Russia has steadily increased from 270 000 to 715 000 by 1993, while the European roe population has grown from 51 000 to 83 000.

In summary, the distribution of roe deer has been determined by both natural and anthropogenic factors. Human interference has become more important only in recent decades when, by either direct or indirect means, numerous populations have been destroyed, resulting in a dramatic reduction and certain discontinuities in the geographical range of roe deer. When such pressures are relieved, roe deer return rapidly to inhabit their former ranges. Fluctuations in deer numbers themselves have a pronounced effect on roe distribution, which expands or contracts with an increase or decrease in the total number of animals. Geographical limits of the distribution are not even stable during a single year, varying from season to season, particularly in Asia. Generally, the northern limit of roe deer distribution is defined by snow depth and the duration of snow lie. The system of mountain ridges across Asia has undoubtedly had an effect on the evolution of modern roe deer forms on either side of this barrier to their movement. The specific ranges of the European and the Siberian roe deer overlap on the right-hand bank of the Volga.

4

Habitat use

4.1 GENERAL PATTERNS

The European roe deer tends to be characteristic of forested areas; it is found in almost all woodland types (coniferous or broadleaved) but prefers a mosaic of mixed woodland in open landscapes (Southern, 1964; Prior, 1968; Stangl and Margl, 1977; Mayer, 1977; Pedroli *et al.*, 1981; Reimoser, 1982). The most typical environment comprises forest of mixed age and composition with a rich shrub understorey, surrounded by meadows and fields (Mottl, 1954; Raesfeld, 1956; Necas, 1963; Pielowski, 1970; Szederjei and Szederjei, 1971; Strandgaard, 1978; Stubbe and Passarge, 1979; Casanova, 1981).

In the north-west of their range, roe deer prefer small mixed forests but also occur in thickets, meadows and fields (Cederlund, 1982), although they avoid open habitats, such as clear-fell areas and bogs, which provide no cover (Cederlund, 1983). Large forests may also be avoided, with deer more commonly found in transitional areas between habitat types, such as at forest edges. However, substantial populations do inhabit the extensive coniferous plantations of Britain, where young stands (5–15 years) or mature woodland support the highest local densities (Henry, 1981; Staines and Welch, 1984), but areas of small extent and the interface between habitat types, such as forest rides, are generally favoured (Welch *et al.*, 1990). In the northern parts of the range in the former USSR, roe show a preference during winter for mixed plantations, glades, open pine forests and clearings covered by young growth. The importance of forest habitats diminishes during spring, and by summer most animals inhabit relatively open areas of shrubby grassland, but as the grass dies back in autumn the deer return to the woodlands (Vereshchagin and Rusakov, 1979). In wooded areas, they show a strong preference for relatively open patches (Baleishis and Prusaite, 1980), favouring spruce and deciduous forest in particular but avoiding lichen-covered pine forests (Padaiga, 1984). Other important habitats for the roe deer of this area include mixed pine–oak–spruce forests, young and mature pine forests, wet meadows with shrubs and hornbeam–oak plantations. Young deciduous forest, such as alder–ash or oak stands, and sphagnum bogs are used only occasionally (Sablina, 1955; Gatikh, 1975).

Further south within their range in the former USSR, roe deer are found in practically all forested zones, spreading well into steppe areas along shelter belts, and are also frequently observed in fields, virgin steppe and domestic gardens (Grechushkin *et al.*, 1975; Gurski, 1975). In mountainous regions they tend to remain in open oak forests, climbing as high as 3500 m above sea level during summer (Dinnik, 1910; Shakhmardanov and Ramazanov, 1980). Deer are also found in hornbeam–oak, beech and chestnut forest, particularly in autumn (Nasimovich, 1939; Arabuli, 1963a, 1967), as high as the lower boundary of the subalpine zone (2200 m above sea level). In winter, animals migrate locally down-slope and may show shifts in home range from northern snow-covered slopes to southern and eastern slopes.

In eastern Europe, roe deer mainly inhabit forest–steppe oak areas including dense coppice, pine forest, flood plain forest, shrubby gulleys and water meadows, but their actual distribution is variable, dependent on a number of factors such as food availability and accessibility of feeding grounds (Novikov and Timofeyeva, 1965; Vatolin, 1975; Prostakov and Obtemperanski, 1986). Aulak and Babinska-Werka (1990a) also noted the importance of cover for roe deer. In Poland, they observed preferential use of the older stands of all forest types, due to the greater amount of cover afforded by the mature forest with well developed ground vegetation, as well as a tendency to prefer mixed deciduous over mixed coniferous woodland. Comparatively recently, roe deer have begun to exploit large open agricultural expanses, largely devoid of cover, in some parts of Europe (Kaluzinski, 1974; Maublanc *et al.*, 1987). Such 'field roe' may often be associated with small pockets of woodland or scrub located in the vicinity, using crops to supplement their diet (Aulak and Babinska-Werka, 1990b), but also may inhabit genuinely open landscapes where herds of more than 100 animals have been observed in winter (Zejda, 1978; Bresinski, 1982). A variety of different crop types may be used by these deer depending on the degree of cover they afford, their nutritive value and the seasonal and annual availability (Zejda and Homolka, 1980).

The Siberian roe deer generally remains within southern forest–steppe and steppe areas where it inhabits birch–aspen woods, pine forests, shelter belts and shrub thickets on banks of rivers and shores of lakes. The highest densities are observed in tall-grass habitats, in birch and pine forests interspersed with meadows, and in clearings and burns (Figure 4.1). They are absent from the dark coniferous taiga but are widespread in mountainous regions, from foothills up to the subalpine zone (to about 3000 m above sea level). In particular, the preferred habitats in these regions include gentle slopes, shrub-grown valleys, medium-density young secondary forests and tall grass meadows near water. An altitudinal change in habitat use is well marked with the arrival of the first snow cover, when deer migrate to areas of shallow snow depth (Zhirnov, 1958; Savinov, 1961; Aizin, 1969; Yelski, 1975; Ziryanov, 1975; Metelski, 1976; Smirnov, 1978; Sokolov and Danilkin, 1981; Dvornikov, 1984; Sludski and Baidavletov, 1984). In Siberia, home range areas change seasonally owing to variation in food availability and relatively long-distance migrations are common in areas of deep snow accumulation (Sabaneev, 1875; Dmitriev, 1938; Antipin, 1941; Ushkov, 1947; Yanushevich and Blagoveshchenski, 1952; Fetisov, 1953; Kucherenko, 1976; Bromlei and Kucherenko, 1983; Darman, 1983, 1986). In Yakutia, the Siberian roe deer occurs mainly in the shallow-snow meadow steppe region; due to the lack of snow accumulation, summer and winter ranges are the same and migrations are not observed (Tavrovski *et al.*, 1971). Until the twentieth century, the main winter feeding places of the Siberian roe were tall grass steppes but, due to the ploughing of virgin lands, the availability of such winter pastures greatly decreased, with a consequent reduction in the number of deer. Recently, soya fields have become increasingly attractive to roe deer in the Amur and Ussuri territories and this may result in the formation of field populations, provided that the animals are better protected so that there is an increase in local density.

In Mongolia, the habitats of the roe deer (Figure 4.2) are of particular interest (Danilkin and Dulamtseren, 1981; Sokolov *et al.*, 1982a). The southern boundary of its range generally coincides with the southern boundary of the forest zone, deviating from this line in only a few places where roe deer are also found in small zones of steppe containing shrubs of various willow and dwarf birch species. Clearly, the roe of this region prefer the open-canopy larch forests of the forest–steppe which occur on gently sloping mountain sides, and deer are especially numerous in river and stream valleys, where high grasses and shrubs predominate. Indeed, in many parts of the Khangai and Khentei, density may reach between 2 and 10 roe/100 ha. In eastern Mongolia, roe can be found among the rushes and shrubs edging lakes and rivers, in small patches of dwarf shrubs which occur in steppe depressions, in the tall grass meadows of woodless valleys among ranges of low altitude hills, and in cereal fields. In the foothills of the Greater Khingan, roe deer inhabit the shrubby sand-hills in the valley of the Khalkhin-Gol river, as well as grass and shrub valleys of other rivers and streams, rush thickets and hilly steppe covered with high grasses (Figure 4.2). Summertime densities in the tall-

(a)

(b)

Figure 4.1 Typical habitats of Siberian roe deer. (a) High-grass, birch forest; (b) forest–steppe.

grass steppe areas may reach the relatively high levels of between 3 and 8 roe/100 ha. It should be noted that, in steppe habitats, the nearest water source is often many miles away and the deer must survive on moisture contained within plants as well as the dew condensing on the plant; in addition, they sometimes drink rainwater collecting in puddles.

(a) (b) (c)

Figure 4.2 Habitats of Siberian roe deer in Mongolia. (a) high-grass, open-canopy larch forest; (b) high-grass, undulating steppe; (c) shrubby valleys of rivers and streams.

(d)

Figure 4.2 *contd* Habitats of Siberian roe deer in Mongolia. (d) patches of dwarf shrubs in steppe depressions.

Towards the south-east, on the Chinese border, roe inhabit low-sward hilly steppe, punctuated by numerous gullies. In winter, migrating roe deer herds travel dozens, even hundreds of miles into the steppes of the plain, to habitats typically of *Procapza gutturosa*. Some groups may remain on the few isolated hills where, owing to the lack of shelter, these animals generally fall prey to poachers.

4.2 FACTORS DETERMINING BIOTOPIC DISTRIBUTION

The distribution of roe deer is clearly determined by a complex of biotic and abiotic factors (Chapter 3), the most important of which will be covered here. Roe deer exploit a broad spectrum of foods (Chapter 5) but prefer readily assimilable plant foods of high digestibility, which are rich in nutrients and water. Availability of such suitable forage is clearly a primary factor in determining the suitability of a habitat. The roe's food requirements are best satisfied by relatively open mosaic landscapes, which may contain small pockets of woodland surrounded by agricultural fields and high grass meadows with occasional shrubs. In woodland habitat, the most favourable conditions are provided by cleared and burnt-out areas covered with young growth, where abundant shoots can be supplemented with the new growth of grasses. Middle-aged and mature forests which lack coppiced zones or open grass patches, and large taiga-type forests do not offer the same variety and abundance of food types and deer will generally avoid such habitats. These preferences for certain habitat types result in an uneven distribution of roe deer, with high population densities in the best of the feeding grounds, despite severe competition for territories. Therefore biotope quality, population density and population productivity are closely related.

Studies have suggested that small forests of between 100 and 500 ha, which consequently have a relatively high proportion of edge relative to surface area, provide optimal conditions (Ueckermann, 1957; Sagesser, 1966a,b; Jurgenson, 1968; Herz, 1971; Bluzma, 1975; Hell, 1979; Baleishis and Prusaite, 1980; Prusaite *et al.*, 1983). Larch and broadleaved forests offer better quality habitat for roe deer than coniferous woodland (Novakova and Hanzl, 1974; Strandgaard, 1978), and mixed forests are preferred over pure coniferous stands (Aulak and Babinska-Werka, 1990a). In mixed-age stands, roe choose young plantations where the availability of young shoots is highest (Staines and Welch, 1984; Timofeyeva, 1965). When assessing a forest biotope for quality, the shrub and dwarf shrub components are of principal importance, and forests where more than 20% of the surface area consists of glades and meadows are generally the most exploited by roe deer (Ueckermann, 1952; Mottl and Janda, 1955; Vincent *et al.*, 1979).

Where deer are permanently resident in agricultural fields, crops are an extremely important food source (Zejda and Homolka, 1980; Aulak and Babinska-Werka, 1990b). During the summer, the range of food and, therefore, of habitats exploited is broad, and depends on the availability of different crop types. Almost all field crops and waste-land vegetation may be consumed (Zejda and Homolka, 1980; Zejda and Bauerova, 1985). Outside the growing season, roe feed largely on any remaining maize, sugar beet and potato, but also use winter wheat, lucerne and clover.

Shelter is as important for roe deer as food, especially in open steppe and field landscapes, and habitats lacking suitable shelter are not used, even if food is abundant. This is particularly evident in the steppe zones of Mongolia where, on the plains, roe inhabit only those areas with patches of dwarf shrubs, cereal crops or depressions containing high grass sward. In addition, roe deer may exist at relatively high density in the woodless regions of the hilly steppe, provided that some shelter is afforded by high grass sward or topographical features. Their distribution in low grass hilly steppe is restricted to areas in and around gullies, and migrants remain in the steppe zone only if adequate shelter is available. Similarly, even within forested areas, habitats that contain a substantial understorey which provides cover may be selected preferentially (Aulak and Babinska-Werka, 1990a). It has been estimated that a forest provides sufficient cover for roe deer if dense thickets account for about 10% of its surface area (Strandgaard, 1978). During the kidding period, places with concealing thickets are vital for does, who search out such sites prior to giving birth.

A further factor which may influence roe distribution is the disturbance caused by human activities (tourists, mushroom pickers, etc.), cattle grazing and the presence of other wild ungulate species. Roe deer density in disturbed areas may be lower than in similar habitats where human activity is insignificant or non-existent (Novakova and Hanzl, 1974; Strandgaard, 1978; Hell, 1979). Furthermore, roe may abandon areas regularly used by livestock or where the density of other ungulate species is high (Vladishevski, 1968; Hell, 1979; Pulliainen, 1980; Pedroli *et al.*, 1981; Danilkin and Dulamtseren, 1981). Disturbance, whether due to predation or human activities, causes an increase in the locomotory activity of roe deer, resulting in higher energy expenditure, and greatly reduces the chance of survival, particularly under extreme climatic conditions.

Snow cover is perhaps the most important of all abiotic factors affecting distribution (Chapter 3). For example, the European roe deer is non-existent or very rare in the north of the Bryansk Region of Russia where the snow depth commonly exceeds 50 cm, whereas density may be as high as 8.3 animals/10 km^2 in the south of the same region where the average snow depth is 29 cm (Vatolin, 1975). Where roe populations are exposed to deep snow overwinter, they may respond by reducing their home range or migrating to regions with shallower snow. Nevertheless, roe deer can overwinter even in places with very deep (1.0–1.5 m) but firm snow if paths and high-quality natural foods are available and they are not harassed (particularly by predators). Firm snow not only enables unhindered movement but also increases the foraging potential, as deer are capable of reaching food at up to 3 m from ground, allowing browsing of previously inaccessible tops of shrubs and low trees. These factors determining the distribution of roe deer play a more significant role in mountain habitats where conditions are more extreme (Metelski, 1975); thus, seasonal changes of home range (migrations) are often more conspicuous in such regions.

Population density may provide an objective indication of relative overall habitat quality, with higher densities generally recorded in more optimal habitats and lower densities in more marginal areas (though the lack of standard and accurate census methods greatly complicates the comparison of results) (Zejda and Danilkin, 1992). In larch forests, and in mixed forests where broadleaved trees are prevalent, density is often as high as 13–18 individuals/100 ha or occasionally even higher (Aulak and Babinska-Werka, 1990c; Vincent *et al.*, 1995). In mixed forests where coniferous trees predominate, density is generally between 10 and 15 roe/100 ha, while in purely coniferous forests a figure of between 2 and 6 individuals/100 ha is more common (Mottl, 1954; Bobek *et al.*, 1974; Bobek, 1977; Bluzma, 1975; Reichholf, 1980; Vincent, 1982; Blant *et al.*, 1982; Reimoser, 1982; Padaiga, 1984; Staines and Welch, 1984; Loudon, 1987). In woodlands generally, the average roe density is directly dependent on soil fertility and the percentage of deciduous trees, and is correlated with the ratio of woodland to surface area (Bluzma, 1975; Padaiga, 1975a; Prusaite *et al.*, 1977, 1983).

Among populations of 'field' roe, density is extremely variable and has variously been reported as 1.1 deer/100 ha in the woodless landscape of southern Slovakia (Hell, 1979), 3 deer/100 ha in Hungary (Szederjei and Szederjei, 1971), between 4 and 13 deer/100 ha in Germany (Stubbe, 1987), 7 deer/100 ha in western Poland (Kaluzinski, 1982), and 12 deer/100 ha in southern Moravia and Czechoslovakia (Zejda *et al.*, 1985). It is determined mainly by the year-round availability of food, but also the extent of shelter provided by small patches of woodland, gullies, shelter belts and areas of high grass sward. In open fields, density is minimal (no more than 1 deer/100 ha), but may be as high as 20 deer/100 ha where adequate shelter is available.

A mosaic of fields and forest is considered to constitute an optimal habitat for roe deer, but moving from the south of its range to the north, average density decreases from 18–20 down to 3–4 deer/100 ha (Sedlaczek, 1964; Baleishis and Prusaite, 1980; Cederlund, 1982), indicating the increasingly pronounced negative influence of a complex of the above environmental factors on roe dynamics with increasing latitude. This general tendency is further illustrated by the gradation of density from not more than 0.1 deer/100 ha in the sub-taiga zone of the north, to 3–6 deer/100 ha in the mixed and broadleaved forest zone further south, and up to 5–12 deer/100 ha in the broadleaved forest zone further south still. Maximum local densities (40–60 deer/100 ha) have been recorded in relatively small areas and are generally seasonal (Muller, 1973; Stangl and Margl, 1977; Mayer, 1977; Reichholf, 1980; Aulak and Babinska-Werka, 1990c).

4.3 ADAPTATION OF LOCAL POPULATIONS TO PARTICULAR BIOTOPES

In the course of time, local populations may slowly adapt to specific local conditions of the environment in which they live, leading to the evolution of 'ecotypes' (Mayr, 1974). There is much debate concerning whether roe deer were originally a forest-dwelling species, or whether they evolved initially in more open habitats, but the lack of vegetational data in the fossil record means that it is difficult to resolve this matter (Lister, 1984). Do the current differences in biotope exploited by different populations of modern roe deer indicate the existence of separate ecotypes?

Many researchers consider roe deer to be primarily a forest species; however, the fact that density is generally low in very dense natural forest, and in some wooded areas roe are entirely absent, might seem to contradict this hypothesis. Moreover, modern roe deer are found throughout the year in habitats which completely lack tree cover such as the steppes and agricultural fields. This could be taken as evidence that roe originally evolved in open steppe conditions but, in contrast, open shelterless habitats are avoided in most regions (Cederlund, 1983). In Asia, numerous populations of Siberian roe deer migrate between forest and steppe biotopes during autumn (or occasionally, from steppe to forest), and on the arrival of spring they repeat the process, but in the opposite direction. Thus, through the annual cycle these animals alternate between forest and steppe habitats and certainly cannot be considered to be specifically adapted to one or the other. It is not uncommon for forest roe deer to leave cover to feed in fields or the steppe, and in some cases does may prefer to drop their kids in a more open, shelterless landscape. These factors highlight the great ecological flexibility of the roe deer within its range and suggest that it should be characterized as a species of the forest edge which is better adapted to living in shrubland and thick, high grass sward than to living in dense forest or in open steppe. It seems likely that the roe deer evolved in a heterogeneous mosaic-type landscape (Raesfeld, 1956; Necas, 1963; Szederjei and Szederjei, 1971).

A variety of morphological characteristics of the roe deer, such as the shorter front limbs, the relatively long neck, the well developed side digits and the abbreviated soft heel of the hoof (Flerov, 1952), as well as the narrow shoulder width, indicate that they are perhaps better adapted to living in more open, relatively humid, high grass and shrub habitats than to dense forest. The jumps of disturbed animals of between 4 and 7 m, and their periodic vertical jumps, which may reach 1.5–2 m in height, also suggest that they are more of a savanna dwelling species.

Indeed, as emphasized earlier, roe deer are more commonly found in the forest–steppe and forest edge. In truly woodland habitats, they prefer areas of relatively thin tree cover with numerous open glades which support abundant browse and grass. In fact roebucks may actually bring about a thinning of the undergrowth during marking behaviour (Chapter 7). Within its home range, each buck may damage hundreds of trees, stunting their growth and creating relatively open bright glades where food availability is higher than under thick canopy forest. The partial destruction of much woodland, particularly the dark coniferous forests, has resulted in a large expansion of the roe deer's range together with a rapid increase in animal numbers. Because roe deer prefer such partially cleared or burnt-out areas which are covered with young growth, Gossow (1977) has classified the roe as an early successional species.

In recent decades, numerous 'field' populations of roe deer which inhabit open agricultural landscapes have appeared in many European countries (Kaluzinski, 1974; Maublanc *et al.*, 1987). Among the factors that stimulated this colonization were high population density in forest habitats leading to intense competition for home ranges, the expansion of agricultural systems which provide a ready food source throughout the year, an intensification of silvicultural practice and an ever growing level of human disturbance. However, it is unclear whether this can be considered as the emergence of a new 'field ecotype' of roe deer. Observations show that, while living in an open landscape practically throughout the year, the field roe deer is still to some extent associated with tree stands, field hedges or small pockets of woodland which may serve as an additional source of food but also provide cover for periods of rest and shelter during adverse weather conditions (Holisova *et al.*, 1982; Zejda and Bauerova, 1985; Maublanc *et al.*, 1987). High densities occur only where large fields have a network of shelter belts or a mosaic of small woods. In genuinely woodless agricultural landscapes roe, if present at all, exist at very low density (Hell and Herz, 1970; Pielowski, 1970; Hell, 1979; Stubbe and Passarge, 1979).

In experimental manipulations (Graczyk, 1978), adults taken from a field population and kept in a forest habitat for several months subsequently returned to the field ecosystem; similarly, adults from a forest population returned to the forest after a period of enclosure in an agricultural environment. In addition, the offspring of the 'forest' roe does, despite being born into a field population, returned to the forest; however, kids of the 'field' roe, which were born and raised in an enclosed forested area, remained within the forest. Thus, it seems unlikely that 'field' roe deer have acquired ecologically determined hereditary characteristics as a result of adaptation to a specific environment, and therefore they do not constitute a separate ecotype. This conclusion was supported by genetic comparisons of field and forest populations which found no apparent differences in the presence of particular alleles, the extent of genetic diversity within populations, and the level of inbreeding, as well as low genetic distances between roe populations of the two habitat types (Kurt *et al.*, 1993). These authors concluded that 'forest' roe deer and 'field' roe deer do not represent distinct ecotypes but rather reflect the considerable behavioural plasticity of the species. Similarly, Markowski and Markowska (1988) were not able to differentiate between field and forest roe on the basis of cranial non-metric traits, although Markowski (1993) has subsequently reported a decrease in the level of fluctuating asymmetry in morphology over time among field roe which he interprets as a gradual adaptation of these populations to their new habitat.

These observations again highlight the enormous ecological plasticity of roe deer and clearly suggest that it is not possible to classify *Capreolus* into defined 'forest', 'field' or 'steppe' ecotypes. The very classification of roe deer as animals of the forest edge and the accompanying lack of precise morpho-ecological specialization point to the fact that they can successfully colonize both open and closed landscapes. Nevertheless, the biotope does have a noticeable impact on the way of life and behaviour of the deer (Turner, 1979). For example, roe living in relatively closed habitats, either throughout the year or seasonally, can be distinguished on the basis of their closer attachment to areas that provide shelter (forest, scrub, rush thickets); when in danger, they almost invariably flee over short distances from open areas towards closed habitats. Adults are motionless comparatively rarely, and then generally only when in the standing position. In contrast, roe deer of the steppe and those found in agricultural fields tend to shelter in relatively closed habitats for short periods only. When disturbed, they are likely to cover much longer distances and may bypass forest patches. In the regions of the steppe where the grass sward exceeds their body height, the deer may lie up in pockets of tall grass or gullies, where they are undetectable while motionless. Indeed, in this situation deer will not bolt, even when approached to within 2–10 m on foot or by vehicle.

The deer of the Mongolian steppes, probably resident in this habitat type for many centuries, may have lost all their affiliations with forest biotopes but this requires confirmation by experiment. These unique steppe populations, which are now on the verge of extinction due to excessive hunting, should be protected as they could provide animals for re-introduction into other steppe regions.

Thus, the distribution of roe deer extends through the temperate and subtropical zones, and includes regions where the climate varies from humid marine to arid continental conditions. They may be found across a wide variety of habitat types, in particular, broadleaved and mixed woodland, taiga, forest–steppe, steppe, plains, highlands and have even spread into semi-desert regions and along the river banks of true desert zones. Despite this apparent diversity, some consistent patterns of their distribution can be observed, reflecting variation in a number of key ecological requirements which determine their overall range.

The information presented in this chapter illustrates the general relationship between deer population density and habitat quality; thus, density may serve as an approximate index of the average ecological capacity of a particular habitat. Average roe deer density is also reflected by the total number of animals in a region (Chapter 3). The total number of Siberian roe deer is nearly an order of magnitude lower than that of the European roe deer, and only during winter in the most preferred areas does the density of Siberian roe climb as high as 10–20 animals/100 ha, and then only in comparatively localized concentrations (Smirnov, 1978; Sokolov and Danilkin, 1981; Danilkin, 1985a; Darman, 1986).

When the habitat modifications which humans have brought about in historical times are considered in relation to the evolution of roe deer population number, it seems that, overall, humans have gradually and regularly improved conditions for the roe over the greater part of its range. The cutting of large, unbroken forests led to the formation of a mosaic habitat structure of different-aged stands interspersed with meadows and fields. In the last 200 years, the steppe shelter belts and changes in agricultural practice have provided the prerequisites for colonization of these basically open habitats, leading to the formation of field populations. Compared with other wild ungulates, the roe deer has proved to be the most flexible in its response to changes in habitat structure that are related to human activity, and it may well be the species that continues to flourish in the heavily managed landscapes of the future.

However, the widespread and heavy impact of recent human activities leading to radical changes in habitats has not always been a positive factor for roe deer. The complete deforestation and ploughing of virgin high grass sward steppe, an important feeding site for the deer, together with a variety of other factors have resulted in a reduction in their total population number in some regions (Belov *et al.*, 1980; Durov, 1980; Shvets, 1980). This has occurred over vast territories, decreasing the availability of feeding grounds with a mosaic habitat structure, and has led to the direct destruction of the best sites such as high grass sward meadows with some scrub. Fortunately, where snow depth remains relatively shallow, the loss of virgin steppe and meadows is somewhat offset by the availability of agricultural areas as an alternative habitat. However, where snow lie is deep, this destruction causes the almost complete loss of the best and, not infrequently, the only suitable habitat.

A further example of human impact on roe deer habitat concerns the populations inhabiting the vast stretches of flood plain. These areas, measuring millions of hectares, have been irretrievably flooded with the construction of a number of hydroelectric power stations. Finally, the shift in forestry practice towards coniferous monocultures which lack the diversity of habitat of deciduous or mixed woodlands has not been particularly favourable for roe deer.

5

Feeding ecology

5.1 DIET AND FOOD SELECTION

In ruminants, the amount of food consumed and the feeding activity rhythm depend largely on the rumen capacity. In roe deer adults, the total capacity of the rumen and reticulum varies between approximately 4 and 5.8 litres (Necas, 1963), and the ratio of its filled mass to total body weight is about 6–8.5% (Nagy and Regelin, 1975; Stubbe and Passarge, 1979) and to carcass weight is 13–19% (Szaniawski, 1973). This relatively small capacity of the rumen indicates that the roe deer has a high rate of food turnover through the digestive tract (Hofmann, 1978a) that depends on the rate of assimilation (Perzanowski, 1978; Drozdz, 1979) and varies according to food quality and quantity, which part of the plant is taken, and possibly also its cellulose content (Bubenik, 1959; Lochman and Barth, 1965). When fed low quality diet, both total mean retention time and rumen retention time are significantly longer; retention time is also longer when deer are provided with only a restricted ration of poor quality food (Holand, 1994). Thus, roe deer may be dependent on selective feeding to enhance survival.

The small stomach capacity and relatively fast digestion mean frequent feeding is necessary, roe deer normally having between five and 11 separate feeding periods in a single day (Bubenik, 1962; Lochman, 1965; Klotzli, 1965; Sokolov and Danilkin, 1981). During periods of optimal food availability (during the growing season), roe may feed at approximately hourly intervals (Lochman, 1965; Hofmann, 1978b). Immediately or a short time after feeding, they begin rapid rumination. In areas subject to frequent human disturbance this feeding rhythm may be disrupted, becoming very irregular, and the deer may only venture out to feed at dusk or during the night, not eating at all in daylight hours. Generally, deer may forage at any time of day

(Figure 5.1) but feeding bouts tend to be longest during morning and evening. The feeding activity rhythm is also greatly influenced by climatic factors, by the amount and availability of food and by the season. For example, on hot summer days the number of feeding bouts tends to be less than on cooler days, and the duration of a single bout tends to be longer in winter than in summer (Sokolov and Danilkin, 1981).

The roe deer may take shoots of a diameter as great as 10 mm but generally concentrates on those in the 2–4 mm range. The foraging height of the European roe deer is up to 120 cm above ground, while for the Siberian roe deer it is up to 150 cm above ground and may increase to as high as 2–3 m when there is deep firm snow (Figure 5.2). When searching for food in winter, roe use their forelegs for digging, sometimes to a depth of 30–50 cm. Siberian roe may even dig up the ground around tall plants in order to cut the stem low at its base and eat it. As noted above, they never take the entire plant this way in summer: this seems to be a winter feeding adaptation that enables the deer to exploit meagre food reserves under adverse conditions.

From our own work and other published data, we have estimated that, within its entire distributional range, the roe deer consumes about 1289 plant species (Holisova *et al.*, 1992). Of this total, most are herbs and forbs (55.2%), while trees and shrub species make up the second most important plant group (24.5%), but grasses are less favoured (15.8%) (Table 5.1). Taken together, these three categories account for more than 95% of the

Figure 5.1 Siberian roe deer grazing.

Figure 5.2 Thick but firm snow cover makes shrub tops as high as 2–3 m accessible to roe deer.

total plant species found in their diet, and therefore, other plant groups are minor food sources.

While feeding, roe deer move almost continuously around their home range. In summer, they rarely consume all the available forage in a particular locality but select particular plants or plant parts before moving on (Maizeret and Sung, 1984). Because roe deer select particular parts of the plant, the contents of the rumen are richer in nutrients than the whole plant (Sokolova and Danilkin, 1977). This style of selective browsing, where only a small proportion of the plant is eaten, means that the plant may well be able to withstand any damage and not infrequently its total biomass increases, allowing it to be browsed several times. However, this very high degree of feeding selectivity may result in a detectable change in the overall biochemical composition of the vegetation of a particular habitat. According to Bubenik (1959), intensive browsing may decrease the total nutrient availability of an area by 54% of protein, 46.5% of calcium and 23.2% of phosphorus.

In selecting particular food items, roe deer use both smell and taste (Esser, 1958; Wagner, 1961; Klotzli, 1965; Stubbe and Passarge, 1979) but they do occasionally eat plants that are generally held to be toxic to humans and domestic animals and therefore the deer must be relatively

Table 5.1 Number of species and the percentage composition of the diet of roe deer

Plant type	European roe		Siberian roe		Species eaten by both		Total	
	No.	%	No.	%	No.	%	No.	%
Trees and shrubs	226	24.7	131	21.9	42	18.7	315	24.5
Dicotyledons	497	54.3	350	58.4	135	60.0	712	55.2
Monocotyledons	146	16.0	97	16.2	40	17.8	203	15.8
Ferns	17	1.9	7	1.2	1	0.4	23	1.8
Mosses	12	1.3	–	–	–	–	12	0.9
Lichens	6	0.7	4	0.7	3	1.3	7	0.5
Fungi	11	1.2	10	1.7	4	1.8	17	1.3
TOTAL	915		599		225		1289	

insensitive to the toxic substances they contain (Prior, 1968; Liepmann, 1972). It may be that the phytotoxins are partially neutralized or biochemically degraded in the digestive tract, but it is also possible that naturally occurring tannins present within food may act as an antidote (Timofeyeva, 1965). Some researchers have noted selective feeding on various *Artemisia* species which may have an antiparasitic function (Dementiev, 1933; Fetisov, 1947, 1953).

Although highly selective, roe deer prefer to feed on a group, or set, of species, rather than concentrating on individual species, and the intensity with which any particular species set is consumed is related to the occurrence and exploitation of other species sets. Food sets are characteristic of various ecosystems and dependent on the habitat structure (Kossak, 1976; Rubis, 1982). Roe are selective in their feeding only when the available vegetation is diverse, particularly during the growing season and in natural undisturbed habitats. In winter, when food is scarce, selective feeding of this nature is hardly possible (Nudds, 1980). Examination of rumen contents has shown that, although the number of plant species taken generally varies between 3 and 42 (Siuda *et al.*, 1969; Cederlund *et al.*, 1980), during periods of snow the diet of the European roe deer in the forest–steppe oak forests of the Belgorod region of Russia includes as many as 90 plant species – but only 17 of these, mainly trees and shrubs, were major food sources (Timofeyeva, 1965). In Hungary, deer living in a forest environment generally consume between 7 and 10 species (average 8.6), while the field roe feed on between 5 and 9 species (average 6.8) (Matrai *et al.*, personal communication). Rather higher figures were reported for the deer of southern Moravia where between 3 and 16 plant species were taken in summer (average 13.6) and between 4 and 27 species in winter (average 14.9) (Holisova *et al.*, 1982, 1984).

On the basis of the frequency of occurrence of particular plant species in the roe deer diet, two categories of food sources can be defined: major and minor.

The **major** sources of food for roe deer are (as noted above) dicotyledonous herbs, trees and shrubs and, to a lesser extent, monocotyledonous species. The dominant plant parts taken include the leaves of small dicotyledonous herbs, and the leaves and young shoots of trees and shrubs, which may account for between 51 and 58% of the total diet (Bubenik, 1959), rising to about 90% during the growing season (Hell, 1979). However, the proportion of these components in the diet varies greatly between habitat types, with field roe feeding much more heavily on grasses (Holisova *et al.*, 1982, 1984, 1986a,b). While the above-ground parts of cereals and other monocotyledonous plants are generally less important components of diet, both qualitatively and quantitatively (Melichar, 1960; Schafer, 1974; Hell, 1979), cereal crops are highly selected by field roe deer. Roe deer also favour the juicy fruits of trees and shrubs, particularly apples, beech nuts, acorns and chestnuts, which they may take from the ground at fruit fall or dig out from under the snow during winter (Mohring, 1963; Necas, 1963; Holisova *et al.*, 1984, 1986a,b). The fruits and seeds of autumn provide high energy foods which can be stored in fat reserves (Necas, 1963; Stubbe and Passarge, 1979), an important source of energy during winter. In addition, the roe diet not infrequently also includes flowers, inflorescences and whole tops of plants (Klotzli, 1965; Borowski and Kossak, 1975; Kossak, 1983; Holisova *et al.*, 1984, 1986a,b). Plant roots constitute a major food source only where they are available in substantial quantities; in particular, sugar-beet roots and potato tubers which may remain following harvest.

It should be noted that roe deer may also eat the needles of coniferous trees (Gunther, 1978; Kampmann, 1983) and indeed some researchers consider them to be a major food; however, they are generally regarded as a minor food source which the deer take during winter only, when the preferred foods of other seasons are scarce (Oszlanyi, 1982). Despite this, their importance to roe has been shown by experiment, where deer may consume needles even when nutrient-rich foods (grain, acorns, leaved forage, dry aspen leaf pellets and clover hay) are available (Kuen and Bubenik, 1980; Prieditis, 1984). It may be supposed that they are required as a source of water (Juon, 1963; Jackson, 1974) or vitamins, as seems to be the case for sika deer (Obrtel *et al.*, 1985).

Minor foods, including plants which are infrequently observed in the diet of some members of a population, or are eaten in only very small quantities, may be taken only when new species are tasted, or during a process of testing particular food items. In many regions, both European and Siberian roe deer may eat a large variety of fungi but generally they account for an insignificant proportion of the total diet. Therefore, fungi are considered as a minor food source, along with mosses, lichens, ferns and *Equisetum*.

5.2 REGIONAL AND HABITAT VARIATION IN FORAGING BEHAVIOUR

The feeding habits of the roe deer across its extensive total range have specific and regional peculiarities. For example, to date we have found that the diet of European roe deer comprises a total of 915 plant species while the Siberian roe deer takes a total of only 599 species (Holisova *et al.*, 1992). However, the list of plants consumed by both roe species is far from being complete. In general terms, the feeding strategies of the European and Siberian roe deer are similar, both taking species from the major plant groups in approximately the same ratios. Nevertheless, their diets differ greatly in specific composition, with only 17.5% of plant species, mainly dicotyledons (Table 5.1), being taken by both European and Siberian roe. This lack of dietary overlap is more indicative of differences in the specific composition of European and Asian floras than of specific feeding choices of the deer.

The total number of plant species consumed varies regionally between 64 and 320. In a given local area, the diet of roe deer comprises between 80 and 130 plant species, but on a larger geographical scale 130–250 plant species may be taken. (European roe deer: Arabuli, 1966; Dahl, 1930; Gusev, 1984; Holisova *et al.*, 1984; Kaluzinski, 1982; Karpenko, 1966; Kossak, 1983; Knorr and Briedermann, 1976; Lavrov, 1929; Manolache and Babulia, 1969; Mottl, 1957; Petrov *et al.*, 1968; Prusaite *et al.*, 1983; Romanov and Romashin, 1982; Sablina, 1955; Timofeyeva, 1965. Siberian roe deer: Averin, 1949; Bromlei and Kucherenko, 1983; Darman, 1986; Fetisov, 1947, 1953; Pasternak, 1955; Smirnov, 1978; Sokolov and Danilkin, 1981.)

In each particular region roe deer favour certain plant species. For example, in the southern Urals, one population showed a high preference for willow-herb during the summer: it was present in 80% of the individuals examined and accounted for between 40 and 90% of the total plant mass volume of the rumen. Indeed, the rumens of two kids shot in August, besides some milk, contained leaves of this plant only. Other preferred plants were meadow clover, *Libanotis sibirica* and *Calamagrostis epigeios*.

However, in another population, deer preferred young aspen shoots in all seasons (Sokolov and Danilkin, 1981).

Towards the north of the southern Urals, in the Ilmen Reserve, the major food sources of the Siberian roe deer are pine, aspen and birch in winter; willow, buttercup, willow-herb and aspen in spring; clover, willow-herb, some umbellifer species, bitter trefoil (*Menyanthes trifoliata*) and paintbrush in summer; but willow-herb, *Polygonum*, umbellifers and compositae in autumn (Averin, 1949). In eastern Siberia, the roe deer prefer aspen, willow and bitter trefoil but also eat wormwood (Fetisov, 1953). Further west in Siberia they feed primarily on vetches (*Astragalus* and *Vicia* spp.), peas (*Lathyrus* spp.), clover and alfalfa (Smirnov, 1978). In the Far East their common foods include aspen leaves and shoots, willow, lespedeza, linden, ashberry, birch, ash, oak, jasmine, grape, hazel, raspberry, windflower and soya (Bromlei and Kucherenko, 1983; Darman, 1986). In western regions of Byelorussia, the European roe deer's main winter foods (84.6%) are buckthorn, hornbeam and oak; further east, oak and willow account for most of their winter diet (70.7%) (Rubis, 1982). Despite these regional differences in food selection, the species set which constitutes the roe deer's diet generally remains constant within relatively local geographical regions, and marked contrasts are evident at the regional level only where the composition of the plant species set changes abruptly.

Roe deer diet differs not only by region but also by habitat type (Necas, 1963; Klotzli, 1965; Bubenik, 1970; Jackson, 1974; Vladishevski and Yelski, 1974; Strandgaard, 1977). For instance, in Lithuania, the main winter foods of deer inhabiting deciduous forests with some spruce are ash and oak, while in mixed spruce–larch forests and pine forests with some deciduous trees, ashberry, buckthorn, aspen and oak are more important (Padaiga, 1965). From the analysis of the contents of 262 rumens of Latvian roe deer, the percentage of dwarf shrubs (mainly *Calluna vulgaris* and *Vaccinum* spp.) in the diet of roe was shown to vary between 2 and 57% in pine forests, 0.3 and 9% in spruce forests, and 0.01 to 0.4% in deciduous forests. Similarly, the percentage of deciduous species (trees and shrubs) varied between 9 and 21% in pine forests, 3 and 15% in spruce forests, and 1 and 8% in deciduous forests. These differences result from variation between habitat types in the frequency of occurrence of the main food sources (Janson and Padaiga, 1975).

Not surprisingly, this variation in diet selection according to the habitat type is also evident for field and forest roe deer, with forest roe consuming greater quantities of browse material but field roe taking a majority of grass species (Stubbe and Passarge, 1979; Hell, 1979; Holisova *et al.*, 1986a,b). Rumen samples of European roe deer taken from a forest surrounded by fields largely contained herb species (57.6%, including

16.2% cultivated species) (Baleishis and Prusaite, 1980); during autumn and winter, beans and soya beans may account for 70–75% of the Siberian roe deer diet in a mixed forest–agricultural habitat (Darman, 1986). Diet differences have been observed between two populations of a single geographical region (Lithuania) according to the proportion of wooded habitat available (7 and 24.8% respectively) (Prusaite *et al.*, 1983). In the first population, 53.2% of the total rumen fill consisted of cultivated plants, 30.2% of herbaceous plants and only 16.5% of woody browse. In contrast, in the latter population, cultivated plants were much less prevalent (10.2%), while herbaceous plants (50.1%) and woody browse predominated. These data show that roe are flexible in their diet selection and are able to adapt to the local ecological conditions.

5.3 SEASONAL FEEDING HABITS

The proportion of certain plant species in the roe deer's diet varies widely between seasons, primarily due to periodicity in the vegetation cycle (Jackson, 1974, 1980; Maizeret and Sung, 1984; Putman, 1986) but also due to changes in nutritional requirements according to the physiological state of the deer (Holisova *et al.*, 1984, 1986a,b). In general, snow cover during winter drastically reduces the variety of available foods and also contributes to the gradual depletion in food abundance; thus, the deer are forced to increase their niche size in order to satisfy metabolic demands (Petrak, 1993). However, this also coincides with a noticeable reduction in the deer's general metabolic rate (Weiner, 1977) and food requirements (Ellenberg, 1978; Kholodova, 1986). Once spring begins, the roe deer's metabolic rate increases, the regrowth of vegetation provides an abundant source of food, and digestion rate increases due to the prevalence of fresh vegetation in the diet. By autumn, the growing season has finished but seeds and fruits become available and are heavily exploited by the deer.

Such seasonal variations in roe deer feeding habits have been observed in nearly all diet studies and most have noted that the winter diet is generally less diverse. For example, in the southern Urals, Siberian roe take between 10 and 26 plant species (average 16) during the growing season but only between 5 and 11 species (average 8) during winter (Sokolov and Danilkin, 1981). In contrast, in the west of the Trans-Baikal region, Siberian roe deer consume almost all the 140 plant species that make up their diet in this region throughout the year, with 93.5% of species eaten during winter, 86.3% in spring, 84.9% in summer and 87.7% in autumn (Smirnov, 1978).

Particularly variable between seasons is the ratio of trees and shrubs to herbs in the roe deer diet. In the Polish section of the Bialowieza Primeval Forest, this ratio was 20 : 80 during spring and summer but 35 :

65 during autumn and winter (Gebczynska, 1980). A more pronounced degree of variation was observed in the Amursk region, where the equivalent figures were 1 : 99 in spring, 0 : 100 in summer, 65 : 35 in autumn and 75 : 25 in winter (Darman, 1986). A similarly dramatic seasonal variation of this ratio was observed in the west of the Trans-Baikal region, where very few trees and shrubs were eaten during spring and summer (ratio 0 : 100 for both) and autumn (ratio 7 : 93) but they formed the vast majority of the winter diet (ratio 80 : 20) (Smirnov, 1978).

Not only do the components which make up the diet of roe deer vary in importance seasonally but also differences in diet composition are observed from year to year for a given season, due to variation in availability (Jackson, 1974; Kuen and Bubenik, 1980; Holisova *et al.*, 1986a,b). In this context, the roe deer should be viewed as an extremely opportunistic species which makes the most of seasonal foods such as mushrooms, berries, fruits and flowers, which are transitory, or of dietary value only for a comparatively short time. However, when available, they are harvested in large quantities and during such periods the roe deer's rumen may be entirely filled with particular mushroom species or fruits, or the berries and leaves of cowberry and bilberry in autumn, early flowering plants in spring, and the leaves of certain preferred plants in summer (Fetisov, 1953; Smirnov, 1978; Sokolov and Danilkin, 1981).

5.3.1 Winter feeding

During winter, roe deer in coniferous forests subsist largely on spruce shoots, fir, pine needles, bilberry (*Vaccinium myrtilis*), blackberry (*Rubus*), cowberry (*Vaccinium vitis-idaea*) and heather (*Calluna*), while in deciduous forests the main sources of food are young shoots, buds and grass (Sablina, 1955, 1959; Novikov and Timofeyeva, 1964; Esser, 1958; Siuda *et al.*, 1969; Henry, 1978a; Yelski, 1979; Helle, 1980). Direct observations of adult roe suggest that they browse between 1700 and 2100 terminal shoots of trees and shrubs per day, each weighing about 1 g (Timofeyeva, 1965). Where beech mast and acorns are locally abundant and persist into winter, deer may consume large quantities of this concentrated high-calorie food (Komarek, 1954; Romanov and Romashin, 1982). In the northern part of their range, roe deer may consume considerable amounts of arboreal lichen (Helle, 1980).

In field habitats, the roe deer's main winter foods are winter agricultural crops and post-harvest remains (Holisova *et al.*, 1986a,b), particularly both the cob and the leaves of maize, alfalfa, soya, wheat and beet, plus any available *Rubus* spp., *Ligustrum vulgare* and other wild species (Holisova *et al.*, 1982, 1986a,b; Matrai *et al.*, personal communication; Darman, 1986). In this habitat, bucks and does have similar diet composition (Holisova *et al.*, 1984) and feeding bout duration (Turner, 1979). In contrast, dietary differences between the sexes have been noted for forest roe deer within a single habitat type; for example, in certain cases, far more does (22.2%) than bucks (0.12%) consume spruce and pine needles (Grigorov, 1976).

The snow depth is an important factor in determining which plant species are exploited in forest habitats during winter, and to what extent, as accessibility to taller plants may increase, whereas low-growing species may be inaccessible (Helle, 1980). When snow is deep, the restriction of food resources may force roe deer to subsist on conifer needles. They may overwinter satisfactorily as long as needles account for no more than 50% of their diet, but when the level of their consumption rises above this level the deer may lose 7–8% of their body weight per month (Prieditis, 1984). In critical situations, roe deer may seek additional sustenance from tree bark (Kiselev, 1979).

5.3.2 Spring feeding

During spring, the main food of forest roe deer consists of young leaves and the buds and shoots of woody browse of species such as *Quercus robur*, *Betula* spp. and *Carpinus betulus* (Gebczynska, 1980), as well as herbaceous plant species (Fetisov, 1953; Siuda *et al.*, 1969). In particular, they prefer early flowering shoots, but comparatively frequently also consume cereals and other monocotyledons (Klotzli, 1965; Timofeyeva, 1965). In the same period, field roe deer prefer the leaves of winter cereals, alfalfa, clover and some weeds, but whenever possible they will supplement their diet by browsing in pockets of woodland such as windbreak belts and small copses (Holisova *et al.*, 1986a,b; Putman, 1986).

5.3.3 Summer feeding

The diet changes considerably by summer, when forest roe deer show a strong preference for dicotyledonous herbaceous plants (Siuda *et al.*, 1969; Smirnov, 1978; Gebzcynska, 1980; Sokolov and Danilkin, 1981), but are less willing to feed on the leaves of trees and shrubs and will rarely take monocotyledonous species (Klotzli, 1965; Kuen and Bubenik, 1980; Kioroglanidis, 1981). At this time, field roe deer prefer the young leaves of maize crops, the leaves and flowers of alfalfa and the leaves of sugar-beet, as well as wheat, although not yet ripe. In late summer, they continue to take alfalfa and wheat but may also consume unripe maize cobs (Holisova *et al.*, 1984).

5.3.4 Autumn feeding

With the arrival of autumn, seeds and fruits become available and, along with the remaining herbs and forbs, are the preferred foods for roe deer during this season. Selection of such high energy forage allows the deer to build up fat reserves (Hewison *et al.*, in press a) which may help them to survive the potentially critical period of the second half of winter and early spring (Hofmann, 1978a). The forest roe deer also particularly favour young shoots (Sablina, 1959; Melichar, 1961), while the field roe exploit unharvested crops of sugar-beet, potato and maize, as well as wild cereals (Holisova *et al.*, 1982, 1986a,b; Stubbe and Passarge, 1979).

5.4 RATES OF FOOD CONSUMPTION

The total annual vegetation consumption of an adult roe deer has been estimated as averaging 750 kg (Weiner, 1975). However, the total amount of food consumed by an individual depends on the sex, age and physiological state of the animal and the season, as well as food availability and the nutritional value of the consumed diet (Necas, 1963; Ellenberg, 1974a, 1978). For the following comparisons, it should be noted that the weight of the rumen contents is strongly dependent on the stage of the feeding cycle at which the deer was shot, so that data reported by different researches may not be comparable.

In the European roe deer, the weight of the rumen contents averages 1600–1900 g (Mottl, 1957; Siuda *et al.*, 1969; Knorr and Briedermann, 1976; Baleishis and Prusaite, 1980; Holisova *et al.*, 1982). In general, the weight of the rumen contents is much higher during winter (1230 g–1530 g) than during spring (700 g–680 g) (Baleishis and Prusaite, 1980; Prusaite *et al.*, 1983). The equivalent figures for the Siberian roe deer of the southern Urals are considerably higher, with an average contents weight of 2500 g, but show the same seasonal trend (averages of 3200 g during winter and 2200 g during spring and summer) (Sokolov and Danilkin, 1981). Similar values have been recorded for deer from the Trans-Baikal region (Smirnov, 1978). Differences have also been observed according to age, with the average rumen contents mass of roe deer kids in January and February much lower (700 g) than that of adults (1137 g).

The average daily intake of the European roe deer varies from 1.5 kg to 4 kg of green vegetation (Mottl, 1957; Necas, 1963; Briedermann, 1974). Under captive conditions, the daily consumption of tree shoots, shrubs and dwarf shrubs was as high as 1.1 kg among kids, 1.7 kg for deer aged between one and three years and 2.3 kg among mature adults (Padaiga, 1975b; also see Prieditis, 1975). Siberian roe have a somewhat greater daily intake of between 2.5 and 3.5 kg of vegetation matter. The European field roe deer may consume as much as 2.15 kg of winter cereals (which corresponds to 390 g dry matter) or 3.5 kg of alfalfa (615 g dry matter) (Pielowski, 1970) per day. However, total consumption varies according to the individual's body weight, with a maximum daily intake of 650 g (as dry matter) and a minimum of 350–400 g. These figures are obtained on captive deer and under natural conditions the values are about 40% higher, indicating the actual average daily intake may be approximately 753 g of dry matter or 84.5 g/kg body weight$^{0.75}$. Of this total, the requirement for assimilable protein varies between 70 and 150 g per day, depending on the season as well as the age and physiological state of the animal.

Similarly, there are large seasonal fluctuations in intake, which is maximal in spring (80 g/kg body weight$^{0.75}$), variable in winter (40–80 g/kg body weight$^{0.75}$), but somewhat lower during summer (60–70 g/kg body weight$^{0.75}$). The quantity of food consumed is directly proportional to its level of assimilation and protein content, and inversely proportional to its cellulose content. The largest quantity of vegetation is consumed during spring (up to 3 kg per day, which corresponds to 250 kcal/kg body weight), when assimilation is high and averages 73.5%. In the latter half of winter, this level drops abruptly and hence the quantity of food consumed also decreases (Drozdz, 1979). Food consumption is especially low when snow cover exists. It must be noted that these data, from experimental studies of captive roe deer, may not reflect exactly the natural situation, as the type of food offered generally differs from natural browse, in both quality and diversity (Drozdz and Osiecki, 1973; Perzanowski, 1978; Drozdz, 1979).

Roe deer kids begin to suckle their mother from birth and continue until September or October, but lactation rarely lasts longer (Borg, 1970). In captivity, European roe deer milk-feed their kids for between 162 and 194 days, each kid receiving an estimated 128–184 litres during this period (Kossak, 1981a). During each suckling bout a kid may take between 30 and 80 ml (Espmark, 1969), or as much as 1200 ml per day (Bubenik, 1965). Roe deer milk is relatively rich in fat and nutrients when compared with that of domestic animals (Stubbe and Passarge, 1979). Kids begin to ingest vegetation 5–10 days after birth, choosing the softest parts of the plant (Bubenik, 1965; Kossak, 1981b; Sokolov *et al.*, 1985c). The number of plant species taken, as well as the quantity consumed, increases gradually with age, with 13 species taken at 11–15 days, 43 species at 40 days and 66 species at 90 days (Kossak, 1981b). At the age of approximately 70 days, ligneous parts of trees and shrubs are eaten for the first time but at a much lower rate than for adult animals (Kossak, 1981b). By 3 months of age, the diet of both European and Siberian roe kids is basically similar to that of the adult deer. Prieditis (1985) reported that, by winter, captive kids consumed an average of 1 kg vegetation (0.4–0.6 kg dry weight) per day, whereas the equivalent figures for adults were 1.3–1.6 kg (0.6–0.9 kg dry weight).

The Siberian roe deer has a daily energy expenditure that greatly exceeds that of the European roe, but this varies dramatically between seasons for both species, with relatively low expenditure during winter (2500–2600 kcal and 1300 kcal respectively) and much higher values during summer (3700–4000 kcal and 2800 kcal respectively) (Ueckermann, 1964, 1985; Drozdz and Osiecki, 1973; Briedermann, 1974; Papageorgiou *et al.*, 1980; Prieditis, 1984; Strothman and Oslage, 1986). However, when these data are expressed per unit body weight, no significant differences between Siberian and European deer in either food consumption or energy requirements are observed (Gromov, 1988a).

The energy requirement of roe deer, although consistently lower in females than in males, is closely related to physiological functions in both sexes and demonstrates pronounced seasonality (Hewison *et al.*, in press a). For example, among females, Ellenberg (1978) has established that the energy requirement during summer is about 50% higher than in winter, and the costs of gestation and lactation may lead to utilization of fat reserves and loss of body weight during this period (Hewison *et al.*, in press a). This increase in energy consumption among pregnant does does not exceed that of barren animals until 4 weeks prior to giving birth, coinciding with the beginning of the growing season, and indeed, the costs of gestation occur mainly in the final stages of pregnancy (Hewison *et al.*, in press a). Average energy requirements of lactating does are approximately 150% higher than for non-lactating does, and during lactation does may have a greater food intake than bucks over the same period (Lochman and Barth, 1965), rising as high as 1300 g of dry matter, containing approximately 120 g digestible protein and 200 g starch (Esser, 1958). Furthermore, energy expenditure increases according to the number of kids at heel; in some cases, food consumption among lactating does may even rise as high as 7.3–7.5 kg of vegetation (2.0–2.3 kg dry weight).

Energy expenditure is particularly high for male roe during the rut, and bucks may experience significant body weight losses prior to, during and even for a short time after rutting, probably due to the metabolism of muscle tissue (Hewison *et al.*, in press a). The level of food consumption declines during this period to the point that intake is less than that of does (Lochman *et al.*, 1961; Sokolov and Danilkin, 1981) and the digestion rate among males also becomes much slower than that of females (Lochman *et al.*, 1974). The minimum daily food consumption for a roebuck weighing about 25 kg is 600 g dry weight vegetation, which contains 35 g digestible protein and about 300 g starch (Becker-Dillingen, 1945).

5.5 MINERAL NUTRITION AND WATER CONSUMPTION

To ensure normal development and metabolic functioning, roe deer (like other ungulates) require minerals and water which are generally obtained directly from vegetation. During pregnancy and lactation for does, and also antler growth for bucks, the mineral demand increases by a factor of 1.5 to 2 (Becker-Dillingen, 1945; Bubenik, 1959; Esser, 1958; Drescher-Kaden, 1974; Briedermann, 1974). Stubbe and Passarge (1979) provide a detailed review of the literature and a description of mineral nutrition in European roe deer.

Where the natural mineral supply is inadequate, supplementary mineral provision may improve conditions for roe deer. For example, in the Zeya district of the Amursk region of Russia, the geochemical profile of the region reveals low levels of phosphorus, copper, cobalt, iodine and, to a lesser extent, calcium. Consequently, productivity of the local roe deer population was low and bucks had unusually weak antlers, even among adults. Through the provision of artificial salt licks fortified with trace elements, the productivity and quality of the population abruptly increased, and buck antlers developed normally to a symmetrical form (Morozov, 1982). In this kind of situation, it is especially clear that the deer are attracted by natural and artificial salt licks, as noted by many researchers (Turkin and Satunin, 1902; Dahl, 1930; Nasimovich, 1938; Averin, 1949; Fetisov, 1953; Arabuli, 1966; Smirnov, 1978). Besides access to minerals and trace elements, deer may visit natural salt licks in order to consume saline soil which is highly palatable, and may serve as a physical aid to digestion and have a positive effect on intestinal microorganisms. Natural salt licks may measure many square metres (Figure 5.3).

Figure 5.3 Natural salt lick visited by roe and red deer in Mongolia.

The chemical analysis of natural licks shows that they are rich in salts of sodium, sulphur, calcium, magnesium and also in other elements and compounds, but their chemical composition differs from region to region. From a geochemical analysis of salt licks of the Sihote-Alin, Yelpatievski and Panichev (1980) suggested that deer will visit natural licks largely to obtain sodium but, to a lesser degree, also for calcium, iron, manganese and probably fluorine. This large sodium requirement means that ungulates will often visit artificial licks of cooking salt. It may be that salt consumption is a response to metabolic disorders which periodically develop in ungulates under the influence of physiological rhythms and environmental factors: salts, in addition to mineral nutrition, act to alleviate these problems. By eating mineralized soils, the deer may also be able to regulate the composition and concentration of ions in the digestive cavity, particularly in spring when they consume large quantities of green vegetation containing more potassium than sodium (Panichev, 1987). Roe deer may also obtain required minerals from alternative sources such as by drinking the water of springs rich in mineral salts or by eating the mud of river banks (Fetisov, 1953).

Free water is undoubtedly essential for roe deer, as for all other animals, however, compared with the water needs of other deer species, this demand is less pronounced in roe. The daily water requirement of roe on a natural diet may be as high as about 1350 ml/10 kg body weight (Bubenik, 1959), although Briedermann (1974) indicated that the roe deer's total daily intake should include 1500 ml of free water. In contrast, when fed dry hay this demand for water increases to 4000 ml (Becker-Dillingen, 1945). During summer, roe deer can obtain sufficient water from fresh vegetation food which is moisture-rich, particularly in the morning and evening and following rainfall. Thus, roe can survive in areas where the nearest natural water source is many kilometres distant, such as in the Mongolian steppes. Lactating does require a higher water intake and they may travel long distances on a daily basis to find free water. Even in winter, the majority of the roe deer's water needs are met by the ingestion of natural vegetation, but this may be supplemented by small quantities of snow, as long as it is not too cold. From a management point of view, the requirement for a considerable amount of water should be taken into account when roe are provided with supplementary dry feeds, as lack of free water may cause functional disorders of the stomach (Borg, 1970).

In summary, the variety of foods taken by roe deer is very diverse, differing between regions across their range and between different habitats. Diet may also vary by season or even during a single day and depends on the sex, age and physiological state of a particular individual, as well as the availability of different food types.

6

Social organization and home range use

6.1 SOCIAL STRUCTURE

The social organization of a roe deer population varies appreciably through the year. Generally, the majority of deer are observed alone or in small family groups during summer but in winter larger groups or herds are more common.

Perhaps 60–90% of all summer observations are of solitary individuals, and groups of four or more animals are extremely rare. At this time of year, pairs commonly comprise a doe and her kid, two kids, or a rutting adult buck with a doe, but less frequently a pair of yearlings, or occasionally an adult buck with a younger buck, may be observed. Three deer together most often constitute a family group of a mother with her two kids, and by the beginning of autumn, such family groups of two or three individuals may account for about a half of all observations of roe deer.

During winter the situation is somewhat different, when mixed groups consisting of four or more animals are most commonly observed. The group size of European and Siberian roe deer inhabiting similar habitat types is typically the same (Figure 6.1) but some differences are noticeable during spring and autumn, due to delayed spring regrowth of vegetation, and the sudden, relatively early frosts and snowfall in Asia.

The social organization of the roe deer population may depend to a large extent on habitat-related factors such as the availability and distribution of food resources and cover (Graczyk and Bereszynski, 1978; Zejda, 1978; Reichholf, 1980; Bresinski, 1982; Maublanc *et al.*, 1987; Sokolov and Danilkin, 1981; Darman, 1986), but intrinsic social factors may also be important. Average group size varies in response to a variety of factors, increasing with habitat openness, population density, falling temperature and deepening snow cover, as well as with decreasing area of feeding

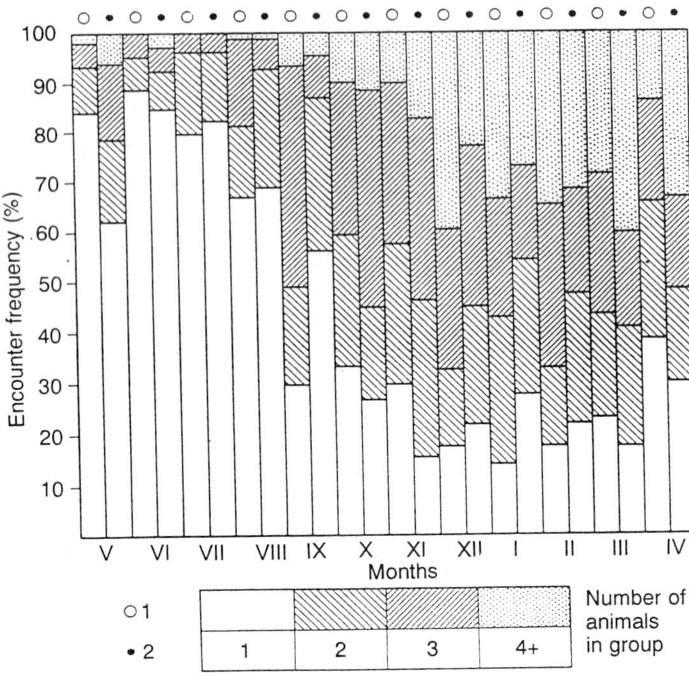

Figure 6.1 Annual changes of group size in forest–steppe populations. (1) European roe deer in Belgorod Region, 1982–1986; n = 2470, totalling 5586 individuals. (2) Siberian roe deer in Southern Urals, 1971–1978; n = 2366, totalling 5172 individuals.

grounds. For example, groups of 10–15 individuals are only very rarely observed in forested areas, but in open habitats (such as agricultural fields) groups often comprise many more deer and frequently may be as large as 40–120 animals. In extreme circumstances in certain parts of Asia, a combination of these factors leads to the aggregation of roe into migrating herds of several hundred deer.

Winter groups begin to form during autumn, generally through the aggregation of two family groups which may be related, or at least occupy adjoining ranges, and possibly some other yearlings. Subsequently, subadult and adult bucks may join these groups, although older bucks in some cases remain solitary throughout the winter. In this season, territoriality and aggressive behaviour between bucks are relaxed; groups containing large numbers of males, and even exclusively male groups of up to seven individuals may be observed among both Siberian and European roe. The core element of the migratory Siberian roe deer populations is also the family unit of a mother and her offspring, accompanied by a single adult male and yearling females. Small groups are formed from the combination of two families and other solitary individuals. Occasional all-male groups are observed during migration, comprising previously territorial individuals, and may be as large as 15 bucks. Large herds of several hundred roe probably occur only during winter, in areas of locally high density, after the mass migrations which follow heavy snowfall.

Large groups are most frequently formed at feeding sites in response to perceived danger and, when disturbed, tend to follow one another. Groups formed in such situations may remain together for several hours, days or, in extreme cases, weeks but more generally, the composition of large groups changes almost continuously, despite an apparent stability (Strandgaard, 1972a). Our observations of the European and Siberian roe deer have shown that herd composition is flexible, large groups often breaking into smaller subunits (commonly family groups) which may recombine, but with a different overall composition.

Small groups are the most stable unit and radiotracking of individual deer has shown that members of a family group generally remain together throughout the winter (Bideau *et al.*, 1983b; Danilkin and Minayev, 1988). Sudden disturbances may lead to the fragmentation of these families but within a few days they reform, generally at feeding sites. Occasionally, kids become detached from their mother and may join some other family or group, however, in most cases, they are met with aggression and so may remain alone throughout the winter, joining other groups only when in danger. Subadults may also be solitary, or perhaps group in pairs, over winter, while individual adult bucks fluctuate in their grouping tendencies, joining and leaving different groups regularly.

The population structure described above is characteristic for the entire winter period, but towards the end of March the number of solitary deer increases, although groups do not dissipate almost completely for a further month (Figure 6.1). Family groups may persist longer and have been observed on 28 May for Siberian roe deer and 19 May for European roe deer. The family unit forms the core of roe deer group structure and serves as the basis for the social organization of the entire population. Adult mother does are often pivotal to the group, acting as group leader in 88% of cases in a forest–steppe population of Siberian roe deer (Sokolov and Danilkin, 1981), and approximately the same leadership pattern has been recorded in a forest–steppe population of European roe. Field populations are also generally led by adults does but the situation is more variable (Bresinski, 1982), as the individual that first responds to danger takes over the leadership role (Graczyk and Bereszynski, 1978).

The above information and other published data for European roe (Sablina, 1955; Raesfeld, 1956; Necas, 1960, 1963; Kurt, 1968a; Nikolandic, 1968; Bramley, 1970, 1972; Zejda, 1978; Reichholf, 1980; Bresinski, 1982; Timofeyeva, 1965; Maublanc *et al.*, 1987; Chapman *et al.*, 1993) and

Siberian roe (Averin, 1949; Filonov, 1974, 1977; Smirnov, 1978; Sokolov and Danilkin, 1981; Sludski and Baidavletov, 1984; Darman, 1986) indicate that the organization of populations of both species is essentially similar. Differences occur only with respect to the dates of change between their alternative seasonal habits, which, in the Siberian roe deer, is about a month later in autumn and slightly earlier in spring, owing to climatic factors.

6.2 SPATIAL ORGANIZATION: HOME RANGE SIZE AND USE

The structure of mammalian populations can be considered from different viewpoints (Naumov, 1963; Petrusewicz, 1966; Schwarz, 1967), but it is the spatial structure (the distribution of a population in space) and the ethological structure (the dynamic system of animal relationships) (Wynne-Edwards, 1962; Naumov, 1967, 1971, 1972, 1977; Shilov, 1972, 1977) that are of the greatest importance. When characterizing the spatial and ethological structure of animal populations, a variety of terminology is employed including area of activity, foraging range, seasonal pasture, individual area, individual range, home range, protected range, hunting range, kidding range and territory. These terms are often poorly or differently defined (Burt, 1943; Hennig, 1962; Barabash-Nikiforov and Formozov, 1963; Naumov, 1963; Jewell, 1966; Kurt, 1966, 1968a; Strandgaard, 1972a; Walther, 1972; Baskin, 1976; Shilov, 1977) and so some revised definitions as they apply to the roe deer are given here:

- The **individual range** is a set of all home ranges of an individual, including its migration routes, over its lifetime (**lifetime range**).
- The **home range** is the total area where an individual (or group) remains for a certain length of time (day, season, period, year). Deer home ranges may overlap widely.
- The **kidding range** is an area protected from other roe deer by a pregnant female, before giving birth until some time after delivery.
- A **territory** is that part of a male's home range with defined boundaries which are defended. These limits are recognized by neighbouring bucks and are generally maintained by mutual avoidance, or by active protection. It should be noted that no reference to marking behaviour is made in this definition, as bucks may frequently mark within adjacent territories.
- The term **territorial** will be used to refer to adult bucks protecting their territory, while young males with no territory will be referred to as **non-territorial**.

In recent years, large-scale studies of individuals marked by either tagging or radiotracking have provided insights into the spatial structure of populations of both European roe (Kurt, 1968a, 1970; Bramley, 1970, 1972; Strandgaard, 1972a; Ellenberg, 1978; Sempere, 1982; Pielowski and Bresinski, 1982; Fruzinski *et al.*, 1983; Zejda and Bauerova, 1985; Cederlund, 1982, 1983; Bideau *et al.*, 1983a; Vincent *et al.*, 1983; Vincent and Bideau, 1987; Sokolov *et al.*, 1986d; Danilkin and Minayev, 1988; Chapman *et al.*, 1993) and Siberian roe (Sokolov and Danilkin, 1981). The spatial and ethological structure of a European roe deer population of the forest–steppe is described below. This example is used as it includes settled as well as seasonally migratory deer, and thus exhibits features typical of both classical and migratory populations (Danilkin and Minayev, 1988). Using this population as a model facilitates a comparative analysis and highlights consistent patterns in the population organization of *Capreolus*. (More information on this population of the Les na Vorksle Reserve, in the Belgorod region of Russia is provided by Novikov and Timofeyeva, 1965.)

The spatial structure of a roe deer population, as well as the overall social organization, changes considerably over the annual cycle but is relatively stable within two periods of considerable duration. During summer, the period of territoriality and reproductive activity, the deer (which are primarily solitary) are dispersed over the population range. In contrast, during winter when the bucks are no longer territorial, the deer are mainly aggregated into groups which are concentrated on feeding ranges. This structure is complicated by the fact that in early autumn and spring, the change between territorial and non-territorial behaviour is asynchronous, so that these two periods overlap. Moreover, in summer, the territorial behaviour of individual deer differs according to sex and age.

6.2.1 Home ranges in summer

(a) Adult bucks

In spring, immediately following snow melt, adult bucks of both European and Siberian roe establish territories which are the same from year to year, defending them against other bucks and marking them with both visual and scent markers (Figure 6.2). Once a buck has established a territory, he may remain there for the rest of his life, and certainly our studies at Belgorod showed this stability, with all five adult bucks remaining in the same territories for 4–6 years. Similar data are reported by Loudon (1978), while Strandgaard (1972a) and Bramley (1970) recorded emigration of only one and two males, respectively. Right of ownership for a particular territory has to be re-established each year in contests with other bucks, particularly young males. Clear evidence of the intensity of such contests is apparent from the scars on the forehead and neck of practically every territorial animal.

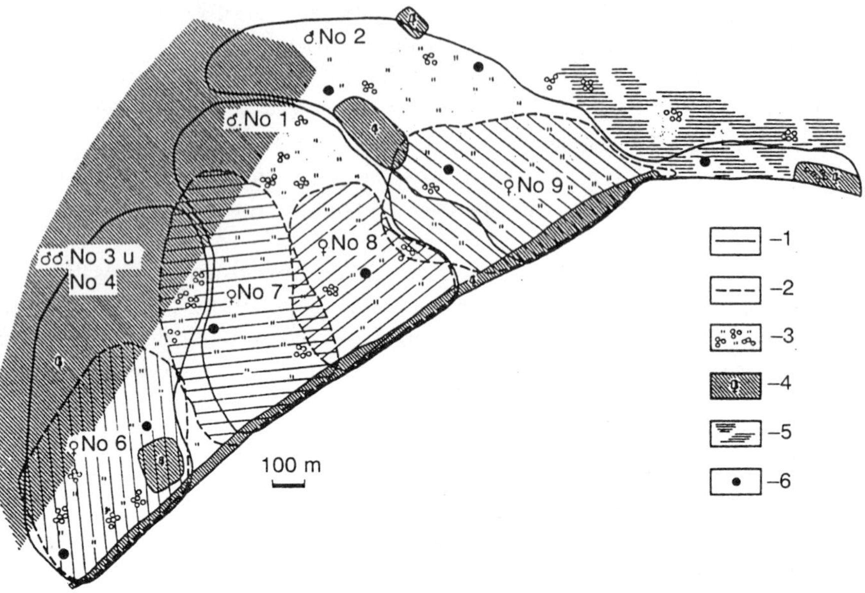

Figure 6.2 Spatial organization in southern Ural population of Siberian roe deer. (1) Summer territories of bucks; (2) home ranges of does; (3) meadow overgrown with shrubs; (4) deciduous forest; (5) bog; (6) observations of non-territorial bucks.

In the Belgorod study population, at a summertime population density of about 5 roe deer/100 ha, buck territories averaged 114 ha (varying from 62 to 157 ha; $n = 9$), which is considerably larger than has been reported for populations in other parts of their range. The largest recorded territory (204 ha) occurred outside the reserve where the population density was at least three times lower, but if this value is included, territory size averaged 122 ha ($n = 10$) (Figure 6.3). Territories of approximately the same size were recorded for Siberian roebucks in the southern Urals (Sokolov and Danilkin, 1981) (Figure 6.2). More generally, analysis of the size of buck territories shows that they are inversely proportional to population density (see Vincent *et al.*, 1995), varying from 2 to 200 ha (Table 6.1).

Territory sizes may be partly determined by summer habitat quality: in rich habitats, population density is generally higher while territories are smaller (Mottl, 1962; Bobek, 1977). Despite the relatively widespread distribution of roe deer during summer, population density is higher at forest edges, in high-grass meadows and near water bodies, where food is abundant and good cover is available. In such habitats, the competition for territory is especially intense, while deer may be absent altogether in

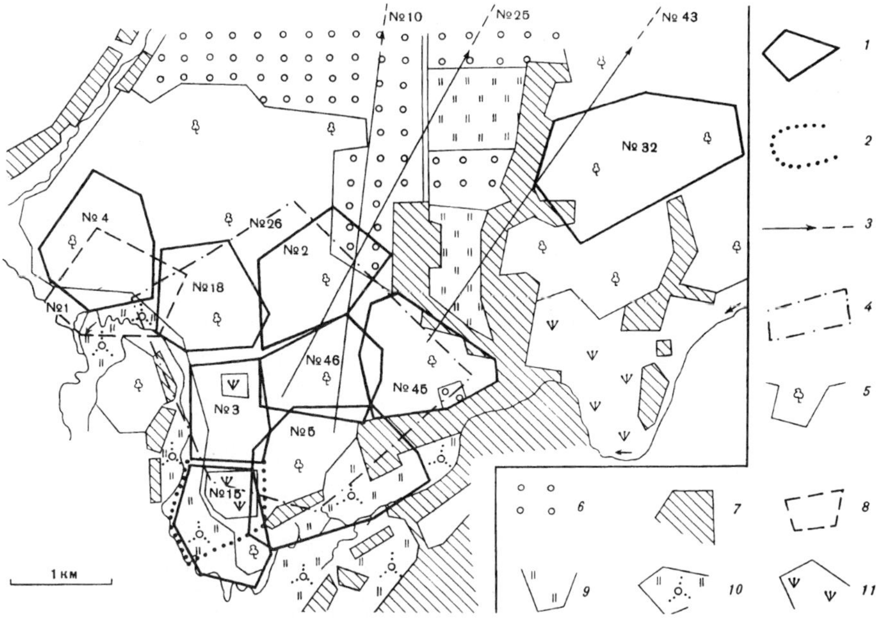

Figure 6.3 Spatial organization of males in a European roe deer population in Belgorod Region during summer (15 April to September). Data from radiotracking and visual observations, 1982–1986. (1) Territories of adult bucks; (2) enlargement of territory buck No. 3; (3) emigration of yearlings; (4) home range of 2-year-old buck No. 26; (5) forest; (6) orchards; (7) buildings; (8) home range of yearling buck No. 1; (9) waste land; (10) meadow with shrubs; (11) fields.

adjacent poorer habitats, or present at a much lower density. However, the detailed vegetational composition of a particular area has little effect on territory sizes which are approximately equal in a given habitat (Strandgaard, 1972a; Sempere, 1982). Further, Chapman *et al.* (1993) found no evidence that roe deer territories were resource based in a coniferous forest in south-east Britain.

In addition to this relationship between territory size, population density and range quality, Bobek (1977) suggested that territory size may also vary with age. Loudon (1978) presented data which contradict this hypothesis and was supported by the conclusions of Ellenberg (1978) from studies on captive roe deer. Our data collected over 4 years do not support Bobek, as adult male territories remained effectively constant over the study period. A single exception occurred when an individual increased his territory size by including a section of a territory previously occupied by another buck which had just died (Figure 6.3).

In the wild, at moderate population density, territories of male neighbours effectively do not overlap (Chapman *et al.*, 1993) (Figures 6.2, 6.3),

Table 6.1 Range in territory size of roebucks at different levels of population density

Population density (deer/100 ha)	Territory size (ha)	
	European roe	Siberian roe
1–2	126–204	100–170
3–10	38–157	35–120
10–20	6–100	18–37
20–30	12–71	–
30–40	7–20	–
40–60	5–10	–
> 60	2–9	–

References: Hennig, 1962; Mottl, 1957, 1962; Kurt, 1968a; Prior, 1968; Bramley, 1970, 1972; Strandgaard, 1972a; Bobek, 1977; Ellenberg, 1978; Janeau *et al.*, 1981; Sokolov and Danilkin, 1981; Sempere, 1982; Sokolov *et al.*, 1982b, 1986d; Cederlund, 1983; Bideau *et al.*, 1983a; Darman, 1986; Lund, 1990; Vincent *et al.*, 1995.

or overlap only slightly (Kurt, 1968a; Prior, 1968; Bramley, 1972; Sokolov and Danilkin, 1981). Boundaries are maintained in the main by mutual avoidance of intrusion into neighbouring territories, although most bucks may venture on short 'raids' into other territories, returning the same or the following day. Territorial limits generally follow visible landmarks such as forest edges, roads, paths, gully slopes, brooks and rivers, and at a constant population density, these boundaries seem to be fixed. However, the mode of their demarcation remains obscure in homogeneous habitats such as large blocks of woodland. At high population density, young males may take over a part of another buck's territory, even sometimes by driving an old buck away (Sempere, 1982). The unoccupied territory of any buck who died in winter may be quickly taken over and occupied by a young buck the following spring. When a territorial buck dies in summer, parts of his territory may be taken over by neighbouring males (Bramley, 1972; Strandgaard, 1972a; Ellenberg, 1978; Danilkin and Minayev, 1988).

At very high density, either in the wild or under captive conditions where space is limited, the ranges of adult bucks may partially overlap due to the joint use of feeding areas (Sokolov *et al.*, 1982b) but the core areas of activity usually remain distinct (Sempere, 1982; Maublanc *et al.*, 1987; Chapman *et al.*, 1993). This partial overlapping of buck territories, accompanied by a considerable flexibility in range size, permits high population densities to form, even when bucks are markedly territorial. However, limitation of resources along with social factors concerning the use of space place a limit on the social carrying capacity of a particular habitat.

Daily variation in the ranges of adult bucks depends on both the territory size and the deer's activity pattern. In the Belgorod region, daily ranges averaged 49 ha (varying from 4 to 97 ha), which constituted only 40% of their total summer range. In Sweden (Cederlund, 1983), daily ranges were even smaller (20.3 ha) and included only 21% of the total territory. These figures illustrate that each buck uses or defends only a fraction of his territory on a day-to-day basis, with as much as 70% of his time spent within a much smaller core zone (Figure 6.4) which seems to be of primary importance for many males (see also Bideau *et al.*, 1993). In contrast, practically the whole of the range is used when the overall territory size is small.

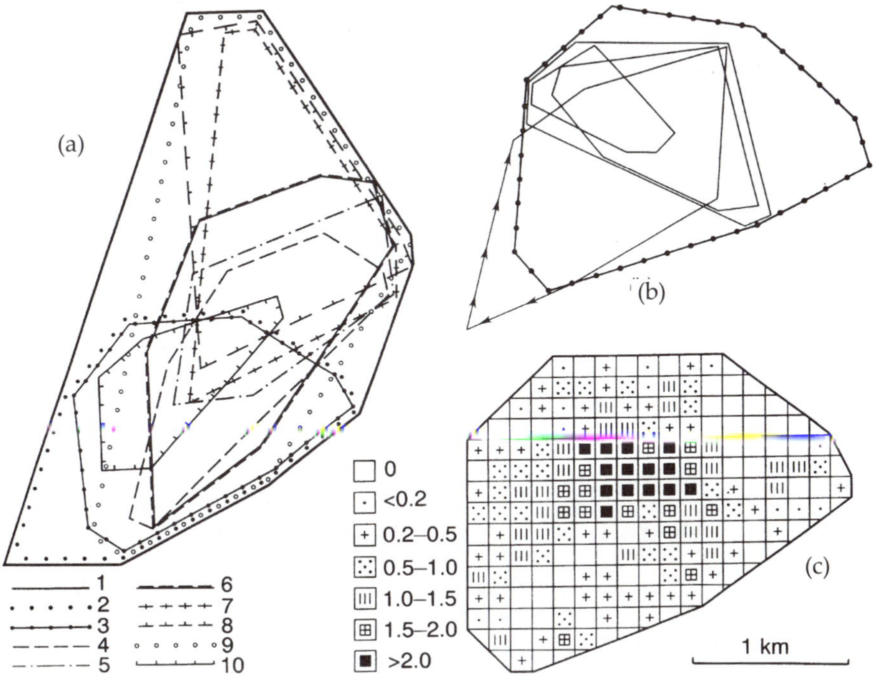

Figure 6.4 Structure of home range of single adult European roebuck (No. 5) in Belgorod Region. (a) Seasonal changes in range: (1) whole observation period (March 1984 to June 1986); (2) summer range (15 April to September); (3) territory (protected part of home range); (4) October range; (5) November range; (6) December range; (7) January range; (8) February range; (9) March range; (10) April range. (b) Daily ranges in period 25–28 August 1984. (c) Frequency of radio detection (in %/ha of buck's territory).

During the rutting period itself, the territorial organization of the roebucks does not change substantially (Bramley, 1972; Sokolov and Danilkin, 1981; Sempere, 1982). At the beginning of the rut, males tend to mate with does within their territory, though no one doe's home range will precisely coincide with a particular buck's territory (Figures 6.2, 6.3, 6.5), and it is possible that the doe may exercise some mate choice between males whose territories overlap with her own. Towards the end of the rut, bucks may undertake 'raids' outside their territory in order to find does on heat. These raids become more frequent but territoriality is still evident, though it decreases progressively over time, especially in migratory populations.

(b) Subadult bucks

Subadult bucks generally show a greater spatial instability than females of an equivalent age, having a more diffuse home range and a higher rate of dispersal, particularly at high density (Bideau *et al.*, 1993). In our study population inhabiting an isolated forest measuring about 1000 ha,

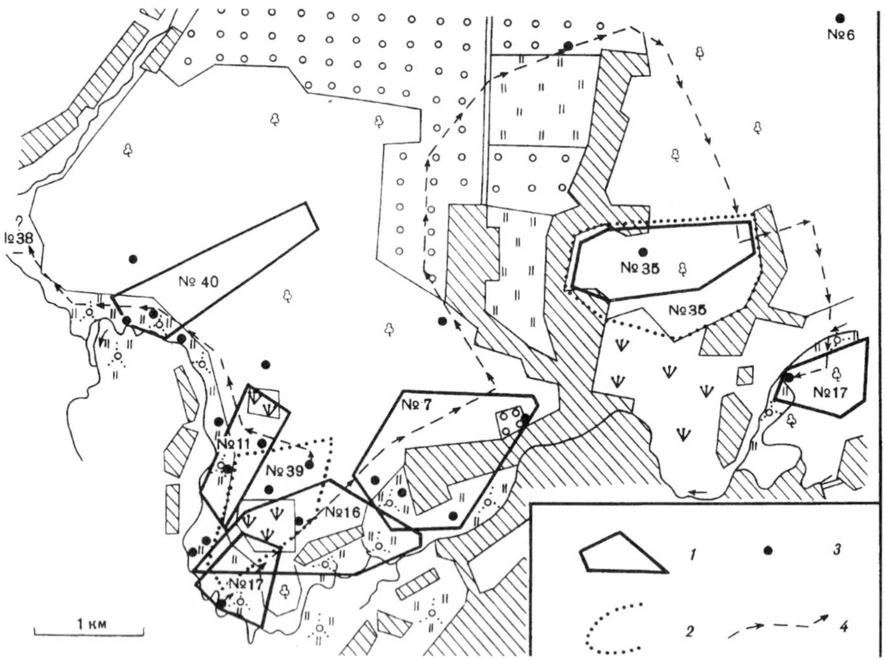

Figure 6.5 Spatial organization of females in a European roe deer population in Belgorod Region during summer (May–August). (1) Home ranges of adult does; (2) ranges of subadults; (3) birth sites; (4) emigration of pregnant does induced by disturbance. Physical features as in Figure 6.3.

subadult buck numbers noticeably fell between April and July; by the rutting period only territorial bucks remained, numbering no more than five or six and ranging widely over the forest. In contrast, most of the males observed outside the woodland, in the river valley or orchards, were yearlings, sometimes accompanied by a doe of the same age.

These observational data fully accord with information obtained from radiotracking. Of the five yearling bucks followed, four emigrated in spring; the fifth occupied a range of about 2 ha which was situated on a river bank at the boundary of two adult bucks' territories, and he entered the forest only rarely. In autumn, only three of these yearlings returned to the reserve: one of these died in winter and a second emigrated again at the beginning of the following summer. The third yearling ranged widely across the reserve, wandering through the territories of nearly all the adult males, but became territorial only in his third year, when he took over the vacant territory of an adult buck who had probably died. Similarly, of the four marked 2-year-old males, three emigrated from the forest but remained in the vicinity for several weeks until finally leaving the 10 km^2 zone of radio observation in July. The area covered by such emigrants and other individuals with unstable ranges was much greater than the summer home range size of adults: it measured 1000–2000 ha.

The emigration of young males from their natal range is largely provoked by the aggressive behaviour of adult bucks who drive them away from their territories. This pursuit of yearlings begins immediately after the family group dissolves, as soon as the maternal bond has been broken. Most young bucks (as many as 58–90%) are obliged to emigrate (Figure 6.3) from their birthplace (Bramley, 1970; Strandgaard, 1972a; Loudon, 1978; Sokolov and Danilkin, 1981; Darman, 1986; Danilkin and Minayev, 1988), and since this may involve large numbers of individuals, the process appears as a seasonal migration. Thus, during autumn, most yearling male migrants return once again to their birthplace, but usually leave again the following spring.

In populations where densities are particularly high and in artificial enclosures, not even 3-year-old bucks are all able to become territorial (Bideau *et al.*, 1993), whereas in low density conditions some yearlings may behave as fully territorial males (Ellenberg, 1978). In densely populated areas, the expelled subadult roe may be continuously replaced by deer immigrating from adjacent zones and adult bucks may have to defend their territory repeatedly. In this situation, many yearling and 2-year-old males remain near their natal range, wandering around the territories of neighbouring males and, in some cases, associating closely with a territorial buck, accompanying him everywhere, at least until the rutting period (Sokolov and Danilkin, 1981; Darman, 1986). However, it is rare that a territorial buck is not openly aggressive toward a younger male and permits him to remain within his territory.

Overall, young males display a more diverse spectrum of spatial behaviour than older bucks (Figure 6.6). They may emigrate (from their birthplace), take over vacant ranges outside the territorial zone of the population, wander about the territories of several bucks, occupy a small territory within the buffer zone separating territories of neighbouring bucks or inhabit a territory of a buck who is not openly aggressive, perhaps associating closely with him. In this context, subadult males may be viewed as an ecological reserve within the population. Owing to their spatial distribution patterns, a population quickly expands into new areas and plugs gaps in the territorial structure resulting from deaths of adult bucks.

(c) Adult does

The spatial distribution of adult does during summer differs from that of bucks (Figure 6.5). By the end of April and the beginning of May, approximately 3–4 weeks prior to giving birth, the does of our study population separate from their groups and occupy kidding ranges which remain the same from year to year and measure several hectares, and are in some cases the summer ranges where they themselves were born. Every April,

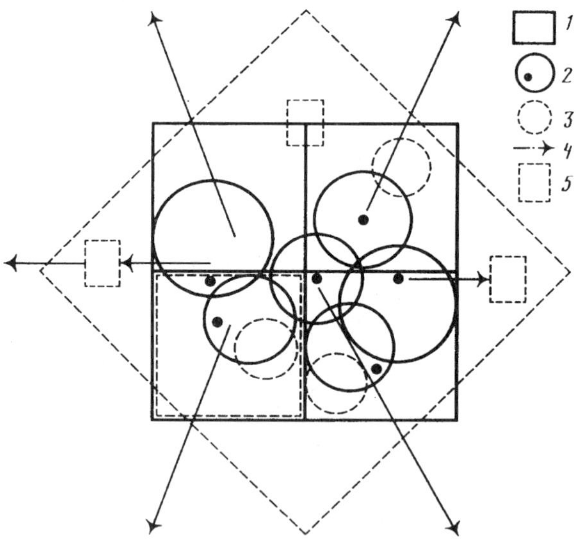

Figure 6.6 Spatial structure of a roe deer population during reproductive period. (1) Buck territories; (2) home ranges of reproductive does (black dot denotes birth site); (3) home ranges of yearling does; (4) emigration of yearlings; (5) various types of home ranges of yearling and 2-year-old bucks. (For explanations, refer to text.)

nearly half the marked does we followed (n = 14) migrated outside the reserve, covering at least 7–10 km, before giving birth. These females returned to the reserve with their kids sometime between October and December. Following some disturbance, two previously settled does moved from their traditional maternity areas only a week or two before parturition (Figure 6.5); having successfully given birth in a new site, they returned to their winter ranges by autumn. The following spring these females returned to these same new kidding ranges, returning to the reserve again by autumn.

A pregnant doe will defend her kidding range, forcing other females and her own yearling offspring to leave through aggressive action, lowering the head and sometimes also using the front limbs. Similar behaviour of does has been recorded in other populations (Kurt, 1968a; Bramley, 1972).

Not all habitats provide suitable maternity ranges. Does may be quite selective over the location of a kidding range, favouring sites with a source of high quality food within the vicinity, good cover and, preferably, a local source of free water. Such requirements are most frequently fulfilled by sites such as forest edges, high-grass meadows and river or stream valleys with some shrub cover. In these habitats, the density of pregnant does is always highest. These preferences for particular kidding habitats are so marked that even at relatively low population density (4–5 deer/100 ha), a shortage of available range space for kidding was observed in some areas.

One or two weeks after kidding, does begin to expand the size of their home ranges, which may increasingly overlap one another as the kids grow. Although ranges of several females, particularly related individuals, may coincide, core areas remain largely separate (Chapman *et al.*, 1993). The majority of mothers remain in their summer ranges until autumn, leaving once the breeding season has finished. The concentration of females into certain preferred habitats means that the territories of some bucks may cover the ranges of several does, while in others no females are present (Figures 6.3, 6.5). However, in the study population at Belgorod, some females left their kidding ranges during June or July, dispersing a few hundred metres to another range which probably provided better cover, and thus ranged within the territory of another buck during the rut who presumably mated them. These occasional changes in range of some does, coupled with the general range expansion among females prior to the rut, may permit males whose territories did not initially encompass any doe ranges to participate in the rut.

In our studies in the Belgorod region, the size of the daily home range among lactating does averaged 24.1 ha (varying from 7 to 34 ha; n = 16), while the seasonal range (occupied from the end of April through to August) averaged 73 ha (varying from 39 to 107 ha; n = 7), which are similar to figures reported by Cederlund (1983) for Swedish roe deer (daily and seasonal ranges averaged 12.3 ha and 94.6 ha, respectively). The daily home ranges of kids are initially several times smaller than those of their mothers, measuring between 1 and 8 ha in June and July, but increasing to between 3 and 14 ha by August. From the middle of September, the sizes of mother and kid ranges are essentially equal, indicating a high degree of synchronization in their activities and spatial behaviour.

As is the case for bucks, the size of a doe's seasonal home range is evidently determined to a large degree by habitat quality. In rich habitats, population density is generally higher and range sizes generally smaller. Seasonally, ranges may vary from 1 to 7 ha during the kidding period up to 180 ha towards the end of the summer season, and in high density populations the ranges generally average between 7 and 60 ha (Mottl, 1957, 1962; Kurt, 1968a, 1970; Nikolandic, 1968; Bramley, 1972; Strandgaard, 1972a; Ellenberg, 1978; Janeau *et al.*, 1981; Fruzinski *et al.*, 1983; Cederlund, 1983; Lund, 1990).

(d) Young females

At moderate density, most young females occupy vacant areas adjoining or overlapping their mothers' ranges, usually not more than 1 km from their site of birth (Bideau *et al.*, 1987). Upon death or artificial removal of an adult doe (Bramley, 1972), the vacant home range is immediately seized by a young female, usually the doe's daughter, who may then kid on her own mother's kidding range (Aylmer, 1977). In contrast, at high population density, many young does leave their parents' ranges in spring (Bramley, 1972; Strandgaard, 1972a; Loudon, 1978). Such migratory behaviour may be stimulated by aggressive behaviour of a pregnant doe, either before or shortly after kidding (Kurt, 1968a), or aggression of a territorial buck directed particularly towards young males.

In the Belgorod population, three yearling does were fitted with radio transmitters in the reserve. Two emigrated in April, one of them accompanying her brother. In autumn only one of the two returned, and next spring she left the study area again. Of the six 2-year-old does that were tracked, four emigrated during spring and two returned to the reserve with their kids in autumn but left again for their kidding grounds the following spring. Once the family groups had dissolved, the two young does that did not emigrate stayed near their mother's home range, remaining in contact with the mother at feeding grounds. Their summer home ranges measured 72 ha and 133 ha. The following spring, both does occupied these same ranges and gave birth there.

Generally, young does show less variability in their spatial behaviour than yearling bucks, either migrating away from their natal area or occupying a range neighbouring their mother's home range. In contrast to immature males, young does rarely range over the territories of several adult bucks; when doing so they are generally accompanied by a buck of the same age, most commonly a brother.

(e) Composite families

From the above, most adult roe generally occupy the same home ranges year after year. Fawning areas may overlap, particularly those of closely related females such as a mother and her kids from the previous year which have settled on adjacent ranges. These ranges of related female offspring may fall within the territory of the same single buck, who may be less aggressive towards his progeny than towards other juvenile deer. Therefore, under certain circumstances, it is possible that a buck may mate with the same does from year to year (Ellenberg, 1978), some of which may be his own offspring. This may lead to a substantial level of inbreeding in forest roe deer populations which is alleviated only when culling rates are high enough to disrupt the family clan structure (Kurt *et al.*, 1993). Thus, during the reproductive period, the core of the spatial structure of a population is formed by the combined home ranges of the composite family, consisting of several closely related does and their offspring, which range within the territory of an adult buck (Kurt, 1968a, 1970; Strandgaard, 1972a; Ellenberg, 1978; Stubbe and Passarge, 1979; Cederlund, 1983; Vincent *et al.*, 1983; Maublanc *et al.*, 1987; Danilkin and Minayev, 1988). This composite family structure may reduce aggressive behaviour between individuals and may facilitate the formation of groups in winter. This pattern of spatial organization during the summer period is shown schematically in Figure 6.6. During winter, the buck may join the group unit which includes his family, thus maintaining the composite family structure, but he is more likely to remain solitary outside the reproductive period.

6.2.2 Home ranges in winter

During the winter period (lasting from October until March or April), the spatial structure within a population of roe deer is fundamentally different from that of summer and, moreover, varies from region to region, reflecting variation in ecological factors – primarily snow depth. This variation, at its most extreme, includes the classical population structure where roe remain settled on a particular range, but also populations which migrate seasonally between widely separated summer and winter ranges, as well as populations which include both sedentary individuals and seasonal migrants, whose numbers depend on snow depth (Figure 6.18).

In the reserve at Belgorod, we observed a clear pattern of deer movements and space use in relation to snow depth (Figure 6.18). When snow cover was less than 20–30 cm, most individuals of the population remained on their summer ranges for the duration of the winter. However, when the snow depth exceeded 30–40 cm, only a few individuals continued to occupy their summer home ranges, and then only when their particular home range occurred in a zone where the snow cover was thinner than the average or where there was a particularly abundant local food source. During periods of heavy snow accumulation (in excess of 40 cm depth) or when the snow surface was not firm and therefore hindered movement, all roe deer left their summer home ranges and migrated to the edge of the reserve or, in some cases, beyond. These individuals tended to establish new ranges where snow was shallower, in the river valley or in orchards with abundant browse. These same sites also attracted roe deer from other adjacent forests, resulting in a three- or four-fold increase in the number of animals within the observation area.

These sites were particularly favoured during the especially harsh winter of 1986–87, when the snow depth was as great as 1.5 m. In this extreme situation, most of the population managed to survive by making use of paths formed by the movements of wild boar and by ranging over open habitats with a relatively firm snow surface, moving slowly with digits splayed to increase the surface area of contact. Due to the depth of the snow, nearly all tree branches in orchards and willow thickets became accessible up to a height of 2.5–3 m, noticeably improving the available nutrition. However, mobility was severely restricted, and predation by poachers and dogs increased.

Roe deer mix freely during winter, ranging more widely (Chapman *et al.*, 1993), and in our study site the winter home ranges of individuals and groups overlapped considerably (Figure 6.7). Daily distance travelled is longest for does in this season, although there may be no seasonal change in the size of the core range area (Chapman *et al.*, 1993). From October to April, in our study population, ranges averaged 508 ha (varying from 124 to 1002 ha; $n = 24$) and were approximately of equal average size for individuals of both sexes and different age classes. This seasonal range clearly contained one or several feeding grounds, where the animals spent most of the day.

The winter home ranges in our study population are among the largest yet reported for the European roe deer, although similar values are reported by Cederlund (1982, 1983) for roe deer in Sweden and by Zejda and Bauerova (1985) for a Czechoslovakian field population (range

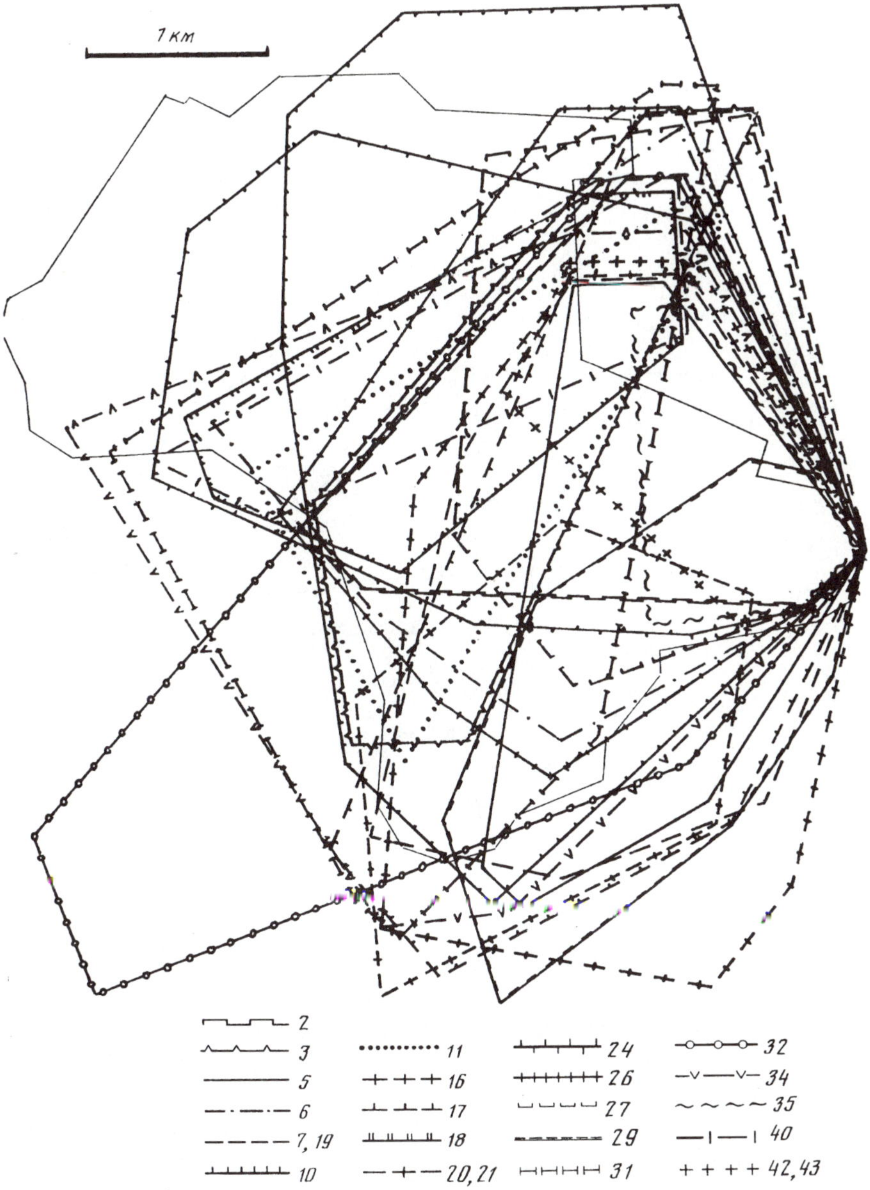

Figure 6.7 Spatial organization of a European roe deer population in Belgorod Region during winter (October–March, 1982–1986). Home ranges of individuals (24 shown) calculated from radiotracking and direct observation, showing high degree of overlap.

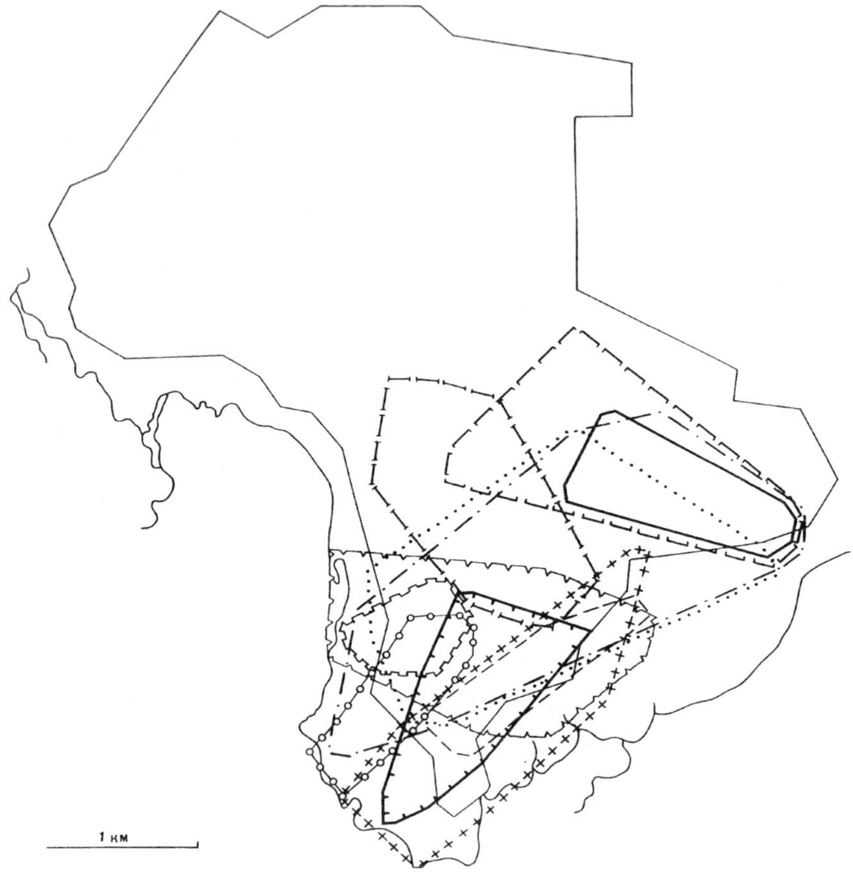

Figure 6.8 Daily home ranges (5 December 1984) of European roe deer in Belgorod Region during winter from 24-hour radiotracking.

40–812 ha, average 200 ha). In all other studies on populations of central and western Europe, ranges are 3–10 times smaller (Mottl, 1962; Kurt, 1970; Ellenberg, 1978; Bideau *et al.*, 1983a). Most probably, these differences may be explained by variation of environmental conditions between populations, with poor conditions forcing the deer to range more widely in search of food, particularly in migratory and field (Zejda and Homolka, 1980) populations. When snow depth was less than 20–25 cm in our study population, ranges were smaller, averaging 103 ha (varying from 30 to 243 ha; $n = 24$) (Figure 6.8). However, this increase in ranging movement in response to a decrease of resources occurs within certain limits and home ranges are smaller when the abrupt deepening of the snow cover limits mobility. Some individuals or groups occupied ranges as small as 2–7 ha on our study site when snow depth was between 50 and 150 cm. A considerable reduction of home range size in especially severe and snowy winters has been observed in some other populations (Cederlund, 1983; Maublanc *et al.*, 1987), and is probably a method of conserving energy through reduced movement. During this period, the basal metabolic rate (Weiner, 1977) and the level of food consumption are also considerably lower.

Over the winter period, even among non-migrating individuals, home ranges may undergo monthly (Figure 6.4) and seasonal shifts, such as changes in configuration and size (Chapman *et al.*, 1993). For example, ranges are often larger in spring than in winter or autumn (Sempere, 1979; Janeau *et al.*, 1981; Cederlund, 1983). At snow melt in spring, both metabolic activity and food consumption increase as the deer begin to forage more widely for fresh browse, and move toward their summer ranges. During autumn, range size may increase to exceed that of the summer period, probably in response to a relaxation of territorial behaviour. The most extreme example of such seasonal shifts in ranging behaviour are the autumn and spring migrations observed in some populations.

6.2.3 Yearly and individual home ranges

The roe deer's yearly range consists of the combined summer and winter ranges, with any routes of migration between them, if they do not coincide. In non-migratory populations in good quality habitats, the seasonal ranges generally are more or less identical; in poorer conditions the winter home range is usually larger than the summer range but may include part of it. In contrast, in migratory populations, the seasonal ranges are entirely separate and distinct for all deer. This yearly home range structure is illustrated in Figure 6.9, and in particularly snowy winters the home ranges of some roe deer varied in approximately this manner in our study population. Roe exhibited both extremes of this yearly ranging behaviour, and some individuals demonstrated both types of behaviour in different years but always with the same fixed summer range.

Yearly range sizes differ across the roe deer's total range. In the southern Urals, the yearly ranges of Siberian roe deer averaged 520 ha (varying from 332 to 828 ha), and similar values have been obtained for European roe deer towards the eastern edge of their distribution. In our study population, yearly ranges of adult bucks (average 385 ha, varying from 189 to 845 ha, $n = 9$) were smaller than those of does (average 459 ha, varying from 237 to 701 ha, $n = 11$). Subadults had the largest yearly ranges, which averaged 598 ha for subadult does (257–825 ha, $n = 5$), while a single 2-year-old buck ranged over 733 ha. Those buck emigrants between 1 and 2 years old that returned to their first winter range (also that of their parents) had ranges measuring more than 2000 ha.

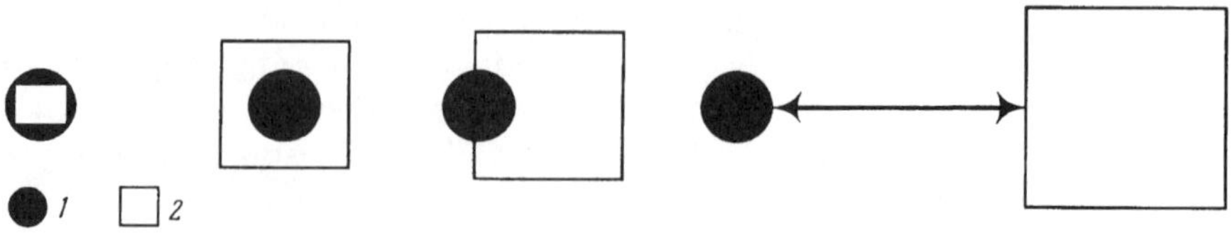

Figure 6.9 Variety of yearly home ranges of roe deer. (1) Summer home range; (2) winter home range. In good quality habitat, seasonal ranges may be identical (left). In poor conditions, winter range may be larger than summer range but may include part of it (centre). Seasonal ranges are distinct in migratory populations (right).

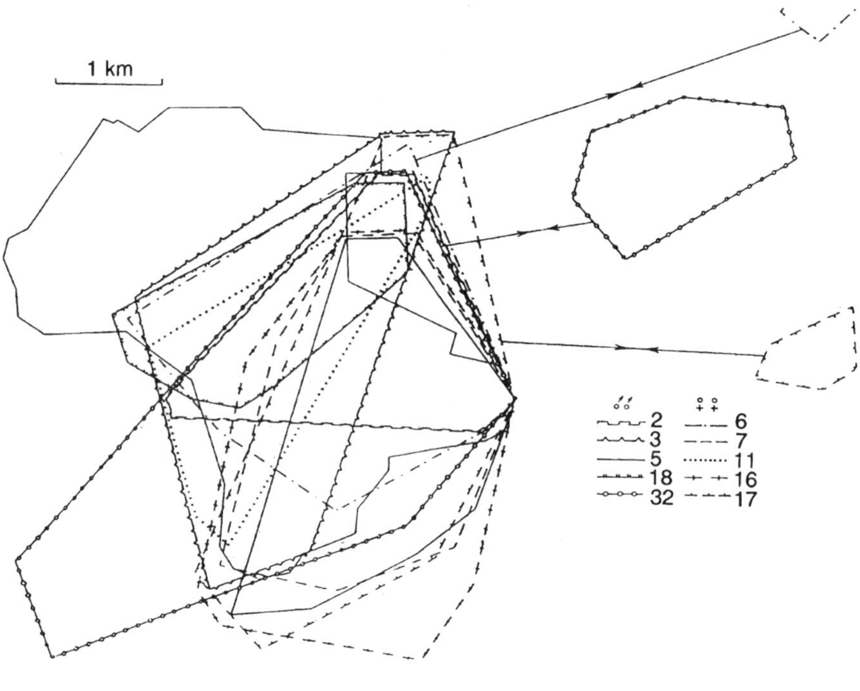

Figure 6.10 Long-term individual home ranges of adult European roe deer in Belgorod Region (1982–1986) showing a high degree of overlap, and no inter-sex differences.

Individual (lifetime) home ranges overlapped widely (Figure 6.10) and were similar in size to yearly ranges, although slightly larger, averaging 593 ha (varying from 304 to 1132 ha, $n = 10$). No differences between the sexes in individual range size were observed in our study site. Similar data have been obtained for a Czechoslovakian field roe population (Zejda and Bauerova, 1985) where sizes of overlapping individual ranges averaged 546 ha (varying from 440 to 755 ha). Somewhat smaller individual ranges were reported from Sweden (Cederlund, 1982, 1983) and also from Poland, both for field roe (Pielowski, 1984) and a forest population (Fruzinski *et al.*, 1983). In other parts of Europe, yearly and individual ranges generally vary between 20 and 100 ha (Mottl, 1962; Strandgaard, 1972a; Bramley, 1972; Janeau *et al.*, 1981; Bideau *et al.*, 1983a; Vincent *et al.*, 1979, 1983), but there is no consistent pattern of differences in range size between the sexes. It should be noted that comparisons of such data between separate studies is difficult as the techniques used are not the same.

In studies on captive roe deer in an enclosure of around 130 ha, Ellenberg (1978) showed that, although the individual ranges of bucks (35-40 ha) were about one third larger than those of does (25-30 ha), they were smaller in territorial bucks (21 ha), particularly in rich habitats (12.5 ha) by comparison with poorer habitats (24–36 ha). Average home range size among does decreased from 23.5 ha down to 14 ha with increasing age; in males, ranges were largest (29–24 ha) during the first three years but stabilized subsequently (18–20 ha).

Population density may also have an important impact on the ranging behaviour of roe deer but the response differs between the sexes: when the density doubled in the enclosure, the yearly home range size of does remained approximately constant; among bucks, range size was halved. Similarly, in a free-ranging forest population where density increased approximately fivefold over a period of 10 years, a reduction of about 30% in the home range sizes of adult males was observed (Vincent *et al.*, 1995). Furthermore, as density increased, does' home ranges progressively overlapped each other (Vincent *et al.*, 1995), whereas bucks continued to defend their territories against intruders, but range size generally decreased due to this increase in social friction. For both sexes, this decrease in range size or increase in the number of deer sharing a particular range may result in a loss in the quantity of food available to each individual. However, where food resources are not limiting even at high density, females may remain in good condition despite extensive philopatry, whereas territorial bucks may suffer body weight loss due to the higher intensity and stress of social interaction (Vincent *et al.*, 1995).

Roe deer do not use their home ranges uniformly but prefer relatively small and generally exclusive core areas close to the centre of the yearly range (Chapman *et al.*, 1993). In Sweden, at least 80% of all animal localization points fell within 400–500 m from the centre. These seasonal geometrical centres of activity for both the buck and the doe living in the same area were generally approximately the same (Cederlund, 1983) but the use of such core areas may be largely separate (Chapman *et al.*, 1993).

This section has highlighted the social organization of roe deer populations into subunits or family groups whose members associate more closely with each other than with other members of the population. Such groups are observed in many mammalian species (Naumov, 1967, 1971) and may have a marked biological importance for population dynamics (Shilov, 1977). In roe deer, it has been suggested that this social organization may play a role in the control of population numbers (e.g. see Bobek, 1977).

6.3 SEASONAL MIGRATIONS

Roe deer exhibit an ecological flexibility such that, despite a strong attachment to a particular fixed home range in the vast majority of situations, under adverse environmental conditions they may undergo temporary migrations between seasonal ranges. This behaviour is especially evident in certain populations of Siberian roe deer, as across much of the range of this species the conditions are such that roe may only survive by adopting a migratory habit.

6.3.1 European roe deer

Results of extensive marking studies of roe deer in Europe have shown that most individuals are subsequently recorded within 1–3 km from the original site of capture. For example, in Denmark, of 146 individuals marked, 102 were re-sighted at a distance of less than 3 km from the capture site and only three animals moved distances of 10 km. For roe deer populations in Switzerland and France, the maximum post-capture dispersal distance recorded was 8–9 km (Blankenhorn, 1975; Sempere, 1979). In a woodland habitat in Poland where 84 deer were caught and marked, 84% of the bucks and 54% of the does were subsequently observed within 1 km; only 17% of does travelled farther than 3 km and the maximum dispersal distance was 2.3 km for bucks and 5 km for does (Fruzinski *et al.*, 1983). Similar data were obtained for a field roe population (Pielowski and Bresinski, 1982).

From extensive tagging surveys of 1361 individual deer, Rieck (1955) found that 74% of kids and 41% of adult roe deer were subsequently recorded within 1 km from the site of capture, while dispersal distances of 10 km were rarely observed (2% of kids, 7% of adults), although distances as great as 50 km were very occasionally recorded (0.3% of kids and 0.6% of adults). Ellenberg (1978) presented data on 1058 deer from several central European countries where the majority were relocated within 2.5 km of release and only a few individuals moved more than 10 km away, with a maximum distance travelled of 70 km. In Austria, of 2734 deer tagged, more than half the adults and 83% of kids were found again within 0.5 km (Reimoser and Zandl, 1987). Some of the longest movements of the European roe deer were recorded towards the north of its distribution, in Norway (90 km: Myrberget, 1965) and Sweden (205 km: Strandgaard, 1972a).

Such data illustrate that most European roe deer are clearly sedentary animals with spatially stable home ranges. However, it is also evident that within these sedentary populations, particularly towards the north and east of the roe deer's distribution, certain individuals undertake relatively long-distance movements which could be referred to as seasonal migrations. Altitudinal migration of European roe between seasons has been recorded in eastern Georgia (Arabuli, 1963b) and in subalpine regions of Switzerland (Robin, 1975). Seasonal movements have also been observed in the central Chermozem region (Prostakov and Obtemperanski, 1986) and Moldavia (Uspenski and Lozan, 1965), in the latter case associated with the summer loss of free water. A more defined route of migration has been recorded in the Baltic region (Figure 6.11), where deer move west to east with remarkable regularity nearly every May and June, crossing the same 15 km long section of the Narva in small groups of two or three. Similar migrations, but on a somewhat smaller scale, have been observed on the Upper Narva, between the shore of the Gulf of Finland and Ivangorod, on the Luga River and near Sabsk-Ivanovskoye (Rusakov, 1978; Vereshchagin and Rusakov, 1979).

Larger migrations of European roe deer have been recorded in Finland. Ranging conditions at this northern limit of its distribution are not favourable for roe, forcing individuals to move quite considerable distances between seasons. In this region, solitary bucks and does with kids favour areas along the sea coast, river valleys and lake shores where snow is shallower and harvested hay may be available for feeding. Some individuals were recorded as far as 700 km from the boundary of the normal range of this species (Pulliainen, 1980). Occasional migratory deer have been noted in other European countries but generally account for no more than 1 or 2% of all individuals (Strandgaard, 1972a; Ellenberg, 1978). In Sweden, one tagged doe migrated with her kid to a previous winter range about 40 km away, returning again the following May, while another doe migrated about 6 km from her summer range (Cederlund, 1982).

Regular autumnal and spring migrations involving the seasonal movement of some individuals over more than 10 km have been recorded through radiotracking studies on the Les na Vorksle Reserve in the Belgorod region (Sokolov *et al.*, 1986d). The autumnal movements occur over a relatively long period and may extend into the winter, depending on snow depth, whereas spring movements are much more synchronous. Adult roebucks move first, heading back towards their territories at the end of March and the beginning of April. One individual crossed a river 200 m wide during flood conditions, and covered a total of about

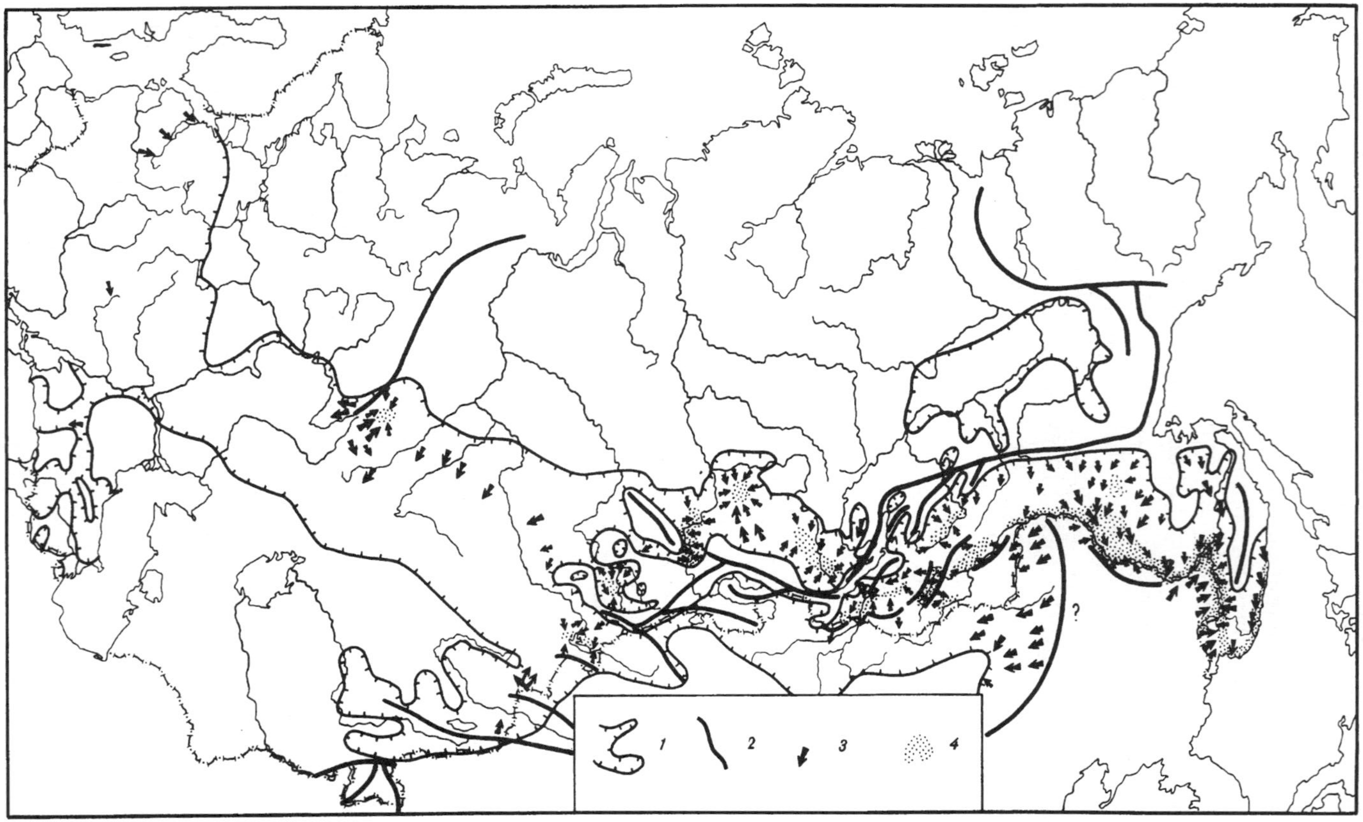

Figure 6.11 Routes of roe deer seasonal migrations. (1) Range limits; (2) main mountain ranges (axes); (3) direction of autumn and winter movements; (4) over-wintering grounds with high deer density.

13 km in 5 days. Adult does do not begin to move to their kidding sites until sometime later, when snow melt is already complete and spring vegetation is available (see also Cederlund, 1982). One adult doe left the reserve each year in the middle of April (between the 10th and the 16th) and covered 10–12 km in a single day, returning to the same home range on the reserve in autumn with her offspring. Climatic factors may affect the timing and synchrony of such migrations: in 1987, when spring was about a month late, the migration period shifted accordingly.

Despite the above examples, the European roe deer probably does not undertake the mass seasonal migrations equivalent to those of the Siberian roe, although some reports indicate this may once have been so. In the middle of the sixteenth century, seasonally migrating deer were killed by the thousand in the Ukraine, and in the 1730s and 1740s herds of 100–300 head congregated on the forest steppe for wintertime (Kirikov, 1959, 1966). In the early twentieth century, in the northern Caucasus, a migration of roe deer from areas of deep snow in the Temni forest of the Stavropol Upland to a valley of the Kuban River was noted (Dinnik, 1910). However, these cases may in fact refer to populations of Siberian roe deer which probably inhabited these areas at the time.

6.3.2 Siberian roe deer

Seasonal migrations are evident over the entire range of the Siberian roe deer (Figure 6.11) and are reported by Sabaneev (1872, 1875), Nikolski (1883), Kirilov (1902), Doppilmair (1926), Troitski (1930), Podarevski (1936), Dmitriev *et* al. (1937), Dmitriev (1938), Averin (1949), Kurazhkovski (1952), Fetisov (1953), Bannikov (1954), Ushkov (1954), Nasimovich (1955), Skalon (1960), Savinov (1967), Tsarev (1969), Subbotin (1973), Pole (1974), Dunin (1975a), Lavov (1975, 1978), Shchetinin (1975), Ziryanov (1975), Belousov (1973), Melnikov (1978), Smirnov (1970, 1978), Sokolov *et al.* (1982a), Yelski (1976, 1983), Bromlei (1984), Iliashenko (1984), Sludski and Baidavletov (1984) and Sobanski (1987). Nevertheless, many regions are inhabited by both migrating and non-migrating populations and in some areas where deep snows do not occur the deer do not migrate at all. Even in those populations where migrations are often observed, during winters of low snow fall the roe deer may fail to migrate and will overwinter on their summer ranges.

In the Far East, however, the summer and winter distribution of roe are clearly distinguished, separated by 400–500 km (Rakov, 1965; Kucherenko, 1976; Bromlei and Kucherenko, 1983), over which the deer migrate twice a year. Most roe overwinter in the valleys of the Amur and its tributaries (Figure 6.11). In this region, migrations during the nineteenth century were on such a scale that, according to Przewalski (1870) and Vetlitsin (1902), a hunter could take as many as 100 head during the mass movement. The supplement to the 1888 Report of the Military Governor of the Amursk Region states that 'there was an unheard of run of roe deer from northern parts toward Manchuria' that winter and about 150 000 animals were taken by hunters during that season. The *Okhotnichiya Gazeta* of the same year (No. 16, p. 184) reported that deep snow caused southbound mass movements of roe deer in December, and over the period of a month about 10 000 roe deer were culled near Blagoveshchensk alone. Mass migrations as marked as these have been observed in various parts of the Far East during this century (Barancheyev, 1962; Rakov, 1965; Kucherenko and Shvets, 1977).

From studying the roe deer migrations of the Far East, Rakov (1965) concluded that these seasonal movements (from October to December and from the end of March to June) are confined to local areas. Furthermore, the 'mass run' of roe deer sometimes observed is merely a concentration of such activity when food is particularly scarce or the snow is deep. When snow conditions are adverse, northern populations leave their wintering quarters earlier and more or less simultaneously, giving rise to 'mass runs' of animals towards higher quality habitats where snow cover is reduced. Early and widespread snowfall, particularly when the snow is wet, may stimulate migration to winter ranges, leading to high local concentrations of deer in small areas. Thus, overpopulation is more likely on these wintering grounds, potentially disturbing the local roe population which may then abandon its summer range.

A mass migration of roe deer in the Amursk region of Russia, in response to a drop in temperature and heavy snowfalls in the north of the region in October 1959, was described by Barancheyev (1962). This is probably one of the most complete accounts of such a phenomenon and it seems appropriate to quote a short extract from his work (pp. 30–32):

> The earliest migrating roe deer appeared in the north of the Mazanovsk region, slightly further north of Ust-Norsk, moving from north to south. Rare roe herds were observed as early as the 5th of October. Mass migration took place from the 15th till the 25th of October when each herd numbered 20–30 or more animals. They moved along three main routes from north-east to south-west. The first stream crossed the Nora River at its mouth, the second and third ones crossed the river 20 and 35 km upstream of its mouth, respectively. They crossed the river by swimming, first in small groups numbering 2 to 4 animals, then in larger groups of 10–15 or more. In one of those days V.S. Samokrutov, a ranger, counted 127 roe deer who crossed the Nora River. When the river was ice-bound, the second wave of migration took place on the 7th and 8th of November. Roe deer moved by night and, according to foot-

> prints, in very large herds. By the 20th of November the migration had ceased. The Zeya River was not an insurmountable barrier for them ... Herds came to the bank and jumped into water without hesitation and swam to the opposite bank. Having crossed the river, the animals took a 40–50-minute rest and resumed their southward journey ... When a continuous sludge blanket flowed along the Zeya River from the 7th till the 10th of November, the roe deer migration was at its most intensive. Large herds numbering as many as 70 animals came to the river, jumped into the water and swam. Most (of these) roe deer were drowned at an ice edge near the right bank since it was impossible for them to climb on to the smooth and slippery ice. No more than 14 to 16 animals survived from a herd of 70. Near the village of Busse on the left bank of the Amur, the highest concentration of the animals was observed at the end of October. The roe deer moved from west to east. Trying to cross the water barrier, the animals jumped on moving ice-floes. Many of them were quickly drowned owing to a dense sludge blanket covering the river. After the river had frozen over, it occurred more than once that roe deer ran into the wood sheds and cattle yards of Busse. The animals lacked almost all fear of man. Their main feeding grounds were soya crops. The migration ceased on the 17th of November, 1959. In spring, numerous roe corpses were found along the route of their mass movement.

Some herds participating in the migration described by Barancheyev were made up of as many as 200, 350 and even 600 roe, and some anecdotal evidence collected by the same author suggests herds of 1000 deer occurred in the Bureya District during November and December 1912. These deer overwintered on the left bank of the Amur, near Blagoveshchensk, feeding primarily on soya crops, and returned along the same routes to their summer ranges between the middle of April and 1 July.

6.3.3 Patterns of migration

For the most part, despite the large scale migration of some Siberian roe deer populations, data on roe movements are scarce and even contradictory, and the ecology of migratory populations remains practically unstudied. Therefore, we recently carried out a study of a large, annually migrating population comprising about 40 000 Siberian roe deer, inhabiting a basin of the Selemdja River in the Amursk region of Russia (Danilkin *et al.*, 1991, 1992b, 1993; Sokolov *et al.*, 1991). Here, 75 roe deer were marked with numbered collars and a further 24 fitted with radio transmitters, following capture at river crossings during the period

1988–1993. The deer were located by radiotracking from a light aircraft every 1–5 days during migration, but two or three times a month during winter and summer. Visual observations were recorded during annual watches performed along a 2 km section of the Nora River, 5 km from its mouth, and included observation during the hours of darkness with the aid of night viewing equipment.

Although it was indicated above that the beginning of roe migration in autumn is generally associated with the arrival of snow and a shortage of food, by contrast, most deer of our study population migrate annually in September when the daytime temperature is above zero and forage is still abundant in these habitats (Tables 6.2 and 6.3). In this case, the onset of frosts is the most likely signal initiating migration (Figure 6.12). The first solitary migrants (Figure 6.13) usually appear at the Nora River observation point within days of the first frost, but large numbers arrive one or two weeks later, after travelling 30–100 km over 8–15 days. Radiotracking of individuals has shown that some roe deer leave their summer range only after the second frost (or sometimes only after a series of them), giving rise to apparent waves of migration (Figure 6.12). Nevertheless, not all members of the population migrate during September, prior to the first snows. Some individuals may not migrate until November, once the snow cover is widespread, and occasional solitary migrants are encountered even during December. Several dozen roe remain on the southern slopes of the foothills of the study site, overwintering practically on the northern limit of the range of this species.

Once they have successfully reached their winter ranges, migrants usually remain there until the following spring. However, in December and January 1990–91, five radio-collared roe deer left the ranges where they had spent approximately the first two months of winter and moved farther west (Table 6.3). These movements coincided with heavy snow-

Table 6.2 Meteorological data for the migratory study population of Siberian roe deer in the Amursk region

Zone	Snow cover		Snow depth (cm)		Total annual precipitation (mm)
	First fall	Snow melt	Average	Maximum	
Northern limit	Oct. 25	May 4	54	69	781
Summer range	Oct. 25	Apr. 27–29	38–43	64–67	695–707
Winter range limit	Oct. 31	Apr. 22–25	36–38	47–48	581–606
Winter core range	Nov. 1	Apr. 20	23–29	32–35	502–558

Table 6.3 Characteristics of the migration of Siberian roe deer in the Amursk region (1989–1992); data are based on a sample of 22 radio-collared animals of both sexes older than one year

	Average (n=22)	Range
Autumn migration begins	Sept. 17	Sept. 11–25
Arrival in winter range	Oct. 10	Sept. 29–Nov. 3
Length of autumn migration (km)	140	103–195
Speed of autumn migration (km/day)	6.1	3.2–26.0
Winter home range area (km^2)	45	9–144
Return migration begins	March 28	Dec. 14–Apr. 13
Arrival in summer range	Apr. 30	Apr. 7–May 25
Length of spring migration (km)	137	108–178
Speed of spring migration (km/day)	3.8	1.1–8.3

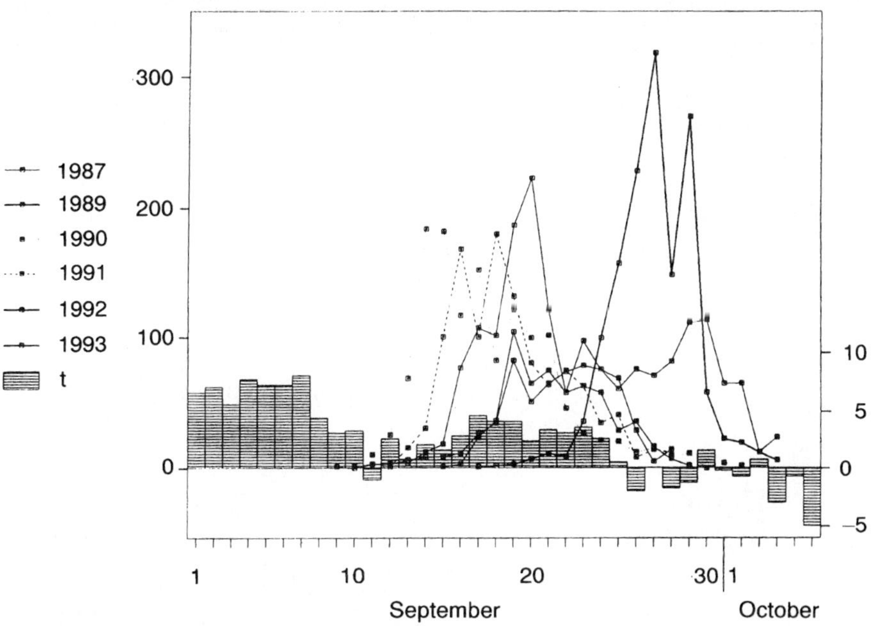

Figure 6.12 Migrational activity of Siberian roe deer population in September–October, 1987–1993. Left axis = number of migratory roe deer crossing 2 km section of river per day; right axis = mean air temperature (t), °C. (1987, n = 648; 1989, n = 1399; 1990, n = 1267; 1991, n = 1059; 1992, n = 1187; 1993, n = 1118.)

Figure 6.13 Siberian roe deer migrating across a river.

falls, again suggesting some influence of climatic factors in their ranging behaviour. The mass return migration of roe deer occurs in spring, towards the end of March and the beginning of April, but migration may not be completed until May. However, a few individuals (Table 6.3) may begin to move back to their summer ranges as early as December or January.

Overall, roe deer are conservative in their ranging behaviour, inhabiting the same home range during consecutive summers, using the same migration routes from year to year and concentrating activity within one or two particular areas each winter. They may even consistently use the same crossing points for traversing rivers (Figure 6.14). For example, one marked individual crossed the Nora River at the same section in each September of three different years between 1989 and 1991, while a further 35 marked deer were observed at the same point in two consecutive years. Furthermore, during the subsequent spring migration to the summer ranges, the deer again use those same routes taken during the autumn.

The distance travelled during migration has been measured for radio-collared roe deer, varying between 103 and 195 km; however, some individuals living to the east of the Baikal–Amur railway cover more than 200 km. The mean speed of deer during the autumnal migration is 6.1 km/day (with a recorded maximum of 26 km/day) but in spring the movement rate is much slower, averaging 3.8 km/day. Accordingly, the estimated duration of the autumnal migration varies from 17 to 27 days, while that of the spring migration ranges between 30 and 47 days.

The intensity of migrational activity varies from year to year, depending largely on population size and climatic factors. In autumn, activity clearly increases following frosts (Figure 6.12), is not affected by rain, but declines somewhat when it becomes warmer. Roe deer may travel at any time of day, but prefer to do so in the morning (Figure 6.15). Indeed, during our observations, individuals crossing the river by night constituted only 39% of the number crossing during daylight hours. The proportion of individuals of each sex and age class migrating across the river at this observation point reflected the approximate sex–age structure of the population as a whole (Figure 6.16). The maximum number of animals observed crossing this 2 km long section of the river by daylight was 318 in a single day, and a total of 1399 deer were observed over a period of 16 days during the autumn of 1989 (Figure 6.12). According to our calculations, about 25 000–30 000 roe deer may follow this particular migration route, with several thousands moving later, during October and November.

Generally, adult females take the lead position within migrating groups, while most adult males migrate alone or in groups of males only. Often bucks are hesitant to cross the river, remaining on the bank for a

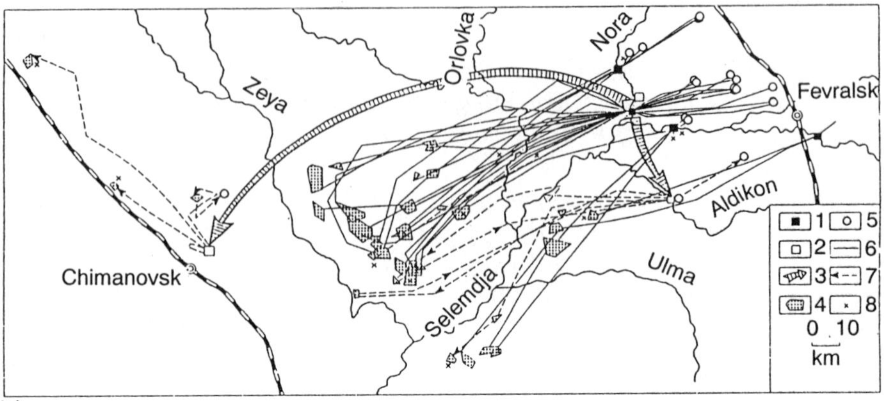

(a)

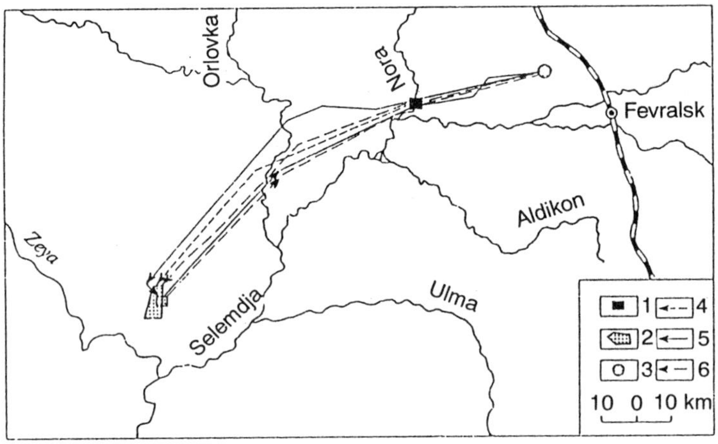

(b)

Figure 6.14 Seasonal routes of migration and home ranges of Siberian roe deer fitted with radio transmitters, in Amur district of Russia (1989–1992). (a) (1) Catching and tagging points; (2) sites of release; (3) translocations for experiment; (4) winter home ranges; (5) summer home ranges; (6) normal seasonal migration routes; (7) migration routes in the homing experiment; (8) site of death. (b) Migration routes of female No. 1 (1989–1991): (1) catching and tagging point; (2) winter home range; (3) summer home range; (4)–(6) routes of migration in 1989/90, 1990/91 and autumn 1991, respectively.

considerable time. Occasionally, dominant individuals may push the weaker yearlings towards the water, or they may wait until other deer have entered the water before darting out of cover to join the swimming group. The attempts of all-male groups to cross are invariably unsuccessful as, having swum several metres, the lead buck often looks back, stop-

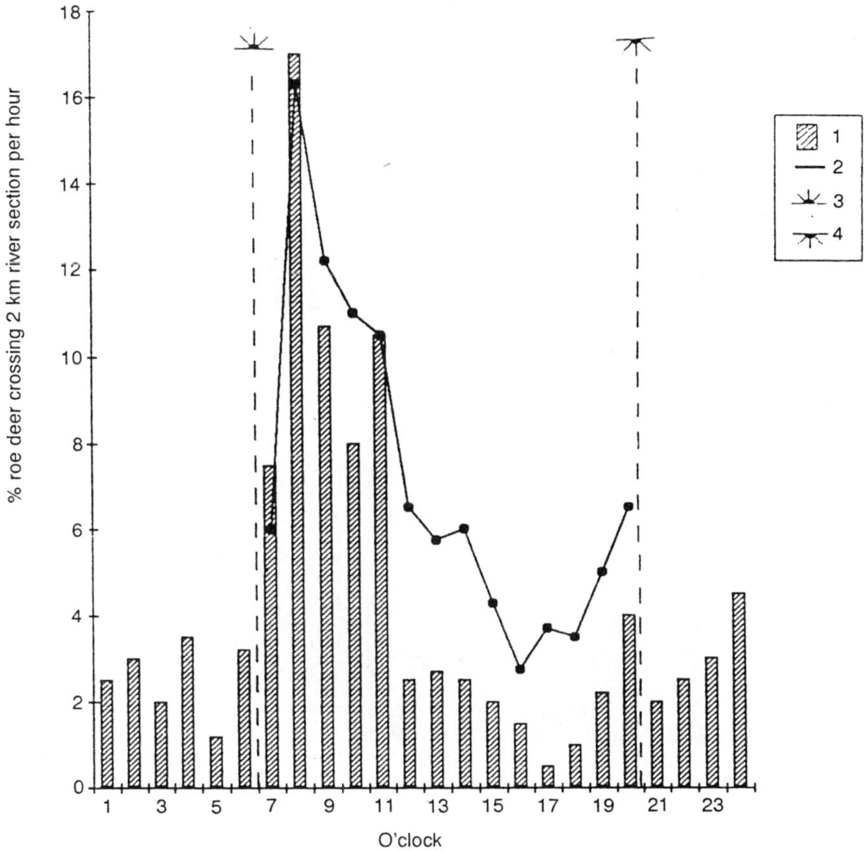

Figure 6.15 Daily activity of migrating Siberian roe deer in September, 1987–1993. (1) Based on nine 24-hour watches (n = 1187); (2) based on daytime observations only (n = 6885); (3) sunrise; (4) sunset.

ping to let the next buck pass. The second buck may then repeat this behaviour, letting the next male pass, and this may continue for some time with the group circling in the water, eventually returning to the same river bank.

At such river crossings, mothers often become separated from their offspring, despite pronounced vocal communication. Kids which have dropped behind their mothers generally give a series of high-pitched bleats and try to join groups of other roe on the bank, but are often driven away. The mother waits for her offspring on the opposite bank, responding vocally to the cries of the kid, and will generally swim back to be reunited. However, when danger is perceived, the mother may flee while the kids will attempt to cross the river alone and, not finding their

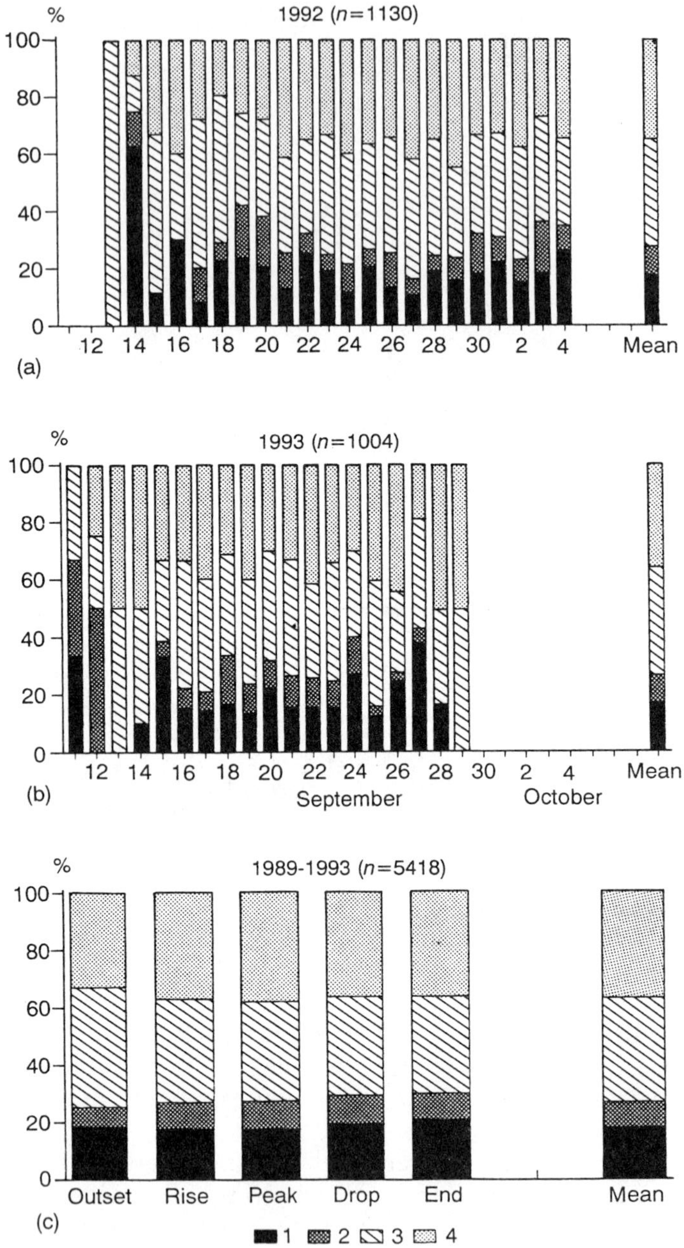

Figure 6.16 Sex–age structure of Siberian roe deer population during the autumn migration. (a), (b) Daily and mean values (September–October) for 1992 and 1993 respectively; (c) cumulative values (1989–1993) for different phases of migration. (1) Adult males; (2) subadult males; (3) subadult and adult females; (4) kids.

mother on the other bank, swim back and forth, gradually becoming exhausted.

Wolves often prey upon kids which have been separated from their mothers; they also prey upon injured deer, and those individuals at the tail end of a wave of migration. Observations of wolf packs were recorded at the river each year, particularly towards the end of the mass migration, where they often succeeded in taking roe deer, either on the bank itself or just as they entered the water.

Some sexual activity was also observed during autumnal migration, when bucks chased and attempted to mate with not only adult and yearling does, but also female kids. The high local density of roe on these migration routes probably leads to high rates of fertilization within the population.

Why does the study population migrate? Clearly, the roe deer migrate between summer and winter ranges, along a particular route. During the winters of 1989 and 1990, deer density in the zone where winter activity was concentrated was as high as 70 animals/1000 ha, while in adjacent areas it was lower by almost an order of magnitude. What attracts roe deer to the wintering range and what distinguishes it from summer ranges?

Firstly, the winter snow depth differs, with about 30 cm accumulating in the overwintering areas but a snow depth of 40–60 cm, or more in some years, on the summer ranges (Table 6.2), making movement and feeding difficult or even impossible. Migratory behaviour may be an adaptation to deep snow cover on the summer ranges, reducing the negative impact of this climatic factor on the survival and reproduction of roe. Secondly, the wintering area is a meadow habitat providing large quantities of shrub birch, a major browse species during autumn when the roe deer population may take as much as 163 kg/ha wet weight, but the availability of birch on the summer range is much lower, or entirely absent. It seems that a combination of these two factors (a limited snow cover and an abundant supply of birch) attract the roe deer to the left bank of the Zeya River where they overwinter at high density.

In terms of behaviour, the low level of aggression between individual deer may facilitate the formation of the very large herds on the wintering range. However, the winter range could also be suitable for roe deer during the summer season, with summer densities on both the winter and summer ranges of our study site approximately equal (23 and 28 animals/1000 ha of feeding grounds, respectively). Why do a proportion of the deer leave these winter ranges which could support them throughout the year, to migrate, sometimes hundreds of kilometres, with all the associated hardships, while some individuals remain settled there year round? The comparison of the quality of summer feeding, protection, climate and other environmental conditions does not suggest any principal ecological difference between the wintering and summer ranges during the summer season. It seems that there is some kind of innate drive which provokes migration towards the summer ranges to breed.

It is not clear how migrating individuals can orientate their migrations or how they locate the same seasonal ranges each year. To investigate this phenomenon, once the migration routes had been well established, a homing experiment was carried out with some deer of our study site. In September 1991, seven roe deer were caught at the Nora river crossing, tagged with radio-transmitters, transported 5–160 km away from the site of capture and released (Figure 6.14a). Following release, each animal remained near the site of release for 4–10 days, despite the fact that prior to capture these deer were in the process of migrating and were expected to continue their movement patterns. Subsequently, all seven individuals began to migrate again, continuing in the same compass direction (i.e. west) as before capture. This homing experiment (Danilkin *et al.*, 1993) suggests that these roe deer may have some kind of migration instinct, particularly in view of the fact that juveniles tend to migrate despite a lack of individual experience of migration. Furthermore, these deer seem to possess some kind of inherent sense of direction and are also able to judge distances migrated. Migration routes are regular, probably due to a process of learning when migrating as a juvenile. When separated from the mother, juveniles may continue to use the same routes by following paths or signs of the migrating population.

6.3.4 Factors stimulating migrations and dispersal

Besides the large-scale movements stimulated by adverse environmental conditions, emigration of juveniles may also be considered along with seasonal migrations. Every spring most yearlings, particularly bucks, leave the home ranges of their mothers and emigrate some distance. This is a response to increased levels of aggression among adults, expelling subadults from their ranges. This dispersal behaviour does not compare with the mass migrations described above, in terms of the number of deer involved as well as the distance travelled, which usually does not exceed a few kilometres (Bramley, 1970; Strandgaard, 1972a; Myrberget, 1973; Ellenberg, 1978; Sokolov and Danilkin, 1981; Fruzinski *et al.*, 1983; Reimoser and Zandl, 1986, 1987). In western Europe, emigration peaks at the end of April and the beginning of May, while in eastern Europe and Asia it peaks in June. Radiotracking of the study population (Danilkin and Minayev, 1988) has shown that in autumn many yearlings return to their birth site and therefore overwinter in the area where they spent the previous winter with their mother. In spring, as they approach 2 years of age, they may emigrate once again, or, if any are available, occupy vacant ranges near their birth site.

From these observations, one may speculate on the origins and development of seasonal migratory behaviour in some roe deer populations. In environments with favourable winter conditions, roe deer typically follow a settled pattern of socio-spatial behaviour with territoriality among males (Sokolov and Danilkin, 1981; Darman, 1986). Historically, such environments may have existed as interglacial zones, steppes and foothills with little snow accumulation. In these areas, increasing population density (by triggering fierce competition for territory) would have caused young roe deer to emigrate long distances in the spring. Those deer which subsequently returned to their parents' range during autumn would have survived in the relatively hospitable conditions, while those which remained on the new (summer) ranges for winter would have died when the snow depth became too great. This migratory pattern – travelling between summer and winter ranges and including established routes and seasonality – may have become more widely adopted within the population as more and more offspring learnt these habits from their mothers each year. Early migration in autumn is favoured because individuals that migrate later may become cut off from the winter ranges by any substantial early snow falls, often starving to death.

In roe deer, migration routes are usually not unidirectional, as in many other migratory species, due to the fact that the deer leave the summer ranges in a fan-like manner, forming numerous streams of migration which converge on the winter ranges in autumn and radiate away from it again in spring (Figure 6.11). Though the roe deer is generally thought of as a sedentary animal, it is capable of quite rapid dispersal and colonization: it took less than 100 years for roe to advance northward over about 1000 km in Sweden; and in eastern Europe both the European and the Siberian roe deer have spread nearly 500 km over the 30 years since the middle of this century, indicating an annual rate of range expansion of between 10 and 15 km on average.

The primary force behind this range expansion, at least for Siberian roe, is seasonal migration. Some individuals, groups or herds have been observed hundreds of kilometres distant from their main ranges and there are numerous cases of migrants remaining on seasonal ranges, reproducing and eventually colonizing the area permanently when conditions are favourable. In Mongolia in 1957, about a dozen roe deer crossed the Kerulen River and settled on the mountain of Tumen-Tsogt, where their descendants currently live in several small patches of dwarf birch and aspen totalling no more than 10 ha, isolated from the main population of roe. Similarly, in the treeless mountains of Matad in eastern Mongolia, 13 of about 100 roe which migrated here during the winter of 1977 remained and successfully produced offspring, but these deer later perished due to the lack of shelter. It seems that migrations of Siberian roe deer from the Urals have encouraged the recolonization of the left bank of the Volga, and now colonization of the right bank has also begun.

Once a new population is well established and density has increased, the frequency of social conflicts increases and more and more young animals are expelled from their parents' home ranges, resulting in the occupation of vacant territories in the vicinity. Nevertheless, as shown above, long-distance movements of young individuals are not rare and they may form the core of a new local population. Schematically, the natural dispersal pattern of roe deer is illustrated in Figure 6.17.

The natural dispersal of roe deer may to some extent be promoted by disturbance. From radiotracking, we recorded two cases in which some form of disturbance caused pregnant does to emigrate from their breeding ranges immediately prior to delivery, one travelling about 23 km and the other more than 10 km (Danilkin and Minayev, 1988). This is likely to become an increasingly important factor in the movements of roe deer populations in the future.

In summary, the social organization of the roe deer population varies appreciably over the course of the year. In summer, most animals remain solitary or form small family groups; in winter, they congregate into larger family groups and herds. In the roe deer, the family is the core unit of most group structures and, overall, serves as the basis for the social organization of the entire population. Most adult roe occupy the same home ranges year after year. During the period of reproductive activity, the core spatial structure of the population is formed from the home ranges of the 'composite family', consisting of several closely related does and their offspring, within the territory of a particular buck. Territory sizes for both European and Siberian roebucks are inversely proportional to population density and vary from 2 to 200 ha. During the rutting period,

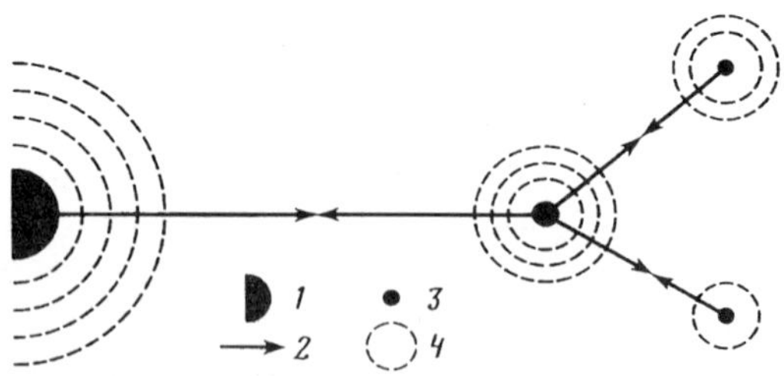

Figure 6.17 Dispersal of roe deer. (1) High density population; (2) seasonal migrations; (3) population formed from migrants; (4) dispersal of juveniles.

the territorial organization of the roebuck population does not undergo considerable change.

In sedentary populations, the home ranges of deer inhabiting relatively rich habitats remain constant from season to season; in poorer environments, the winter home range is usually larger than the summer range but probably includes part or all of it. The spring/summer emigration of juvenile deer is due to social factors and is probably more pronounced at high population density. In migratory populations, the winter and summer ranges are entirely distinct. Seasonal migrations provoked by adverse environmental conditions occur in both Siberian and European roe deer populations, and increase in intensity with snow depth (Figure 6.18). Deep snow cover inhibits free movement and may make it difficult or even impossible for the deer to obtain sufficient food. Snow may also act as a stimulus for such migrations to begin: in mountainous areas where the snow cover forms earlier, migration commences correspondingly early. The deer generally move away from mountain ranges, or at lower altitude, migration is usually southward, towards areas of lower snowfall (Figure 6.11).

Routes of migration are consistent from year to year and such migratory behaviour may be learnt and thus transmitted from generation to generation (Jurgenson, 1969). Most populations of roe deer in areas of low snow cover are sedentary (though some populations may include both migratory and sedentary individuals), indicating that migratory behaviour is not necessarily a characteristic feature of the genus *Capreolus*.

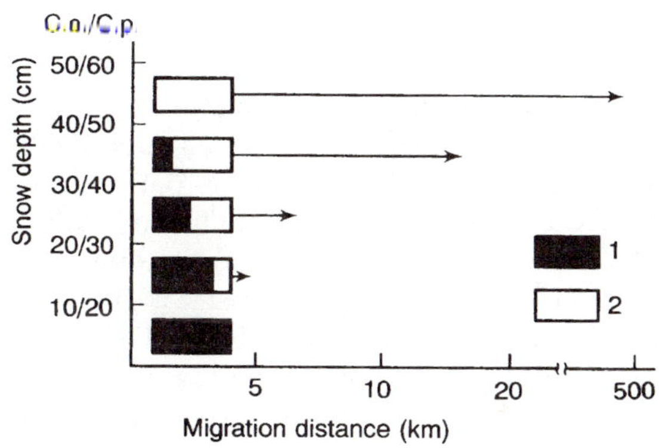

Figure 6.18 Migrational tendency of roe deer populations in relation to snow depth in terms of proportion of populations migrating or remaining sedentary (rectangles). (1) Sedentary behaviour; (2) migration behaviour. Arrows indicate migration distance (*C.c.* = European roe deer; *C.p.* = Siberian roe deer).

7

Behaviour

7.1 GENERAL ACTIVITY PATTERNS

The roe deer, due to its selective feeding strategy and consequently relatively short bouts of rumination, is a highly active animal (Cederlund, 1989). Patterns of activity are determined by a variety of factors, varying between years and seasons and with the time of day, and also depending on sex, age, level of stress as well as climatic and other environmental factors. For example, the level of activity changes periodically within the course of the day as the animal alternates between periods of movement and grazing activity and periods of rest and rumination, generally four to 12 times a day. Both individual deer and groups are most active at dawn and dusk, when activity is the most predictable and is of the longest duration (Mottl, 1957; Bubenik, 1960, 1962; Lochmann, 1965; Eiberle, 1967; Vincent *et al.*, 1979; Turner, 1980; Cederlund, 1981; Janeau *et al.*, 1981; Sokolov and Danilkin, 1981; Sempere, 1982; Lund, 1989). This is particularly marked at dawn when the arrival of daylight may act as a cue which synchronizes daily activity patterns (Turner, 1980).

From a purely energetical point of view, the pattern and duration of activity is determined by the energy budget (Turner, 1978, 1979). In this context, the relative length of time spent on feeding, rumination and resting strongly depends on the forage diversity and quality, and therefore varies between seasons (Cederlund, 1989). During spring, when fresh, easily digested forage is readily available, rumination takes up less time than grazing activity, whereas the reverse is true during all other seasons. The proportion of time spent in phases of rest is the highest during winter and early spring, and this is related to conservation of energy at a time when only poor quality forage is available and inactive bouts are therefore longest (Lochman, 1965; Cederlund, 1989; Jeppesen, 1989). In summer, when high quality, easily digestible forage is still readily found, inactive bouts for rumination are shortest. On average, in spring and autumn the total daily activity time is practically equal to the total time spent resting, but activity falls off somewhat (to around 40%) in winter, correlating with shortening day length (Jeppesen, 1989). Radiotelemetry studies have shown that the roe deer's daily level of activity in winter is generally reduced, with six daily peaks of activity during this season compared with nine during summer, but these patterns may vary from day to day for a single individual (Berg, 1978).

In contrast, Sempere (1982) suggested that the number of activity periods per day is relatively stable (average of 6.7) throughout the year. However, his conclusions were not supported by observations on captive animals which, as previously observed, indicated an average of four to eight periods of activity during periods of snow in winter, but eight to 14 during summer (Ellenberg, 1978). The number of activity periods was approximately 13% lower among does compared with bucks. Ellenberg suggested that a reduced level of activity in roe deer during winter is not only a response to relatively low quality forage and a slower rate of digestion but also a physiological adaptation to reduced food availability. These seasonal peaks in activity in spring and autumn may correspond to shifts in foraging strategy between low quality winter browse and high quality herbaceous foods not available during winter (Cederlund, 1989).

From October to April, field roe deer have between four and seven daytime activity periods accounting for approximately 33.4% of the total time, of which 76.6% is spent feeding, 7.8% on grooming in a standing position and 4.1% on other activities. The remainder of the time is spent lying, about a third of it ruminating, 13% sleeping and 4.6% grooming. The average number of individuals occupied in foraging activity at any one moment in mid winter (54%) is lower than in spring (74%). In winter, the number of activity periods for bucks is one more than for does, but the total duration of activity is approximately the same. By day, field roe deer are more active than forest deer, spending less time ruminating and more time grazing, probably due to the differing quality of forage between these habitats (Zejda *et al.*, 1985).

The activity rhythms of group members are highly synchronized, particularly at dawn and dusk (Mertens, 1984), and are probably coordinated by the behavioural tendency to follow a leader. The level of synchronization is highest during winter and lowest in late spring (Jeppesen, 1989). The synchronization of activity within family groups occurs only when fawns are at least 2½ or 3 months old (Figure 8.6). Bucks tend to graze and rest for longer periods, spending less time moving than does. With increasing group size, the time spent grazing becomes longer and locomotor activity shorter but the number of grazing and resting periods remains the same. The duration of periods of rest and time spent on social interactions are approximately the same in groups of varying size (Buttner, 1980).

Climatic factors play an important role in rhythms of feeding behaviour and activity. Roe deer feed less during hot weather than on cool and rainy days, and feeding bouts are longer when ground frost occurs. Turner (1980) observed a negative correlation of the level of population activity with both wind speed and air temperature which he attributed to problems of thermoregulation. Rainfall only reduced population activity during the dawn period.

During spring and summer, roe deer are most active by night and at dawn and dusk; during winter, they are generally most active at dawn only (Bideau *et al.*, 1983a). As energy requirements increase, locomotor activity among adult does decreases but among bucks it increases (Turner, 1978, 1979). Thus, prior to fawning, does may reduce activity levels by avoiding energetically expensive behaviours, while activity may be highest among males during the rut (Turner, 1980). According to Sempere (1982), the mean speed of movement in the European roe deer during the period of territoriality (1.26±0.19 km/h) is higher than when the deer are no longer territorial (0.57±0.05 km/h), with a maximum recorded speed of 6.2 km/h which occurred during the rutting period. These high-speed movements during summer are due to the patrolling activity of bucks along territorial boundaries, while the lower movement rates observed during winter are related to conservation of energy. In general, high levels of population activity are observed during the rut due to increased frequency of social interactions among all age and sex classes of deer (Turner, 1980). At other times of the year, differences in the duration of active–inactive cycles between deer of different ages or sex may be related to variations in basal metabolic rate (Jeppesen, 1989). Similar tendencies have been observed for the Siberian roe deer (Sokolov and Danilkin, 1981).

Visual observations and radiotracking data show that when danger is perceived, roe deer may rise from their resting place and move a short distance but may lie down again in the vicinity. The average number of times per day that roe stand or move has been estimated as 22, varying from 12 to 30 (Gonzalez and Angibault, 1987). This number seems to correlate directly with the level of anthropogenic disturbance, as well as the population densities of other large mammals. Where disturbance is very marked, especially during the hunting season, the normal daily activity pattern breaks down. In such situations, roe deer are active by night and at dawn and dusk only, becoming easily unsettled, and this possibly affects digestion efficiency. However, Turner (1980) found only limited influence of human disturbance on the level of activity of a field roe deer population: there was a tendency for the level of activity to increase when disturbance was low but increasing, but any further increase in disturbance produced a decline in population activity as the deer sought deep cover.

7.2 VOCAL COMMUNICATION

Vocal signals play an important role in the social interactions of both the European and the Siberian roe deer. The most detailed descriptions and attempts to classify particular types of vocalization can be found in studies by Meyer (1968), Stubbe and Passarge (1979), Smirnov (1975b, 1977), Danilkin (1977), Sokolov and Danilkin (1981) and Sokolov *et al.* (1987). Based on this literature and our observations of free-ranging and captive roe deer, six main types of signal can be distinguished (Table 7.1).

(a) Squealing

Squealing (or whistling) is a harmonic sound of two subtypes: a soft, short call (140–200 ms) and a loud, long call (230–590 ms), each of which has a particular social meaning (Table 7.1). Soft squealing is associated with individual interactions at close distance and is most common either during mother–offspring contact, or in various other situations where it is used either as a standard call or as an expression of uneasiness. Most commonly, it is used by newborn kids, who have lost visual contact with their mother, to signal hunger. Less frequently, such sounds are produced by does whilst searching for their offspring.

Similar squeals may be produced by subordinate deer during contact with dominant individuals (or even with humans) in order to express uneasiness and, possibly, submission. This type of submissive call is commonly heard from juveniles and does, and less frequently from antlerless adult bucks (in winter). When approached by a dominant animal, the subordinate individual bends its legs slightly, extends the neck while lowering the head, often urinating, and gives between one and 10 soft squeals. This may be followed by nasal contact and/or escape.

This sequence of behaviour of subordinate individuals closely resembles that of does on heat when approached by an adult buck. These squeals are rather louder at the beginning of the rutting period, when being chased by males, particularly aggressive bucks. Later in the course of the rut, when does seem to be less disturbed by these interactions with bucks, this call is heard less frequently, if at all. During the final stages of courtship and also during mountings, Siberian does may emit pairs of soft, short (55 ms) squeals, forming a short series of two or three calls which follow at 462 ms intervals (Sokolov *et al.*, 1987).

The second type of squealing vocalization is associated with extremely alarmed deer which may produce a very loud, long and shrill whistle which is clearly a sign of fear or great disturbance. This sound is typically

Table 7.1 Types of vocal and non-vocal communication in the roe deer

Type (*phonetic*)	Nature of sound[1]				Sex–age class	Season	Context
	H/N[2]	Group	Duration (ms)	Frequency (kHz)			
Squeal	H	European	144–175	1.4–2.5	Does; kids	May–Sep	Mother–offspring
(*pee*)		Siberian	365–489	1.0–2.1	(rarely bucks)	(rarely other)	Unease, discomfort
							When chased at rut
(*pee-ye*)	H	European	365–489	1.0–2.1			
		Siberian	237–583	0.8–2.1			Extreme fear
Rasp	N	European	397–510	1.0–2.7	Bucks (more	May–Oct	Excitement at rut
(*choof*)		Siberian	397–510	1.0–2.7	rarely does)	(rarely other)	Bucks aggression)
(*sh-h-h*)							Mother–offspring
Bark	N	European	145–185	1.4–2.5	Adult bucks (rarely	All year	Unease
(*baao*)		Siberian	216–292	0.7–2.3	immature does)	(esp. spring–	During exploration
						autumn)	Alert
(*boh-boh*)							Discomfort
Whine	H	Siberian	151–159	0.5–0.6	Bucks	Jul–Aug	Discomfort at rut
Scream	H	European			All	All year	Pain or fear
		Siberian					
Other					All	All year	Unease
stamp							Alert
scrape							Conflict

[1] After Sokolov *et al.* (1987)
[2] H = harmonic, N = non-harmonic

produced by kids when in apparent danger. The reaction to such calls is particularly pronounced in lactating does during the first weeks after birth, but occasionally bucks nearby may also be drawn to the sound. In such situations, the adult will run quickly towards the source of the sound and display aggressive behaviour towards any intruder.

(b) Rasping

Rasping (or panting) indicates a state of great excitement in the animal. Depending on the situation, this call may be soft and short, loud and shrill, or intermediate. The former type is most commonly given by a buck when approaching a doe on heat, but becomes louder when the doe is chased. Roe deer may also produce rasping sounds when displaying aggressive behaviour. Bucks may produce such sounds during combat display and fighting, either with another buck or with a perceived opponent (humans, or even shrubs and trees), when chasing or tracking other bucks, and also during periods of frenzied marking activity. Lactating does may also be heard rasping when protecting a newborn kid from predators, humans or even another doe. This type of vocalization is first recorded in kids at 3–4 months of age during play.

Very soft rasping, audible to the human ear at no more than 20 m, may be produced by a mother when approaching her offspring, and also by dominant deer when moving towards a subordinate, probably indicating that contact is desired. As the animal becomes more excited, the call may become louder and occasionally develops into another type of vocalization such as squealing or barking. We have recorded such a change in both captive and free-ranging roe deer of both species.

(c) Barking

Barking is the most typical and intensive call, generally consisting of a series of between three and 30 sounds, with a recorded maximum of 168, of which the first one is often the loudest and longest and can be heard up to a distance of 3 km. On average, each separate bark is from 140 to 330 ms in duration, with intervals of between 0.3 and 1.4 seconds, and a frequency in the range of 1–2.5 kHz. The dynamic spectrum of barking is very similar to that of the Cervinae, especially Bukhara deer (Nikolski, 1976). The normal bark is generally produced from a standing position, but this is transformed into a very short sound when the deer is moving. Roe deer kids do not use this call initially, only beginning to bark at 6–8 months of age, suggesting that this sound may develop from another type of vocalization.

Barking is often a response to some form of disturbance, particularly at dusk or at night, but rarely during the day. Such disturbance may include any kind of noise or scent, but also visual contacts with humans or potential predators. When the source of the disturbance is not identified, the deer may move in an arc, barking and stamping, in order to position itself downwind to assess any potential danger by sense of smell.

In spring, deer seem particularly sensitive to disturbance as barking is frequently provoked simultaneously in several members of a group, often eliciting further vocalization in nearby animals. During this season, roe deer are especially alert to such calls during exploratory behaviour. While barking may sometimes alert other members of a group in a similar way in summer, commonly other deer react only briefly to such a signal before resuming feeding or resting when no danger is perceived, even if the bark is repeated.

Fighting may also be followed by a spell of barking of the victor, or both combatants if the outcome of the fight was inconclusive, and highly active bucks may bark while marking a territory. Some researchers have suggested that roebucks may bark for no apparent reason and consider such vocalizations to be acoustic marking of a territory (Meyer, 1968; Smirnov, 1975a, 1977). Our own long-term observations show that barking is always a provoked act but the cause cannot necessarily be ascertained by an observer each time. Thus, roe deer do not possess a particular call for territory marking; but barking, together with rasping and non-vocal communication, may play a role in territoriality.

A comparative study of free-ranging roe has shown that Siberian roe deer bark more frequently than European roe, and bucks bark more frequently than does. In summer, in the southern Urals, barking was recorded in 24.5% of observations of Siberian roe, of which 70% involved bucks. In similar circumstances, European roe deer of the Belgorod and Kursk regions of Russia barked in only 7% of all cases, but again this call was heard primarily from bucks (67.8%) rather than does. From November to March, both species barked relatively little (4% of all cases for Siberian roe and 1% for European roe), while barking was most intensive among both bucks and does during spring (29% of all cases for Siberian roe and 9% for European roe), when the deer were still mostly in groups. The lower rate of barking among European roe may be because these deer are habituated to human contact due to the relatively higher level of anthropogenic activity in this part of the roe deer's distribution, while such activity is more disturbing for the Siberian roe deer which is less accustomed to such contact.

(d) Whining

Whining (or mewing) involves a series of three to 20 sounds and resembles the whining of a dog. It has been recorded in some Siberian roebucks during the rutting period when the buck stops to rest, either standing or lying down, after chasing a doe for a long time without success. Its importance is not clear, but the nature of the sound seems to express a mixture of discomfort (squeaking) and sexual excitement (rasping).

(e) Screaming

Screaming (or moaning, howling) is a very intensive vocalization and probably includes components of several different calls. It is given by a wounded or captured animal and it seems likely that it expresses fear or pain.

(f) Non-vocal signals

Signals of non-vocal origin which may be used as a form of communication include stamping, when a roe deer moves slowly, alternately raising each foreleg to beat it against the ground. This is usually a response to an unidentified or perceived disturbance and may be an element of exploratory, defensive and agonistic behaviour. It is likely that stamping expresses indecision prior to choosing a course of further action, but it may also warn neighbours of danger and may possibly be used to intimidate a potential opponent. Stamping has been recorded in roe deer when coming into contact with foxes, dogs and humans, and also during confrontations between displaying adult bucks.

A similar signal may be made by scraping the ground with the forelegs, producing areas of bare ground. Such behaviour is common during snow lie, to obtain food and prepare resting sites; however, in summer, scraping may be a signal used in display behaviour in confrontations between bucks. Generally, such signals are used to induce an opponent to charge.

Thus, during the first months of life, roe deer seem to be limited to only one type of vocalization – squealing (soft and loud) – from which the other types of call may subsequently develop. These forms of communication, with the exception of whining, are the same in bucks and does and, by and large, are also similar in the European and Siberian roe deer, carrying the same functional meaning. However, European roebucks have not as yet been known to produce a sound equivalent to the whining which has been recorded in some Siberian bucks (Sokolov and Danilkin, 1981) nor have pairs of squeals been recorded in European does. The squeals produced when kids lose contact with the mother are somewhat different, being of a slightly higher pitch in the European roe (Sokolov *et al.*, 1987), although they may become identical with age. Nevertheless, we cannot yet come to a final conclusion regarding the differences between the vocal signals of these two roe deer species, as the only comparative work to date included only a few animals and differences were only detected in a small proportion of these individuals. It seems likely that their equivalents will be found in the European roe. The primary difference between species may in fact concern the frequency and duration of vocalizations, as is observed in kids during their development.

7.3 CHEMICAL COMMUNICATION

From spring until autumn, roebucks define their ranges with olfactory and visual marks (Raesfeld, 1956; Hennig, 1962; Turcek, 1962; Kurt, 1966, 1968a; Stark, 1966; Meyer, 1968; Prior, 1968; Smirnov, 1978; Sokolov and Danilkin, 1981). Scent marking involves a secretion from the glands of the head and skin, which swell greatly in summer. The secretion is applied by rubbing the forehead, cheeks and neck against trees, shrubs and tall grasses (Figure 7.1). Visual marking includes the fraying of trees with antlers (Figure 7.2), which may then be marked with scent also. These scent and visual marks may attract bucks from some distance. The typical behavioural sequence generally begins with the buck sniffing the tree which may retain the scent of a previous marking secretion. He rubs his forehead against the tree (having passed it between his antlers), frays the bark and sniffs again at the frayed spot; then he rubs his forehead or cheek against the tree and sniffs again. This procedure is generally repeated many times. Roe deer preferentially mark small trees of less than 8–12 cm in diameter, fraying them at heights from 15 to 100 or 120 cm. When highly excited, the buck not only rubs himself against the tree but may also butt it, sometimes bending it to the ground, breaking branches or even the trunk. In such situations, it is especially noticeable that what was initially marking behaviour becomes an act of aggression and the object being marked becomes a simulated opponent.

A further form of marking involves scraping the ground with the forelegs, producing obvious patches of bare ground measuring around 1 m in length and 0.5 m in width. These scrapes often occur at the base of trees and such spots are probably also scent marked with secretions from the interdigital glands (Figure 7.2). In rare cases, an excited buck that is preoccupied with marking activity may bark, but it seems unlikely that this signal is really associated with the act of territory marking and more probably this is merely the expression of a state of high excitement. Unlike bucks, does do not display specific marking activity, though they probably do perform some elements of this behaviour (Kurt, 1968a).

Towards the west of their total range, European roebucks begin tree rubbing as early as February, visual marks appearing once the antlers have been cleaned of velvet, but scraping activity only begins in March.

Figure 7.1 Scent marking: roebuck rubbing forehead, cheek and neck against a tree to mark it. (Photograph by V. Semenov.)

However, in northern and eastern areas of their range these behaviours commence later: olfactory marking is observed from the end of March onwards and visual marking does not start until April. The Siberian roe deer of the Urals do not begin marking their ranges until May but in such regions, where snow is deep and lies for a long time, marking activities do coincide with the onset of the growing season.

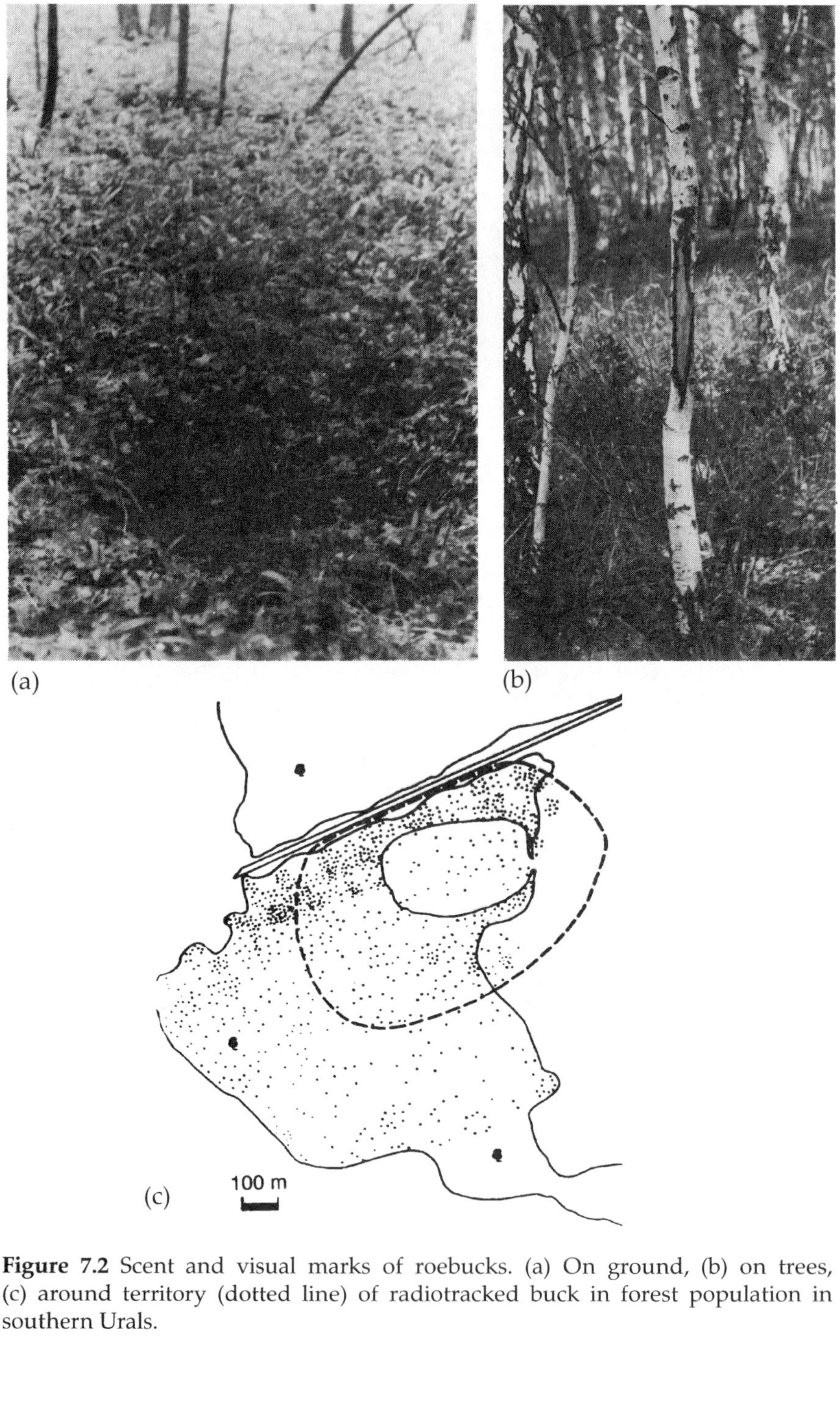

Figure 7.2 Scent and visual marks of roebucks. (a) On ground, (b) on trees, (c) around territory (dotted line) of radiotracked buck in forest population in southern Urals.

Adults begin to display marking behaviour 2–3 weeks earlier than juveniles, setting the limits of their exclusive territories. Lack of marking on a particular area indicates that a potential range is available, allowing juveniles to find and establish a territory without extensive confrontation. Buck marking activity has two peaks – one at the beginning of the territorial period and one immediately before the rut – which also coincide with displays of aggressive behaviour. The decline in marking activity at the beginning of summer occurs when most territories have been claimed, their limits established and marked, and when most young bucks have been expelled from already occupied areas. Thus, adult bucks can retain their territories with minimal effort, enabling them to conserve energy immediately prior to the rut. Following the rutting period, marking behaviour ceases (Ellenberg, 1978; Sokolov and Danilkin, 1981; Sempere, 1982).

Roebucks mark their entire territory (Figure 7.2), but are especially likely to do so at boundary intersections, paths (sometimes every other tree), and isolated trees. In some cases, where trees are rare in a particular habitat, every tree may be marked. Despite the fact that roe deer generally confine their day-to-day activities to the centre of the home range, it is the periphery of the territory that is the most intensively marked. Adult males mark not only their own ranges, but they may also do so within the territories of neighbouring bucks during raids, in particular, concentrating either on trees previously marked by the resident buck or on other trees in the vicinity. During such raids, the bucks become particularly excited at scrapes and latrine areas. Marking activity within an occupied territory is also occasionally observed in yearling and 2-year-old males, sometimes in the presence of the resident buck. In this case, the subadult marks a single tree, which may already bear the mark of the resident buck, and is followed by the adult male, who then sniffs and remarks the same tree. The resident buck then often tries to charge the intruder who, when approached, assumes the threat posture without engaging in combat, before running to another marked tree. This sequence is repeated many times.

During the period from spring to autumn, each territorial buck may make hundreds of scrapes and mark thousands of trees (Figure 7.2), many of which may die but will remain targets of marking behaviour for many years to come. The intensity of marking depends on the season, the time of day, the weather and the arrangement of ranges in a particular population. For example, during rainy and cold weather and by daylight, marking behaviour is very rare and may not occur at all. At the beginning of the territorial period, bucks may apply as many as 500 or 600 olfactory marks per day, but by the middle of the period this may fall to between 40 and 150, and during the final stages only 10 marks per day may be used (Ellenberg, 1978). Olfactory marks are often renewed, some of them frequently but others only rarely. Our observations of a single Siberian roebuck during August showed that he marked a total of 457 trees, of which one particular tree was marked eight times, 50 trees were marked four times each, 132 trees were marked three times each, 93 trees were each marked twice and 181 trees were marked once. In this case, isolated trees and those at territorial boundaries were marked the most frequently.

External disturbances, particularly intruders, have a pronounced effect on marking activity. For example, in Siberian roebucks, a rate of marking of about 100 rubs per day when undisturbed increased approximately fourfold when a known human was present within their range, and when fresh footprints of an intruding young male were found the marking rate became as high as 400 rubs per hour. An excited roebuck can mark as many as 370 trees in a period of three hours. When no intruders are present in the territory, a buck will alternate between marking behaviour and feeding, but in the presence of an intruder the buck devotes all his time to marking (Sokolov and Danilkin, 1981).

The rate of marking also greatly depends on the vegetation structure and is less intense in thick forest compared with more open stands. Scrapes are particularly prevalent in clearings, and in steppe and field habitats. Under captive conditions, the marking activity of bucks is almost invariably associated with aggressive behaviour, and in addition to trees the deer may also mark fences and various other artificial constructions. In captivity, dominant bucks may display marking behaviour even in winter (something never observed in nature) but in this case the marking is performed without first sniffing the mark site.

Thus, the marking behaviour of the roe deer across its range is noticeably seasonal, coinciding with the onset of the growing season, and is associated with aggressive activity and territory protection, related to the level of blood hormones (Barth *et al.*, 1976; Sempere, 1982). It is clear that marking acts a signal, indicating to other roe that a particular range is already occupied, thus often preventing direct confrontation. In some cases, however, despite intensive marking activity of a territorial buck, the boundaries of his range may be violated relatively frequently. Nevertheless, as has been observed in other species, the individual that has marked the area still seems to hold some kind of psychological advantage. Once within an occupied territory, an intruder becomes uneasy and the territorial buck generally prevails in any combat.

Other substances which may signal the presence of a roe deer on a particular range, and possibly indicate the age, sex and physiological status of the animal, include urine, faeces and probably also saliva deposited on the ground and on vegetation during feeding and licking. In addition, scented secretions of the sexual organs, metatarsal and interdigital glands and secretions of glandular complexes in the skin of the metatarsus and metacarpus may play some kind of role in roe deer communication. Indeed, roe deer are able to locate other animals by following the scent of footprints, sniffing the ground continuously. In this case, the deer are responding mainly to the scent of the interdigital gland secretions which are deposited on the ground and also to the secretions, produced by glands contained in the skin on the metatarsus and metacarpus and by the metatarsal glands, which are deposited on vegetation and at snow level. The secretions of the sexual organs are particularly important during the rutting period, when a buck chasing a doe on heat keeps his head and neck parallel to the ground and at the level of her genitals. If he loses sight of the doe, he will resort to tracking her once more, following the doe's footprints.

Roe deer, particularly bucks, also show significant interest in urine odours, often responding to urine produced by does on heat by raising the upper lip to bare the teeth in the typical rutting gesture ('flehmen'). During summer, bucks may also become very excited after sniffing the spot where another buck has recently urinated.

Initially, at the first meeting of two deer, naso-nasal contact (probably, the animals respond to the scent of saliva or nasal skin glands) is followed by sniffing the body, head, metatarsal glands and genitals, most attention being paid to the genitals.

It is clear from the behaviours described above that chemical communication plays an important role in the life of the roe deer. Histological studies of skins of both the European and Siberian roe deer indicate that gland secretions can be deposited on either ground or vegetation (Meyer, 1968; Sokolov and Danilkin, 1981). Perhaps the most important elements in the chemical communication and behaviour of roe deer are the sweat glands and sebaceous glands (which swell in summer) of the head, neck and anterior body. Also important are specific glands such as the metatarsal glands and the interdigital glands of the forelegs and hind legs, as well as gland complexes in the skin on distal sections of the metatarsus and metacarpus, whose structure allows them to be considered as specific glands (Figure 7.3).

Experimental manipulation has shown that roe deer are able to distinguish mature individuals from immature deer on the basis of the scent of the metatarsal gland secretion. Furthermore, when adult bucks are artificially presented with the scent of the forehead skin secretion, they become aggressive and display marking behaviour. These observations suggest that the secretion of the metatarsal gland may facilitate individual recognition, and show that forehead skin glands play an informative role in territorial behaviour (Broom and Johnson, 1980). Gas chromatographic analysis of these secretions shows that territorial bucks produce more highly scented substances than non-territorial individuals and lac-

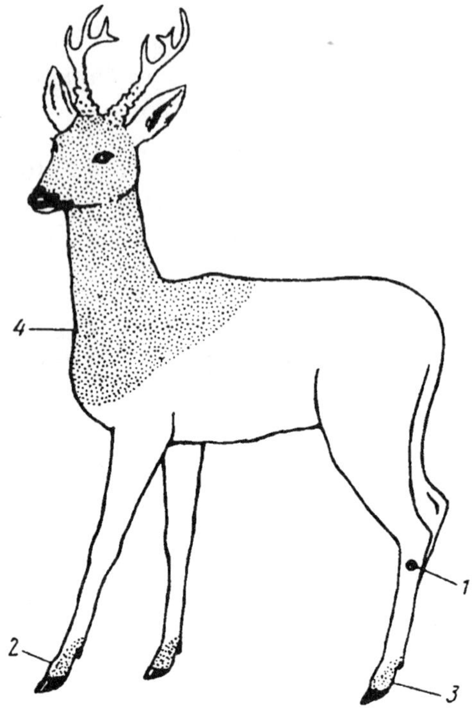

Figure 7.3 Location of zones containing sebaceous glands and sweat glands. (1) Metatarsal gland; (2) anterior interdigital gland and distal glandular complex of metacarpus; (3) posterior interdigital gland and distal glandular complex of metatarsus; (4) skin glands of head, neck and chest which swell in bucks only during summer.

tating does tend to produce a greater quantity than subadults or does without kids (Ellenberg, 1978). Moreover, the secretion of the interdigital gland differs between individuals, also indicating a possible role in individual recognition.

The presence of a specific skin gland on the head of male roe deer, in front of or between the antlers, seems to vary (Schumacher, 1939; Schaffer, 1940; Raesfeld, 1956; Necas, 1963; Pielowski, 1970). Meyer (1968), despite a detailed histological analysis of the skin of the head of European roe, was not able to locate it, and we also did not find such a gland in Siberian roe deer (Sokolov and Danilkin, 1977, 1981). Examination did not reveal any specialized portions of thickened skin, indicative of a hypertrophied well-defined gland complex, as in the metatarsal and interdigital regions. In winter, the skin of the head and neck is of uniform thickness among bucks, but a thickening from the nose to the back of the head is regularly observed by summer (Figure

7.13). This may be accompanied by a swelling of the sebaceous and sweat glands during summer, which then diminish in size severalfold during autumn and winter when the secretions, produced in small quantities only, may not function as agents of communication. These glands are also slightly bigger in summer than in winter among does. Kids posses only poorly developed glands, while in yearling bucks they are nearly as big as adults. It may be that roebucks do not possess a specific forehead gland for territory marking, and that in fact the skin of the head, neck and anterior body carries nothing but numerous sebaceous and sweat glands which swell seasonally, accompanied by changes of skin thickness. It seems likely that this localized area of thicker skin may have initially served only as a means of protection in male confrontations. However, the seasonal thickening of this region and the swelling of the glands associated with it provoked the production of a larger quantity of secretion and thus, in the course of evolution, a new function developed related to marking and communication behaviour.

7.4 SEXUAL BEHAVIOUR

Rutting behaviour of wild ungulates varies widely (Adolf, 1954; Sludski, 1956; Walther, 1958, 1963; Savinov, 1962; Ustinov, 1967; Lott, 1991) but from comparative studies and a review of the literature it seems that the sexual behaviour of European and Siberian roe deer is identical. In the roe deer, behaviour during the rutting period entails some particularly vigorous exchanges, the buck chasing hard any doe on heat, while rasping. The doe generally tries to escape, creating the impression that the buck is attempting to drive her off (Figure 7.4). Indeed, some observers report that rutting roebucks are extremely aggressive, attacking does (occasionally even killing them) and forcing them to escape (Turkin and Satunin, 1902; Dahl, 1930; Flerov, 1952; Averin, 1949; Heptner *et al.*, 1961, Cherkasov, 1962). However, others have suggested that the level of aggression toward the doe is exaggerated, and that he does not drive her away but pursues her in order to maintain visual contact (Smirnov, 1975a, 1978).

Both points of view are correct to some extent. From our long-term observations, certain males, particularly young bucks, may be rather aggressive towards the females the first day of the rutting period, even to the extent of butting them with their antlers. Typically, as the buck approaches a doe he lowers his head with antlers pointing forward, often rasping, signifying a threat and provoking a defensive reaction in the doe. She will turn her head towards the buck, cowering, and urinate, often giving several submissive vocalizations before moving away. The buck may take up the chase immediately, or may first sniff the doe's urine on the ground and perform a 'flehmen'. The chase is generally long

Figure 7.4 The rut. Bucks are often observed chasing does on heat.

and continuous, and describes large circles before the tiring doe resorts to running around trees, shrubs or raised ground to avoid the buck. By repeatedly following the same trajectory many times, roe deer often beat runs in the shape of a ring or a figure of eight, characteristic of rutting behaviour (Figure 7.5). Finally, exhausted, the doe will abandon her attempts at escape, stopping and allowing the buck to mount her. Subsequently both participants seek out cover, lying down to rest. A tired doe may even lie down during the chase, but in this case the excited buck will butt the doe, forcing her to get up, and will then mount her. When the doe succeeds in escaping, the buck will quickly track her footprints and resume the chase.

Once the buck has successfully mounted for the first time, the behaviour of both deer changes considerably. The doe becomes less afraid of the buck, making fewer attempts to escape, and further copulations occur frequently, almost immediately the buck approaches. The deer engage in naso-nasal contact, the buck sniffs and quickly licks the doe's

Figure 7.5 Typical run patterns beaten by roe deer during the rutting period.

head, flanks, back and genitals, and then mounts her (Figure 7.6). The attempted copulations of young, inexperienced males are invariably unsuccessful; sometimes they try to mount while standing beside or in front of the doe, before assuming the correct posture by moving their hind legs around the female step by step. On average, a single mounting lasts 22 seconds for yearling and 2-year-old bucks (varying from 8 to 120 seconds), but only 6 seconds for adult bucks. Two or three servings may

(a)

(b)

Figure 7.6 Mating. (a) Buck sniffs and licks doe and (b) mounts her.

be separated by only a few minutes but subsequently the deer generally rest for an hour or more. Upon ejaculation, the buck falls heavily on to the doe and almost invariably she then lies down, generally in the most secluded spots. The buck often loses sight of the doe between copulations, relocating her by tracking.

In most natural situations, a doe on heat is usually chased by a single adult buck and only rarely by two males; in exceptional cases, three or four bucks may be involved. As soon as the doe ceases to run, it is the dominant male that remains with her, assuming a threat posture to keep the others at not less than 30 m. Under these circumstances a dominance hierarchy amongst the males is clearly apparent, with the subordinate buck dominating the third male who, in turn, drives off the fourth (generally the youngest) buck, and the whole group circulating around the rutting pair. Having mated the doe, the dominant male lies down to rest and the doe may then be covered by the second and even the third buck, with as many as 27 copulations observed over a period of several hours. If the doe escapes, the buck group quickly follows, maintaining the hierarchical order.

Under captive conditions in enclosures of limited size, it is not uncommon that the doe is mated repeatedly by all the males present. For example, in an enclosure situated in the Moscow region, the reproductive activity of one particular Siberian roe doe helps to illustrate the situation. Two roebucks in the pen (a 3-year-old and a yearling) failed to cover the 2-year-old doe during her first day on heat, despite a protracted chase. On the second day, they again became involved in a long chase, and only succeeded in mounting the doe once she became exhausted and fell whilst running, having circled a tree more than 40 times. The older male mounted her after he had forced her to stand by butting her, and within two minutes she had also been covered by the yearling. Then, after a 3-hour rest, they subsequently mounted the doe 18 times in the following 5½ hours before darkness, with the yearling gaining two thirds of the copulations and the adult only one third. The doe showed a much stronger tendency to run when the adult male approached and he was obliged to enter into a prolonged chase in order to have an opportunity to copulate. In contrast, the doe was clearly less intimidated by the yearling and she did not run on any of the 12 occasions he attempted to mount. A further seven copulations were recorded the following morning. Subsequently, after a substantial period of rest, the doe ceased to permit the males access. When the female population density is high relative to that of males and mating opportunities are therefore frequent, the opposite situation may occur where one buck alternately covers two does on heat during a single day (Holmes, 1973). Some observations have indicated that a doe prefers to maintain an association with a single buck, mating with him for several years successively.

Once a doe is fertilized and no longer at the peak of oestrus, both the doe and the buck behave differently. As the buck approaches, the doe cowers and urinates. The buck spends a long time (about a minute) sniffing her urine and performing the 'flehmen', and then begins feeding or, more rarely, resumes chasing the doe. In the latter case she urinates a second time and the buck, having sniffed the urine again, becomes disinterested. This sequence is generally repeated again one or two hours later and the buck may remain with the doe for one or two days more. Towards the end of the rut, raids into the territories of other bucks, obviously with the purpose of finding a sexual partner, become more frequent. During this period, bucks respond especially readily to squeaking-like noises and are often taken by hunters imitating the sound (Kerschagl, 1952; Kurt, 1968a; Bayern and Bayern 1975; Ellenberg, 1978; Stubbe and Passarge, 1979).

Generally, roe deer rut for between 2 and 5 days. During this period the buck practically ceases to feed (Hewison *et al.*, in press a), becomes less vigilant and will not leave the doe, even when danger is apparent. Those does which were not fertilized may come into heat a second time, with an interval between periods of oestrus varying from 9 to 84 days, suggesting that there is no clear cycle of oestrus activity.

Radiotracking studies have shown that the system of territories is generally maintained during the rutting period (Sokolov and Danilkin, 1981; Sempere, 1982), largely due to the fact that the doe's home range commonly falls within the territory of a single buck which follows her without attempting to interfere with her normal patterns of movement. However, we have observed four cases in the wild of a buck running to block the path of a doe, driving her back from the boundary of the buck's territory. In captivity, a buck chasing a doe on heat also often tries to restrict her movements to an area where there are no rival bucks present. The system of territoriality also diminishes the number of contacts and potential conflicts between bucks. Nevertheless, rutting pairs often do cross into another buck's territory, provoking investigation and possibly some tracking behaviour in the neighbouring adult male, but he will join them only in exceptional cases. More commonly, the violation of the territory border triggers an aggressive response in the second buck, who intensifies the level of marking activity.

In roe deer, the rut takes place in summer when the behavioural patterns described above are observed. In addition, the roe is the only species of ungulate to exhibit embryonic diapause. How have these characteristics evolved and to what extent are they interrelated? A hypothesis to explain these phenomena was proposed earlier which can now be expanded a little (also see Danilkin, 1990). It seems likely that the immediate ancestor of modern roe deer rutted during the autumn, as is the case in closely related species of the Cervidae, and a shift towards summertime territoriality evolved subsequently. The drive to maximize reproductive success within a territorial system may have provoked bucks into early sexual activity, even pursuing females on their territories before they apparently came into oestrus. Such premature activity has been observed in modern wild populations (Sokolov and Danilkin, 1981). This premature sexual behaviour, including vigorous long chases by excited and aggressive bucks, may stimulate sexual responses in does. This stimulation is clearly apparent in the wild and, furthermore, once exhausted following a long chase, a doe is unable to prevent the buck from mounting. An earlier rut may have been selectively favoured because reproductive activity (the rutting and birth periods) is thus concentrated into the climatically milder summer months, possibly enabling roe deer to inhabit more northern ranges with relatively large seasonal shifts in environmental conditions (Loudon, 1982). It has also been suggested that this would promote female choice at a time when high quality resources are available, and so male–male competition is at its most intense (Sandell, 1990). Gradually, a shift from an autumn to a summer rut may have occurred but, with the evolution of delayed implantation, the birth period did not change. Thus, the fertilized ovum, under a mechanism of physiological control as yet not well understood, does not develop normally until the middle of winter. This hypothesis seems to explain the relationship of the evolutionary, ethological and physiological aspects of this phenomenon.

7.5 AGGRESSION

Aggressive behaviour is observed in most mammals and plays an important role in their social interactions. This phenomenon has been widely studied but a precise definition of the term and an evaluation of its causes are still subjects of debate. In ethology, the term aggression is generally used in a narrow sense, describing the behaviour of one individual towards another aimed at establishing a certain hierarchical status, claiming territory, food or some associated object, or ensuring protection, which involves demonstration or actual combat. Demonstrative behaviour, along with territorial, defensive, feeding and other behaviours, is often classified separately to actual acts of aggression. However, aggression is a part of an intricate complex of behaviours defined as agonistic behaviour, which includes demonstration, threat, attack, submission and escape, and it is rather difficult to separate aggressive behaviour as such from this complex. Undoubtedly, aggression is caused primarily by an individual encroaching to within a certain distance, provoking visual, acoustic and olfactory contact, but it may also result from a conflict between various types of activity. The level of aggression depends on heredity (for example, a strong hereditary component of aggression is found in fox terriers and fighting cocks), hormonal status and level of nervous activity (Krushinski, 1960; Lorenz, 1967; Showen, 1972; Tinbergen, 1974; Hinde, 1975; Baskin, 1976; Shilov, 1977; Panov, 1983).

7.5.1 Aggressive interactions in roe deer

Roe deer are aggressive mainly during the period of reproduction. The relationship between a doe and her offspring of the previous year is broken 2–4 weeks prior to the birth period. This rupture is caused by an increased level of aggression in the mother when occupying her kidding range, which she protects from other deer before giving birth. Subsequently, does express clearly pronounced aggressive behaviour in the first 1–2 weeks after kidding, demonstrating aggressively to almost any intruder who approaches her offspring, particularly if the kid gives a vocal signal (Kurt, 1968a; Espmark, 1969; Sokolov and Danilkin, 1981). In this situation, the mother assumes a threat posture, ears pressed back, lowers her head to the ground, stamps her forelegs and makes hissing sounds before suddenly rushing towards the intruder, chasing for between 50 and 100 m, or running past the source of the disturbance before turning to repeat the attack. If the intruding roe doe does not run away, the two opponents approach each other, their heads and necks vertical but held slightly to one side. At a distance of 4 or 5 m, they assume threat postures before charging to butt each other, retreating to repeat the attack several seconds later. Invariably, the victor in such aggressive encounters is the doe with kids present in the near vicinity. Equally, at this time of year, does may sometimes display threatening behaviour towards adult bucks, while in all other seasons they tend to avoid any contact.

As the offspring develop, the level of aggressive behaviour in lactating does decreases and the home ranges of two or more does may begin to overlap. From the end of August, when the breeding season is over, there are practically no traces of intolerance remaining; does with neighbouring ranges, who may be related, may form female groups. This cycle of aggression levels among does, intensifying markedly in the last weeks of pregnancy and reaching a peak immediately after birth, is probably under hormonal control.

The level of social intolerance between related does is clearly less pronounced than that between unrelated deer. Thus, philopatric subadult daughters may stay near their mother's home range, producing kids there a year later, resulting in a high degree of range overlap and, thus, a composite family structure. Members of a composite family are often aggressive as a unit towards others trying to join their group, particularly kids. However, kids which have lost their mother and subadults are readily adopted by a related doe. A dominance hierarchy system among female roe during winter is only observed at high population density or if food resources are limited (Geiger and Kramer, 1974a,b; Espmark, 1974) when, under such extreme conditions, the offspring of dominant females fare better. As shown by Ellenberg (1978), progeny of dominant mothers are in better condition and have a higher chance of survival, commonly becoming migrants.

Roebucks are generally more aggressive than does. The onset of male aggressive behaviour commences when the antlers are cleaned of velvet and marking activities begin (in captive males, it is even earlier) (Figure 7.7). In the west of their total range, such behaviour is recorded 1–2 months earlier than in the north and east, coinciding with snow melt and the start of the growing season. The synchronicity of aggressive and marking behaviour with the acquisition of hard antlers at the beginning

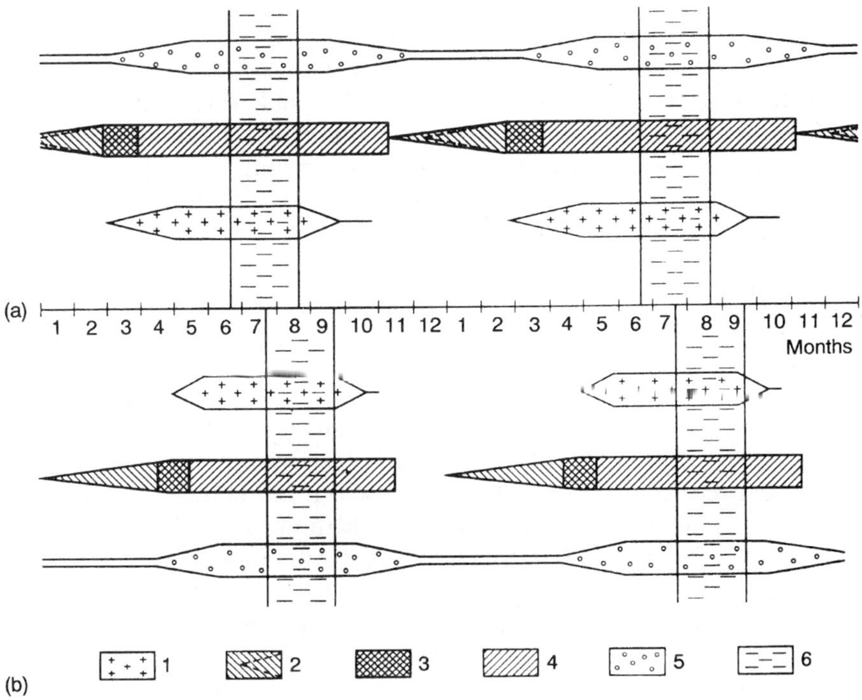

Figure 7.7 Annual cycle of sex-related characteristics in (a) European and (b) Siberian roebucks. (1) Territorial behaviour (aggression, marking); (2) antler growth commences (dotted line gives variation across range); (3) shedding of velvet; (4) buck in hard horn to casting; (5) formation of a 'skin shield'; (6) sexual behaviour.

of the territorial period illustrates the advantage of manifesting such characteristics of territoriality as early as possible, despite the considerable energy expense. Males capable of demonstrations of aggressive behaviour early in the year may hold an advantage regarding the taking and occupation of a territory.

Most aggressive encounters take place during the taking of territories prior to or at the beginning of the rut (Sempere, 1982), while aggressive contact between adult deer and younger animals is most frequent at antler cleaning and during the mating period. The rise in the level of male aggression post-rut, during autumn, observed in captive bucks (Ellenberg, 1978) has not yet been adequately explained, although the author supposes that such behaviour results from a degree of instability just prior to the onset of winter which may provoke migration to winter ranges. However, an alternative hypothesis might be that the rutting period actually occurred in autumn in the ancestors of modern roe deer, as is the case for other closely related species, and such behaviour, particularly the autumnal surge in sexual activity of juveniles, represents a trace manifestation of previous breeding patterns.

Adult bucks are especially intolerant of non-territorial deer and will chase them on sight or track them once detected. When the dominant buck approaches, the non-territorial animal retreats and leaves the territory, and only rarely does such a situation result in combat. Conflict between neighbouring territorial bucks which have been in proximity for a number of years are also comparatively rare and, despite mutually threatening behaviour, contact between these bucks usually does not develop beyond demonstration. Fights are most commonly observed over contested ranges or during raids into occupied territories, when the intruder fails to detect the territory holder approaching.

The demonstration and fighting behaviour of the European roe and the Siberian roe are similar (Hennig, 1962; Kurt, 1966, 1968a; Meyer, 1968; Smirnov, 1978; Sokolov and Danilkin, 1981). Once two rivals have observed each other, they approach to about 30–40 m and assume an aggressive posture, each facing the other. If the territorial buck does not charge the intruder immediately, the deer approach even nearer at a slow, deliberate pace to 4 or 5 m before stopping again. At this point, they stand side by side and then walk in parallel, their necks erect, heads held slightly to one side away from the opponent, while constantly holding each others gaze (Figure 7.8a). Then suddenly, as if controlled by a signal, both bucks turn and retreat. This demonstration behaviour lasts from 2–5 to 40 minutes, alternating confrontation with activities such as rubbing the head and neck against, or butting, trees and shrubs (Figure 7.8c,d) and scraping the ground with their forelegs (Figure 7.8e). Occasionally a buck may take some grass or leaves as if feeding, but

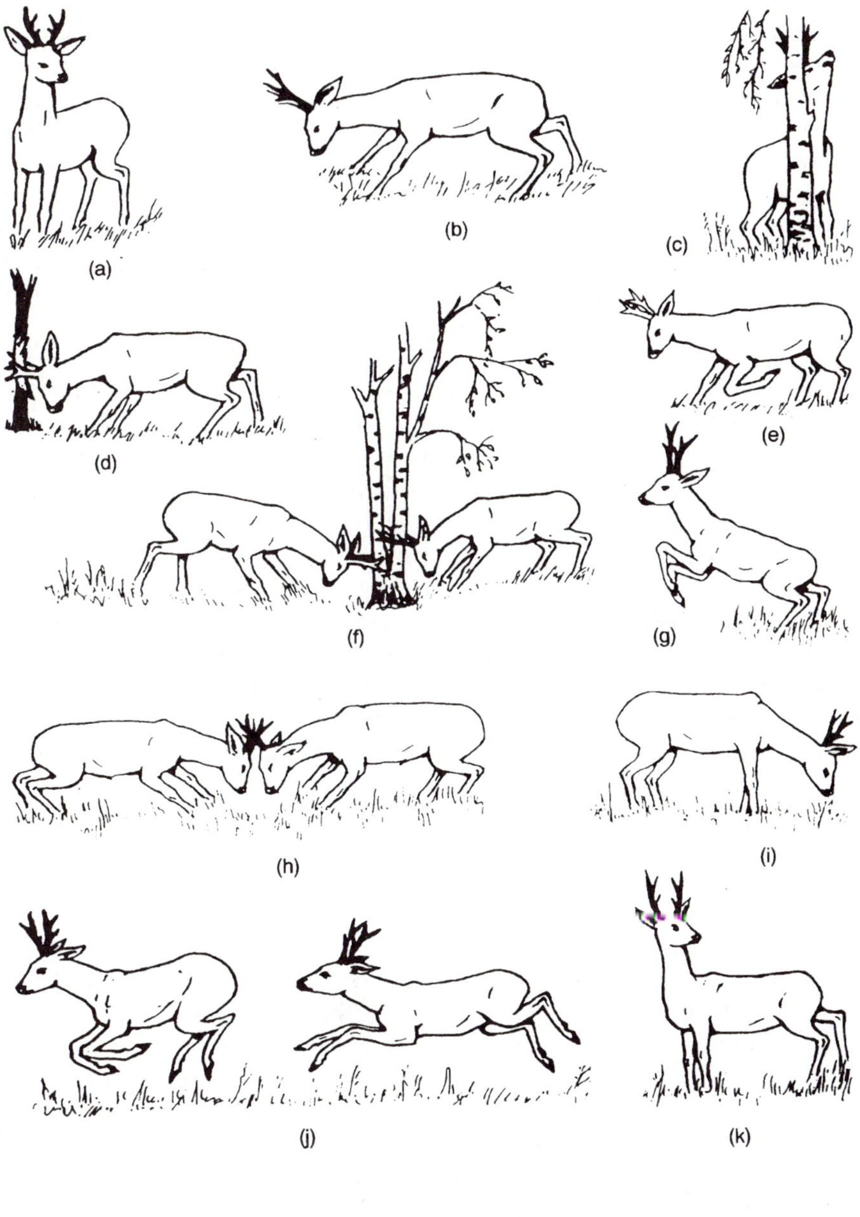

Figure 7.8 Agonistic behaviour in roebucks. (a) Pre-fight posture; (b) threat posture; (c) rubbing and (d) butting trees; (e) ground scraping; (f) simultaneous tree butting; (g) stand on hind legs; (h) clash; (i) submissive posture; (j) flee; (k) victorious posture.

seemingly he does not ingest it.

These demonstrations gradually develop in to a threat posture (Figure 7.8b), typified by a curved back with the neck lowered, the ears pressed back, and the antlers levelled at the opponent. The hair is generally on end, the whites of the eyes are visible, and froth may appear at the mouth, particularly in captive roe. Then one or other buck generally darts towards his opponent. The deer do not actually clash at this point of their fight, stopping at 0.5–2.5 m and then retreating while watching each other closely. During the intervals between these aborted charges and between actual clashes, the bucks repeat the above behaviour patterns, butting shrubs and small trees and scraping the ground. Occasionally, both bucks will charge and butt the same tree or shrub which is violently beaten, with the bucks' antlers nearly touching each other (Figure 7.8f). In this highly excited state, the deer give occasional hissing sounds to which the other responds by charging forward, and occasionally they may rise up on their hind legs in threat (Figure 7.8g). Finally, after a series of these threat patterns, the two bucks charge each other and make full contact. Each of the combatants attempts to drive the other back (Figure 7.8h). When they are well matched, the pair may spend several minutes struggling on a single spot, with a high degree of strain particularly on the muscles of the neck. Once the antlers become interlocked, each buck tries to twist the other's head, while searching for an opportunity to spike his opponent with an antler in the head or neck. After such a clash, the deer separate slowly. If the outcome was inconclusive, they repeat the above pattern of demonstration, threat and attack. The number of actual clashes per fight varies from one to eight but the number of aborted charges may be as high as 22.

The defeated buck is obliged to make an escape (Figure 7.8j), and the victor usually chases him, generally to the boundary of his territory, but occasionally even farther. Following a successful fight, the buck keeps his head erect (Figure 7.8k), often barking and butting trees and shrubs, and may stamp the ground for some time. The defeated buck assumes a typical submissive posture (Figure 7.8i) and often rubs or touches trees with his head and neck, but does not actually butt them. In unresolved contests, both bucks back away slowly, often mimicking feeding behaviour, and then commence to bark with the head erect, both adopting the behaviour of a victor, butting trees and shrubs as they leave, probably to mark them.

Thus, conflicts between roe deer are largely ritual and their demonstrations of strength, like those of many other mammals, follow a certain system of visual signals, consisting of typical elements or postures which indicate the level of motivation of the animal in relation to external stimuli. During demonstration behaviour of opponents, certain tendencies indicating a willingness to attack (aggressiveness) or a disposition to run away can be identified. A change in the balance of these tendencies dictates the outcome of a particular fight (Panov, 1969, 1983; Tinbergen, 1969, 1970, 1974; Hinde, 1975). The ritualization of combat is also illustrated by a type of behaviour often observed when bucks, rather than attacking their opponent, butt shrubs and trees instead.

It should also be noted that the order of the above elements of the agonistic behaviour of roe deer is not strictly fixed but variable. At first contact, bucks may immediately begin to threaten or alternatively they may first rub and butt trees, only then proceeding to demonstrate to the other. Some aspects of this behavioural pattern such as marking may be omitted altogether if the interaction takes place in open habitats such as fields or meadows. As buck population density increases and accordingly the size of individual territories decreases, encounters and conflicts become more frequent. The fact that almost all free-ranging adult bucks caught or shot carry some wound scars indicates that combat takes place much more often than is indicated from direct observation.

With rare exceptions, all adult males are intolerant of young bucks, actively chasing them and driving them out of the territory. Interestingly, adults are especially aggressive towards young bucks in the morning and in the evening, i.e. during the periods of peak activity. In contrast, during the middle of the day non-territorial bucks are sometimes able to feed and even to lie down only a few metres from the dominant buck which drove them off that morning and is likely to do so again once evening comes. There are also inter-individual differences in the level of aggression expressed by bucks. Some adults become extremely excited and vigorously chase any non-territorial deer on sight, whereas other bucks merely follow slowly behind an intruder, occasionally marking trees on the way, until he is beyond the territory boundary.

The aggressive behaviour observed in bucks is linked to the system of territoriality in roe deer. It is more difficult to explain why sometimes an adult shows a high degree of tolerance towards a particular young male who may become his companion throughout the territorial period in the absence of any aggressive behaviour. Strandgaard (1972a) may have provided the answer with his discovery of a clear relationship between the degree of physical development of bucks and the occurrence of yearling dispersal. In spring, those young bucks in particularly good condition begin to display typically territorial behaviour, including aggression and marking activities, triggering a response among territorial bucks. In contrast, young bucks in poor condition, with poorly developed antlers, do not claim territories and some adult bucks may be quite tolerant of them, only displaying aggressive behaviour towards these individuals the following year.

A similar relationship between physical development and the tendency to disperse was recorded by Ellenberg (1978). He also observed that territorial bucks are less aggressive towards the progeny of does living within their territories, which are tolerated until they become markedly territorial, than towards those of does inhabiting the territories of neighbouring bucks. Thus, the aggressive behaviour of territorial bucks forces young bucks in good condition to disperse, resulting in the dissociation of family groups and possibly limiting the potential for inbreeding. This dispersal may play a role in population regulation and newly formed populations may tend to be founded from larger than average individuals from the source population.

Following the reproductive period, the level of aggression among roe deer decreases markedly and may even disappear altogether. During winter, large groups of free-ranging deer may feed together without noticeable antagonism. By the end of winter adult males, who are territorial in summer, may temporarily form single sex groups. In Siberian roe deer, the members of such groups do not become aggressive until the end of April, despite the fact that all the adult bucks have fully developed antlers, although still in velvet (Sokolov and Danilkin, 1981). Nevertheless, antagonism between roe may be observed in winter where the deer are at high density such as at feeding racks and on supplementary feeding grounds. In Sweden, Espmark (1974) recorded 1048 aggressive contacts at a feeding rack and found a strong relationship between the number of aggressive acts and group size. In Switzerland, Geiger and Kramer (1974a,b) found, from observations of tagged roe on a winter supplementary feeding ground, that aggressive interactions were more common among bucks than does and that the rank of a particular animal was related to the distance between the feeding ground and its territory. The dominance hierarchy was approximately linear and relatively stable throughout the winter, with adult bucks dominating all other individuals. Once the antlers were cast, the individual quickly lost rank and, inversely, those individuals that cleaned their antlers of velvet earliest immediately rose to a higher social status. The behaviour of does at feeding points was determined by a principle of priority, with those individuals which appeared early at the beginning of winter gaining an advantage. Buck kids ranked higher than their sisters and, by the end of winter, they sometimes dominated all female members of the group, including their mothers.

In pens where establishment of territories is impossible owing to limited space availability, an even more marked linear hierarchical social structure may be observed throughout the year. Generally, one adult buck with particularly well-developed antlers dominates other adult and subadult bucks, who in turn are dominant over adult and subadult does, with kids ranking lowest in the hierarchy. In the absence of external stimuli, the dominant buck is not especially overtly aggressive, and a slight lowering of the head with the ears pressed back is usually sufficient to induce a submissive reaction in another individual, who retreats or runs away. Nevertheless, displays of aggression occur practically daily in such situations, causing substantial disturbance in the population, even resulting in the death of subordinate deer.

In experimental manipulations (Sokolov and Danilkin, 1981), the dominant buck lost his position in the hierarchy immediately after his antlers were sawn off, and following antler casting in the autumn even does may dominate over bucks. When captive bucks come into hard horn, combat becomes frequent, violent and repeated over several days, with the victor becoming dominant and aggressive, chasing any defeated bucks for a substantial time each day. These subordinate bucks may occasionally assume a threat posture but the approach of the dominant buck almost invariably forces a retreat, although the retreating buck will continue to display and perform marking behaviour at the opposite end of the pen. Gradually, with time, the level of marking and aggressive behaviour of the subordinate individual during these interactions diminishes, and he responds by retreating and running away only.

When bucks inhabiting the same cage are separated by a net partition, both display behaviour typical of a dominant animal, spending a considerable time in demonstration and appropriately aggressive postures, sometimes butting each other through the net. In this situation, it seems that both the dominant and the subordinate bucks have taken possession of a territory, deriving some additional stimulus which triggers aggressive and marking behaviour. A similar situation is observed in free-ranging populations, where the behaviour of an intruder in another deer's territory changes from the typically aggressive conduct displayed in his own territory to tentative exploratory behaviour, and subsequently becomes defensive when the territory owner is encountered. Such behaviour patterns have been observed in other mammals, with aggressive behaviour primarily expressed in familiar surroundings, and the outcome of a particular encounter decided by adjustments in the balance of these behavioural patterns (Shilov, 1977).

Thus, the behaviour of roe deer may differ greatly according to environmental conditions, varying from a strictly linear hierarchy to a clearly territorial structure during the rut, with antagonism practically non-existent outside the territorial period. It seems possible that the evolution of social behaviour in *Capreolus* followed these lines, developing from a gregarious habit towards a more solitary way of life.

7.5.2 Aggressive interactions with other species and humans

A distinctive feature of the behaviour of wild ungulates, including roe deer, is an innate vigilance and fear of predators and humans. The usual behaviour of roe deer when a predator is encountered involves a passive defence, becoming immobile or escaping. Roe deer also clearly avoid contact with other large ungulates (moose, red deer, fallow deer and wild boar) and usually move away as they approach, but are generally indifferent towards small mammals and birds. Aggressive behaviour towards other species is rarely observed in free-ranging roe deer, but does do display some level of aggression during the first one or two weeks after giving birth when protecting their offspring which are unable to escape. During the territorial period when the bucks are highly agitated, they may display threat postures to foxes, small dogs, hares and large birds, but rarely to other species of ungulate or to humans – though occasionally such threats do result in attacks on humans, even in wild populations (Holmes, 1973).

In captivity, the behaviour of roe deer is greatly altered. Hand-reared kids often form a very strong attachment to their foster parent and do not develop a fear of people or domestic animals. Under such conditions, does remain placid in their interactions with humans throughout their life but, in contrast, bucks become aggressive during their second year and, lacking almost all fear, may sometimes attack people, demonstrating the whole pattern of specific display and combat postures and reactions, as if fighting another deer. When the noises made by a human resemble even vaguely those of a roe deer, a buck may become highly excited, sometimes provoking an immediate attack in a highly agitated individual which will continue to pursue such intruders when they make an escape.

Such aggressive behaviour of bucks towards humans, as towards other roe deer, is markedly seasonal, being strongest in spring and summer but subsiding appreciably in autumn. By winter, most bucks do not show any aggressive responses, but the level of aggression rises again once their antlers are clean of velvet in spring, particularly during morning and evening when activity is highest. Experiments have shown that, in captivity, roebucks respond more readily to a foreign stimulus (a human being) than to another roe deer, and that they are especially aggressive towards individuals with whom they are most familiar. Recognition is probably based on a combination of the voice, clothes, smell and even the facial features of a particular individual. When a buck is aggressive towards a human who then leaves the enclosure, the dominant buck persists in his aggressive behaviour by directing it towards a subordinate deer. If the human subsequently reappears, the focus of this aggression generally switches back or, if the dominant buck is highly excited, he may concentrate his activity on any moving object. Indeed, although a buck usually exhibits only exploratory behaviour when confronted with a set of moving objects, these same objects may elicit an extremely aggressive response when he is provoked to a state of excitement (Sokolov and Danilkin, 1981).

Thus, aggressive behaviour of roe deer is essentially dependent on the individual's physiological state and predisposition, strongly influenced by hormonal factors (e.g. blood testosterone levels) or whether he possesses a territory or not and his dominance status, and also on the season, time of day and other external stimuli (Barth *et al.*, 1976; Schams and Barth, 1982; Sempere, 1982).

7.5.3 Morphological adaptations for combat

In many species of mammal, a system of intraspecific antagonism which minimizes physical damage has evolved where, for example, dominance is established through display rituals. In this aspect of behaviour, the roe deer seems to be similar to other cervids in that the strength of an individual may be determined from his appearance – the size of the antlers, body and neck – without any physical violence. Nevertheless, fighting between adult bucks, often associated with breeding behaviour, is not uncommon and sometimes even results in death (Dementiev, 1933; Averin, 1949; Severtsov, 1951a; Kirikov, 1952; Sokolov, 1959; Cherkasov, 1962; Holmes, 1973; Stubbe and Passarge, 1979). As outlined above, the focus of these disputes is not a particular doe but actually a territory.

Darwin (1859) observed that the success of a male in a confrontation often depends on the development of special structures, either weapons or means of protection, and even a slight difference between two individuals in this respect may give a significant advantage. The roe deer possesses antlers, a hard shield of skin on the forehead and neck muscles which thicken during the period of maximum aggressive interactions. It is generally accepted that the antlers of ungulates may serve as both offensive and defensive weapons in male–male conflict (Severtsov, 1951b, Bubenik, 1966a, 1968; Geist, 1968; Lincoln, 1972) and their development is under hormonal control (Chapman, 1975; Jaczewski, 1976; Jaczewski *et al.*, 1976). The use of antlers for defence against predators and as a means of obtaining food and their function in thermoregulation are probably characteristics of secondary importance (Lincoln, 1972). Experimental removal of antlers clearly highlights their primary function as a means for determining social rank and as weapons used in intraspecific conflicts, particularly for obtaining or defending territories once the antlers are fully developed (Figure 7.7).

Antlers develop faster, attain a larger overall size and are cleaned of velvet two or three weeks earlier among adults than in subadults (Figure 7.9). From the time a buck starts to lose antler velvet to the moment when he is finally in hard horn is relatively short, taking only 2–5 days. During this period, the buck repeatedly rubs his antlers against trunks and branches of trees, leaving pieces of velvet and skin behind which may be consumed immediately by licking the branches (Figure 7.10). The antlers become bloody and seem to be highly irritating, causing the buck to continue rubbing them – sometimes for 40 to 50 minutes at a time. There is considerable variation in the date when bucks come into hard horn, and those doing so first gain a marked advantage, facilitating the acquisition of a territory and probably giving the psychological advantage of social dominance on a particular range.

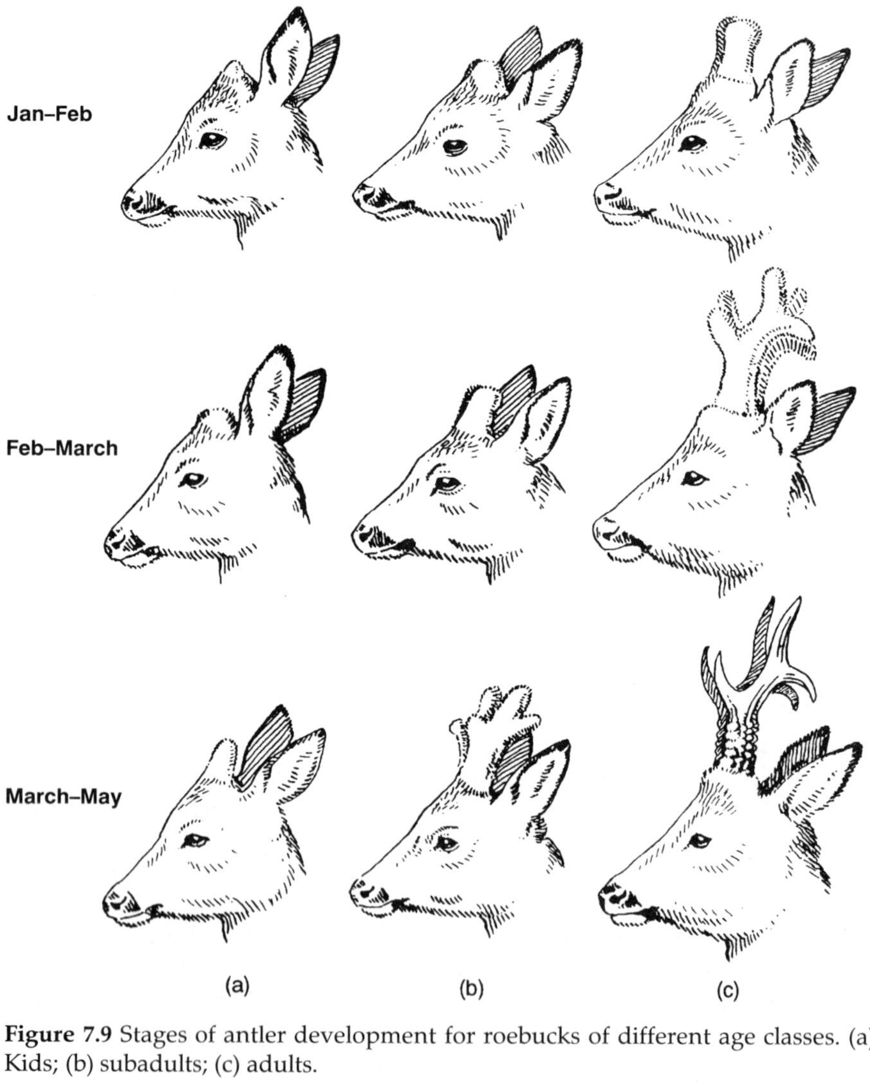

Figure 7.9 Stages of antler development for roebucks of different age classes. (a) Kids; (b) subadults; (c) adults.

Well-developed antlers are one of the most important features determining the outcome of a particular contest. In combat, the antlers generally interlock, cushioning the impact of a head charge and diminishing the possibility of penetrating the skin. Nevertheless, the antlers may occasionally pass either side so that the blow falls on the head – mostly on the neck, forehead and cheeks. Examination of adult buck skins during summer shows damage mainly on the head and neck (Figure 7.11); moreover, the smaller the antlers, the more numerous are the scars on an individual's body (Figure 7.12). The maximum number of wounds found on one animal was 27 (the minimum was 2), while the most serious wound observed measured 8 mm deep by 7 cm long.

The number and severity of skin penetrating wounds are minimized due to the thickness of the skin in the most exposed regions during head-to-head contact. The thickness of the skin on the head and anterior parts of the body is related to age and seasonal factors, which are especially pronounced in the larger Siberian roe deer (Figure 7.13). In summer, the head and neck skin of adult bucks is 2.5 times thicker than that on the back and flanks, creating a type of shield which covers the particularly vulnerable parts of the body to the scapulae or even farther in the spinal area (Figures 7.13c, 7.14). The skin is thickest in the region just behind the antlers (up to 10.5 mm). During autumn and late winter, the skin is 2–3 and 3–5 times thinner, respectively, than in summer (Figure 7.13b), becoming of a similar thickness to that observed in does (Figure 7.14). In contrast, the thickness of the skin in does on different parts of the body (Figures 7.13, 7.14) does not vary significantly between seasons, and at its maximum is 2.5 mm. The skin shield begins to develop at an early age in roebucks, thickening noticeably in July and August at 2–3 months of age, but then during winter, at 9–10 months old, the skin becomes twice as thin again. Among yearlings during the rut the shield is already pronounced, with the skin on the head and neck generally 3–5 times thicker than during the previous winter (Figure 7.15).

The thickening of the skin in roebucks coincides with the intensification of aggressive behaviour (Figure 7.7) and, conversely, the skin shield thins again once aggressive activity declines. The skin shield protects those regions which are the most vulnerable to antler blows during combat and it therefore becomes thickest during the period of most intense fighting, protecting vital organs from damage and so reducing mortality. In addition, the considerable swelling of the skin glands in this same region of skin also indicates it is important for scent marking of territories.

Similarly, a further remarkable morphological character which exhibits pronounced seasonal variation is the thickness of the neck in roebucks. While the circumference of the neck remains essentially constant in does throughout the year, among bucks it may measure several centimetres more in summer, compared with other times of the year (Figure 7.16),

(a)

(b)

(c)

Figure 7.10 (a)–(c) Siberian roebuck cleaning his antlers of velvet.

Figure 7.10 (d) Velvet pieces left on the tree are eaten by the buck.

Figure 7.11 Usual location of scars suffered during butting in adult Siberian roebucks.

largely due to the thickening of the skin and muscles of the neck. For example, during summer the neck muscles of Siberian roebucks are 3.5–5 times thicker (3.5–4.6 cm) than those of does (0.8–1.0 cm) (Figure 7.17), while in winter the muscle is approximately half the thickness (down to 1.3–2.3 cm). This feature has probably evolved as bucks with strong neck muscles will gain a significant advantage during rutting combats when much strain is placed on the neck. However, in addition, it is also probable

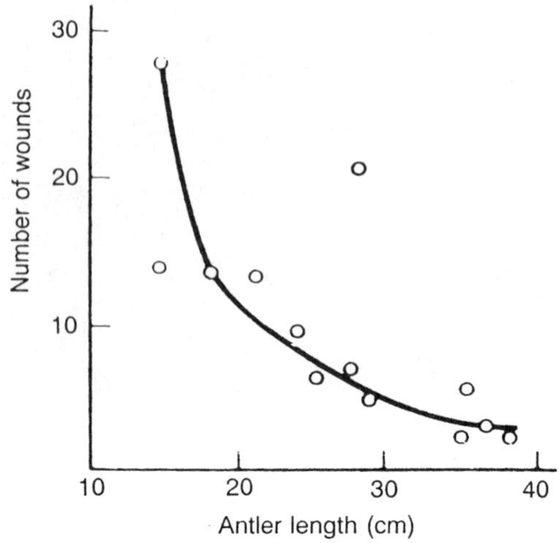

Figure 7.12 Relationship between antler length and number of wounds sustained in adult Siberian roebucks.

that a powerful thick neck provides a visual indication of the strength of an individual and is an important feature in display behaviour.

The features associated with rutting behaviour in roe deer described above (aggressive activity, the possession of antlers, the thickening of the skin shield and the neck muscles, etc.) are not apparent in winter when roe become significantly more gregarious. It is possible that, initially, the immediate ancestors of *Capreolus* were not so aggressive but were more gregarious in habit, and therefore may not have possessed such exaggerated secondary sexual characteristics. With the evolution of territoriality, roe deer have developed specific behaviour and morphological characteristics which differ from those of other species of deer that have a stronger tendency towards group living. The seasonal variability in morphology of these secondary sexual characteristics is compared with the annual cycle of territorial behaviour in Figure 7.7.

7.6 ALARM BEHAVIOUR

Exploratory behaviour is usually defined as a set of reactions aimed at providing a better perception of the immediate environment or investigating a source of disturbance, generally implying a higher level of sensitivity to stimulation in the animal and a readiness for action. Exploratory actions may be closely connected with defensive behaviour; a defensive reaction quickly follows when a stimulus or disturbance is particularly strong. There is often a conflict between exploration and fear, which may

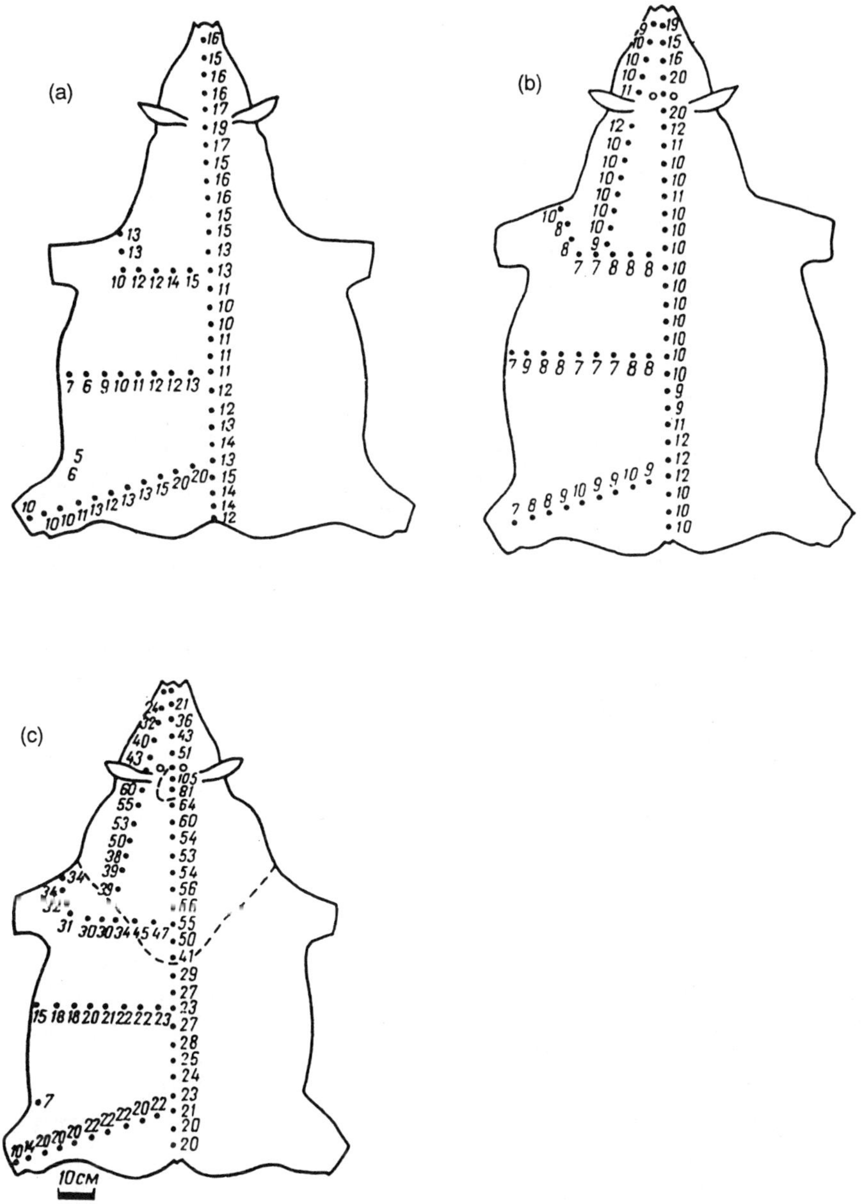

Figure 7.13 Skin thickness (mm x 10^{-1}) in adult Siberian roe deer. (a) Doe (July); (b) buck (March); (c) buck (July). Dots show measurement points. Dotted line (c) delineates region of thickest skin.

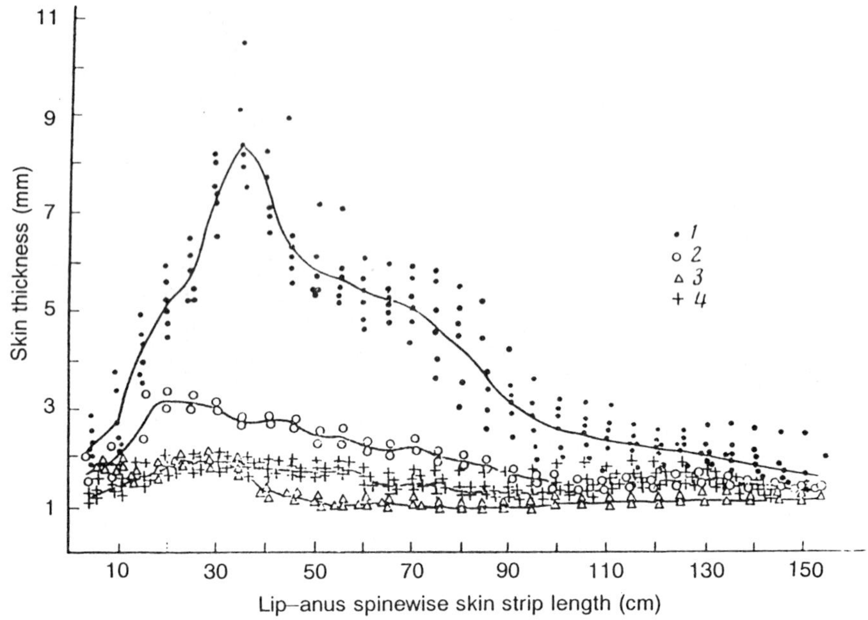

Figure 7.14 Seasonal changes of skin thickness in adult Siberian roe deer. (1) Bucks in June, July and August; (2) bucks in November; (3) bucks in March and April; (4) does in July and August, November and March.

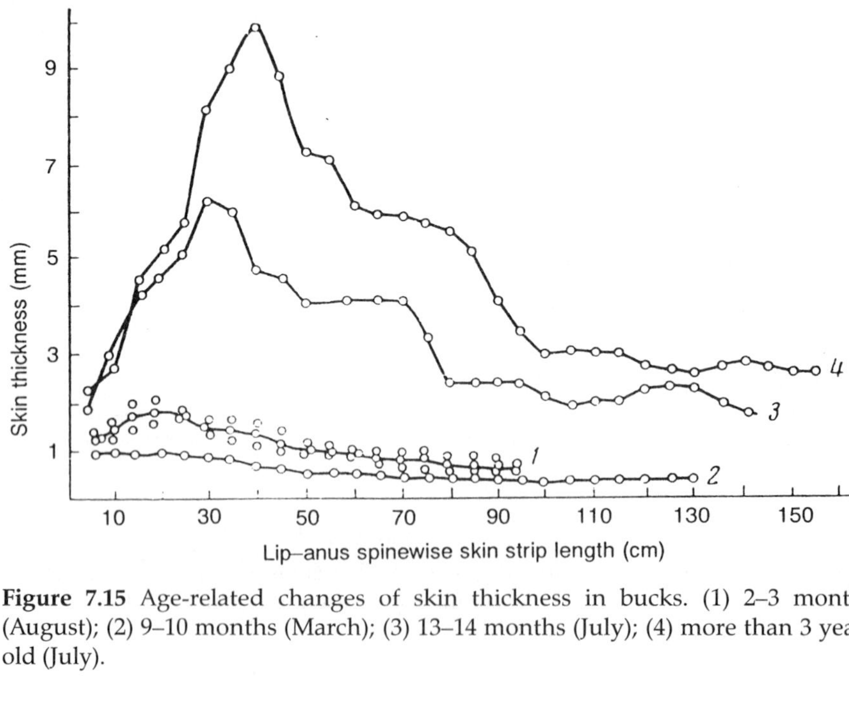

Figure 7.15 Age-related changes of skin thickness in bucks. (1) 2–3 months (August); (2) 9–10 months (March); (3) 13–14 months (July); (4) more than 3 years old (July).

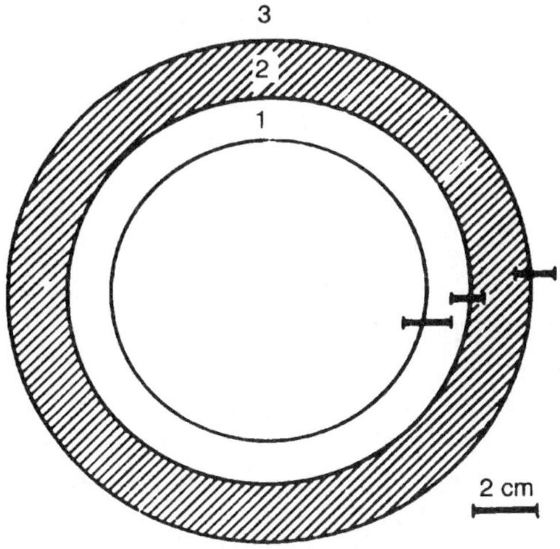

Figure 7.16 Seasonal changes in thickness of neck of Siberian roe deer. (1) Does (all seasons); (2) bucks in winter; (3) bucks in summer.

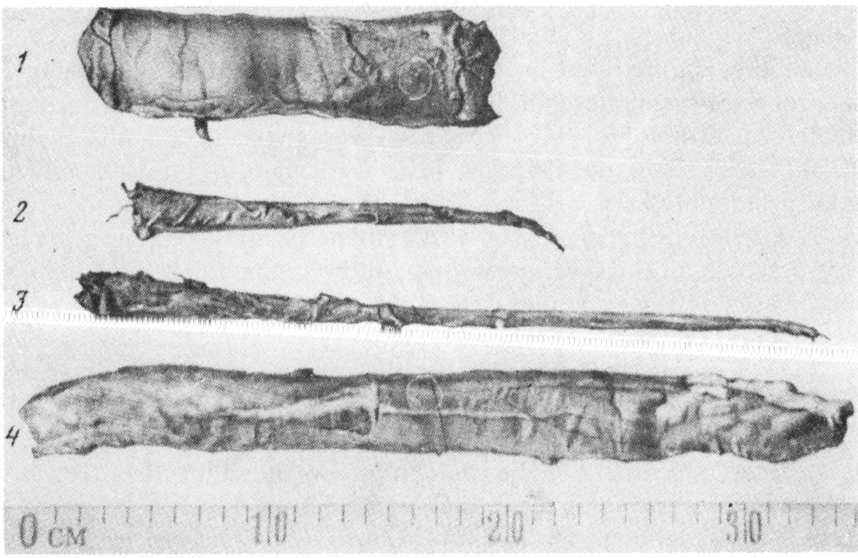

Figure 7.17 Thickness of cervical muscle (sterno-mastoideus) in Siberian roe deer. (1), (4) bucks; (2), (3) does.

result in a displaced reaction (Pavlov, 1938; Sokolov, 1966; Showen, 1972; Hinde, 1975).

In the roe deer, exploratory and defensive behaviour comprises a series of separate elements ('unit reactions' is the term coined by Krushinski,

1960), but not all of these elements are necessarily displayed in a given situation and the order in which they appear may differ. During periods of activity, exploratory behaviour accounts for between one quarter and one half of the roe deer's time budget (depending on the situation, the source of the disturbance, etc.). Even when no source of disturbance is apparent, roe may periodically raise the head in vigilance, searching for sources of danger with both sight and smell. However, their vision is somewhat limited and it is often possible to approach a deer to between 20 and 50m when downwind.

Alarmed behaviour – moving quickly away from perceived danger – is often the first type of reaction expressed by deer in response to some disturbance and is not preceded by any vigilance behaviour. However, once deer have fled approximately 10–30 m, they will often turn and assume a posture of vigilance, only then running away a substantial distance if there is clearly a threat. This behaviour is more common in closed forest habitats and may help to explain why groups of forest-dwelling roe deer rarely consist of more than 10 individuals. In contrast, roe deer in groups tend to bunch together in response to disturbance, becoming vigilant.

When in a **vigilance posture** (Figure 7.18), the head is held high above the ground, the muscles taut, the target hair erect, ears pricked and orientated towards the disturbance, eyes wide open, as the animal sniffs the air. The roe deer may then lower the head again, as if to feed, only to jerk up suddenly to observe once more. It can often be clearly seen that the deer moistens the nostrils by licking them, possibly to improve olfactory perception. This posture is frequently accompanied by urination and defaecation. From studies of time budgeting, it has been shown that the vigilance posture is assumed at intervals of 36.2 seconds on average but the degree of vigilance varies from day to day, and even over the course of a single day, becoming less pronounced towards dusk. The relative level of vigilance in roe deer is comparable with that of most other deer species but is much higher than in moose and fallow deer (Pelosse, 1976). Within a group of roe, when a single animal adopts a vigilance posture the neighbouring animals are immediately alerted, provoking the deer to regroup more closely and subsequently to adopt vigilance postures as well. If the disturbance is not identified, the deer generally become calmer gradually or proceed with exploratory behaviour. Vigilance behaviour may also include vocal signalling or stamping.

When some potential danger is perceived, roe deer often **freeze** on the spot. Kids usually lie down, pressing their heads to the ground, while adults stand completely still, their heads lowered, or may also lie down, particularly in open steppe habitats.

Figure 7.18 Vigilance posture.

Disturbances may provoke **barking** in a single individual or several animals at the same time. While barking in such situations, roe deer often raise their forelegs high and stamp down on the ground (as opposed to the scraping movements performed in other social contexts) as they move towards or away from the disturbance. For example, a group of six Siberian roe deer lying down responded to the arrival of an observer 70 m downwind. Firstly, a single doe stood up, assumed a vigilance posture and moved 2 or 3 m away from the disturbance, stamping her forelegs on the ground. The whole group then followed her and also became vigilant, while the doe proceeded to bark 37 times in succession. After 18 minutes, during which time the observer remained stationary, all members of the group lay down again, one after another. A second disturbance, when the observer moved once more, provoked the buck to stand and adopt a vigilance posture, shortly followed by two does. The buck then moved, stamping his forelegs, and finally ran after bounding once, followed by the entire group, but stopping after 200 m and becoming vigilant once more. Barking and stamping indicate the state of an animal responding to a disturbance; they serve to warn other group members of danger and possibly to frighten away a potential threat.

For precise identification of the source and type of a particular disturbance, roe deer often rely on the **sense of smell**. Hearing and vision allow the roe deer to perceive the existence of some disturbance but more precise information is obtained only by smell; thus in such situations the deer attempt to move downwind. The olfactory organ is the most important of the roe's sensory organs, with a total surface area in the nostrils measuring up to 90 cm^2 (against 2.5 cm^2 in humans) and the number of olfactory cells may be as many as 300 million (against approximately 30 million in humans) (Kolb, 1979). Kurt (1965) has estimated that of 42 different elements of social behaviour in the roe deer, 26 are triggered by olfactory perception, 13 are provoked by sound and only three are caused by visual stimuli, highlighting the importance of the sense of smell.

The range of detection of a given object by sense of smell varies with climate and habitat. Under windy conditions, roe may detect the presence of a human at 200–400 m in open country and at 100–200 m in forest, but when there is no wind these distances decrease by between 20 and 80% and deer may be approached relatively easily from upwind.

Bounding almost invariably occurs once the roe deer has spotted some potential danger. Moving a few metres away from the source of the disturbance, the deer performs a high vertical jump and lands, clicking its legs together. When moving through tall grass habitats, roe deer bound every several dozen metres. The bounding of a single animal immediately triggers vigilance in others and may cause them to run; however, it is more likely that during summer, when roe are essentially solitary, bounding enables the deer to ascertain the location of a potential threat and choose a direction of movement.

Roe deer may sometimes signal the presence of danger to other individuals of the group by **running past in close proximity**. In this case, the animal which perceives danger does not run away, even once the source of the disturbance has been precisely identified. Instead, it runs towards one or more other members of the group which may be lying down or unaware of the danger. Usually the animal runs past in close proximity, or occasionally may even jump over resting deer, thus provoking vigilance or causing the whole group to run. Such behaviour may be considered as a warning signal in some cases and is mostly displayed by kids of between 5 and 10 months, which run to their mother, or by subadults. For example, instead of running away, an adult buck which had detected the presence of an observer turned towards the potential threat and ran past another 2-year-old buck lying not far away. When this produced no reaction, the buck remained vigilant for about one minute and then ran once more towards the second buck, jumping over him and causing him to get up, become vigilant and finally to follow.

When one individual in a roe deer group starts to run, this generally provokes a similar reaction in the other group members. After covering between approximately 30 and 250 m, the group often stops, adopts a vigilance posture and, if the source of danger is still perceived, resumes its escape. If the deer are not being chased, they will generally run no more than 300 m in forest, but will continue for a much longer distance in open fields. In forest and forest–steppe habitats, an **escape response** is generally provoked when an observer or a vehicle approaches within 20–150 m, but this distance may be much higher in areas where disturbance is high.

In general, kids are the most responsive to potential danger, provoking the escape response in many cases. Adult deer are almost invariably the last to run, but when they do so they immediately become the group leaders, particularly does. When no real danger exists, the group leader quickly abandons the escape and the juvenile deer copy this example. In this way, adult roe express a more controlled pattern of alarm behaviour which juveniles gradually learn, thus reducing the number of unnecessary escape responses manifested within the group.

In situations of danger, roe deer noticeably display a '**follow the leader**' type of behaviour pattern, with group members following a single lead animal in a chain, with the target flared (Severtsov, 1951a). This behaviour may maintain ties between juveniles and their family group and allow young deer to learn from the more experienced adults.

The behaviour of the European and the Siberian roe deer is practically identical and does not seem to be an obstacle preventing the hybridization of these two species, either in the wild or in captivity. Thus, differences in behaviour patterns have probably not played any significant role in *Capreolus* speciation.

8

Reproduction

8.1 PUBERTY

Sexual activity is first displayed by roebucks at an age of 4–6 months, when both testicle mass and the diameter of the testicular tubules increase slightly (Figure 8.1), hormonal levels rise (Figure 8.2) and the first stages of spermatogenesis, possibly giving rise to primary spermatocytes but rarely to spermatozoa, are recorded. However, spermatozoa disappear from the epididymides towards the end of December, reappearing again only in the following May or June. As these gametes are effectively immature, this first period of sexual activity can be considered as non-reproductive for the vast majority of male kids. Reports of successful covering of a mother or another doe by a kid seem to be exceptional (Stieve, 1950; Wandeler, 1975; Tsaplyuk, 1977; Sempere, 1982; see also Stubbe and Danilkin, 1992). Nevertheless, this precocious spermatogenesis emphasizes the fact that roe deer experience puberty unusually early for an ungulate species.

Roebucks actually reach puberty at the beginning of their second year, indicated by behaviour as well as histological studies of their testes and epididymes. As they approach the rutting period, yearling bucks experience an abrupt rise in the level of testosterone (Figure 8.2). In addition, the weight of the testes and the diameter of the testicular tubules increase significantly (Figure 8.1). The epididymis tubules may be filled with spermatozoa at a density equivalent to that found in adults, indicating that yearlings can be considered as fully mature. However, the testes are often more poorly developed and the testicular tubule diameter is generally smaller among this age class in comparison with mature bucks (Stieve, 1950; Otskheli, 1964; Bramley, 1970; Pole, 1973; Sempere, 1982). Furthermore, yearling bucks are physically inferior and are generally expelled from the territory of an adult. Due to this enforced non-territori-

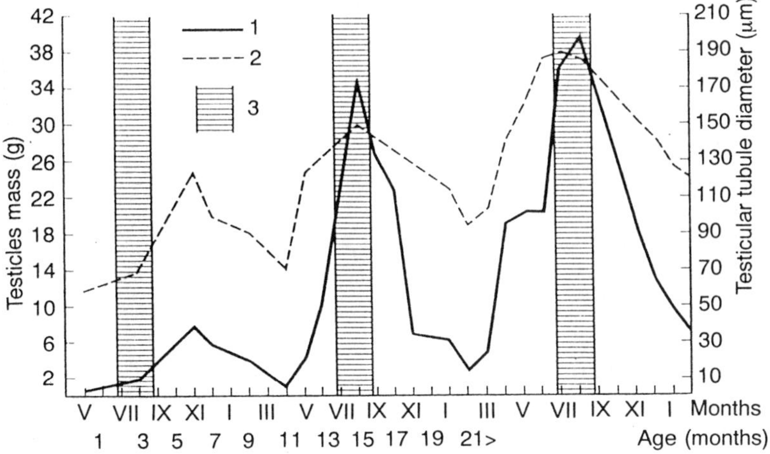

Figure 8.1 Age-related and seasonal changes of (1) testicle mass and (2) testicular tubule diameter in Siberian roe deer (adapted from Tsaplyuk, 1977). Hatched bars (3) denote rutting period.

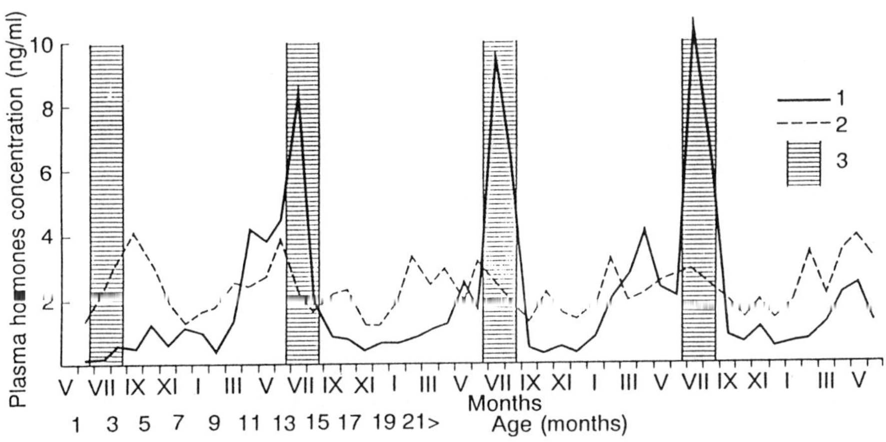

Figure 8.2 Age-related and seasonal changes in levels of (1) blood plasma testosterone and (2) luteinizing hormone in European roebucks (adapted from Sempere, 1982). Hatched bars (3) denote rutting period.

ality, they are unable to participate in the rut in the wild, although under captive conditions yearlings may successfully mount does. From this point of view, they can be considered as a kind of reproductive reserve within the population, only becoming involved in breeding when the number of mature territorial bucks is low.

Does may attain puberty as early as their first year, if the appropriate threshold body weight is reached (Loudon, 1987; Hewison, 1993, in press a). In August, when 2–3 months old, female kids may carry numerous small ovarian follicles, which are probably associated with the approaching oestrous cycle. Towards October these bodies generally become atrophied but some may persist and mature, making fertilization possible at this time (Tsaplyuk, 1977). However. only relatively few kids actually breed, probably due to their social immaturity or the low occurrence of spermatozoa in bucks during the autumn, following the rut, making fertilization unlikely. Despite this, corpora lutea of pregnancy have sometimes been recorded in European roe deer kids, indicating that successful fertilization has taken place. In a single British population at Pickering, North Yorkshire, Hewison (in press a) found almost half (48%) of the 232 kids sampled were carrying such corpora lutea. However, in the other eight populations of Hewison's study, as well as in other reported work (Georgii, 1974; Wandeler, 1975; Chapman and Chapman 1971), these indications of successful fertilization among kids have generally been observed in less than 10% of cases. Similarly, corpora lutea have been found at low frequency among kids in Siberian roe deer populations of Kazakhstan (Tsaplyuk, 1977), and we have also observed them in a single female kid from the southern Urals.

Despite this low but widespread occurrence of fertilization during their first rut, cases of roe deer kids actually giving birth when one year old are rare (Neubert, 1970; Borg, 1970; Ellenberg, 1978). Even in the population at Pickering in the above study, where corpora lutea were commonly found among females of this age class, successful implantation of a fetus occurred in only 8% of all kids and did not occur at all in 13 of the other 14 populations where these data were available (Hewison, in press a). This suggests that social constraints and the abrupt fall off of spermatogenesis among bucks following the rut are not the only factors limiting reproductive activity among female roe deer kids. Thus, as is the case for males, most female roe attain puberty during their second year, giving birth at the age of 2 years; however, the proportion of yearlings breeding within the population and/or the average litter size they produce may be somewhat lower than among older age classes (Gaillard *et al.*, 1992; Hewison, 1993, in press a).

8.2 MATING

The mating periods for European and Siberian roe deer overlap, taking place from June to August and from July to September respectively, with only a few female kids coming into oestrus later on in the year (Table 8.1). The onset of the rut tends to occur later, moving across the range of

Table 8.1 The dates of the rut across the roe deer's geographic range

Location	Dates of the rut: Start	Finish	Reference
		European roe deer	
France	July 21	August 15	Sempere, personal communication
Austria	July 20	August 20	Kerschagel, 1952
Bulgaria	July 12	August 15	Petrov *et al.*, 1968
Czechoslovakia	July	August	Hell, 1979
Germany	June 30	August 17	Ellenberg, 1978
Germany	June 17	August 29	Stubbe and Stubbe, 1985
Denmark	July	August 20	Strandgaard, 1972a
Switzerland	July 20	August 15	Sagesser, 1968
Poland	July	August	Kaluzinski, 1978
S. Ukraine	July	August	Steklenev, 1980
Ukraine – left bank	July	September	Karpenko, 1966
N. W. USSR	July	September	Vereshchagin and Rusakov, 1979
Byelorussia	July 15	August 15	Sablina, 1955
Latvia	July 15	August	Kalninsh, 1950
Georgia	July	August	Otskheli, 1964
		Siberian roe deer	
S. Urals	July 25	September 6	Filonov, 1974
S. Urals	July	September 22	Kiselev, 1979
S. Urals – 1972	August 16	September 18	Danilkin's data
S. Urals – 1973	August 7	September 11	Danilkin's data
S. Urals – 1974	August 4	September 9	Danilkin's data
S. Urals – 1975	August 9	September 14	Danilkin's data
S. Urals – 1976	July 22	September 8	Danilkin's data
Kazakhstan	August 5	September 20	Zhirnov, 1958
Kazakhstan	August	September	Pole, 1973
Kazakhstan	June 19–23	September	Sludski and Baidavletov, 1984
Kirghizia	August	September	Aizin, 1969
E. Siberia	August	September 8	Fetisov, 1953
Trans-Baikal	August 5	September 17	Smirnov, 1978
Yakutia	August	September	Yegorov, 1965
Far East	August	September	Bromlei and Kucherenko, 1983
Far East	June 26	October 30	Darman, 1986
Mongolia	July 19	September	Danilkin's data

the species, both from west to east and from south to north. Even within a single area the exact dates depend on altitude above sea level and may also vary from year to year. In the European roe deer the interval between giving birth and the next oestrus is comparatively constant, measured as 67.3 ± 1.9 days (Ellenberg, 1978), but yearling females generally come into oestrus somewhat earlier than older does (Kaluzinski, 1978; Eberle, 1986). The onset of oestrus may well be synchronized photoperiodically with a day length of 15.75 hours seemingly critical (Lehmann, 1960b).

Successful fertilization of does from December to April–May is practically impossible due to pronounced annual physiological cycles of bucks such that the mass and volume of the testicular tubules of adult males is 5–10 times lower in winter than in summer (Figure 8.1). Testes mass is highest in July, i.e. during the breeding season, and lowest from December to March, but some gonadal growth is detectable as early as February. In the Urals, an area harbouring the largest subspecies of Siberian roe deer, the testes of adult males weighed within the range of 3.5–54.3 g (Sokolov and Danilkin, 1981); the equivalent range for deer of the Tien Shan subspecies in Kazakhstan was 2.9–44 g (Pole, 1973). Similar data have been obtained for the European roe deer (Stieve, 1950; Short and Mann, 1965, 1966; Bramley, 1970; Steklenev, 1980; Sempere, 1982).

Ziegler (1848) found that spermatozoa disappear from the testes in November or December and, therefore, any mating occurring in late autumn or winter in roe deer was unlikely to be productive. Later this was supported by Bischoff (1854), Keibel (1899) and other researchers. However, from histological examination of the sexual organs of European roe deer, Stieve (1950) demonstrated that, despite a rapid reduction in testes size and a cessation of spermatogenesis after the rut, fertilization remains theoretically possible until the end of December due to the persistence of live sperm in the epididymides (see also Sempere, 1982). Nevertheless, there are only a few spermatozoa in the epididymides later than September or October when immature sex cells are prevalent, and it is still unclear whether successful fertilization is actually possible at this time (Short and Mann, 1966; Arabuli, 1966).

From studies of the seasonal dynamics of the histological structure of testes and epididymides in Siberian roe deer, Tsaplyuk (1977) noted that epididymides cease to form among adult males in May, but only some of these tubules may contain a few spermatozoa at this time. (In European roebucks, free spermatozoa have been recorded in the sexual organs as early as March; Sempere, 1982.) The level of spermatogenesis increases to a moderate rate by June but becomes extremely intense during July; by August, the concentration of spermatozoa in the epididymides is especially high. However, the rate of production of primary spermatocytes begins to decline as early as August when spermiogenesis becomes prevalent. Towards the end of September spermatogenesis stops entirely and the epithelium perishes on some sections of the epididymides. By the end of November all spermatogenic cells and solitary spermatozoa remaining in the tubules begin to degenerate and no live spermatozoa are found in the sexual organs of Siberian roebucks by the end of December. In January, resorption intensifies in the testes and epididymides and continues into February, but at the same time replacement of the epithelium begins in the epididymides. Mitosis of the spermatogonia takes place in the testicular tubules in March, with the formation of spermatogenic cells in April.

It should be noted that activity of the gonads in *Capreolus* is stimulated in winter, some time after the winter solstice, and is linked to increasing day length, whereas in other cervids this occurs at the beginning of summer, coinciding with a decrease in daylight hours. What are the hormonal changes which drive these cycles? Following the rut, blood testosterone levels fall abruptly among roebucks to a minimum in winter; they rise again in spring and reach a maximum by the start of the next rutting period. The dynamics of pituitary hormone (LH) concentration is identical to the profile of the annual cycle in the level of testicular hormone (testosterone) but is in advance of the testosterone cycle by about a month (Figure 8.2). Castration essentially does not modify the plasma LH cycle profile. Seasonal variations of testicular activity and prolactinaemia are strictly synchronous, while the exocrinal and endocrinal functions of testes are very nearly so. However, the processes involved in spermatogenesis accelerate very rapidly while endocrinal functions remain checked throughout the spring; therefore, it seems unlikely that androgenic and spermatogenic functions are fully coordinated with each other. Nevertheless, age-related and seasonal variations in roe deer cycles are quite similar (Figures 8.1, 8.2), suggesting that testicular activity is controlled by both the hypothalamus and the pituitary glands (Short and Mann, 1966; Sempere, 1982), and histological studies of the pituitary gland support this (Cadariu *et al.*, 1977). Activity of the gonadotropic beta cells becomes noticeable in May, intensifies in June as spermatogenesis begins and, having reached a maximum in July, then starts to decline once more. Similar changes were found in the delta cells, with the maximum activity recorded during August. These phenomena seem to be directly associated with the reproductive cycle. Thus, roebucks are physiologically capable of reproducing only during a single annual sexual season from May (or possibly March–April; Sempere, 1982) to December, though it is unlikely that successful coupling can occur at the very beginning or end of this period (Figure 8.3).

Stieve (1950) suggested that roe does have two mating seasons per year. Although this seems not to be the case (Short and Mann, 1966; Aitken, 1974), the belief that roe deer have one main rut but with several additional periods of reproductive activity is still widespread. Tsaplyuk (1977) has rather convincingly demonstrated that the Siberian roe does of

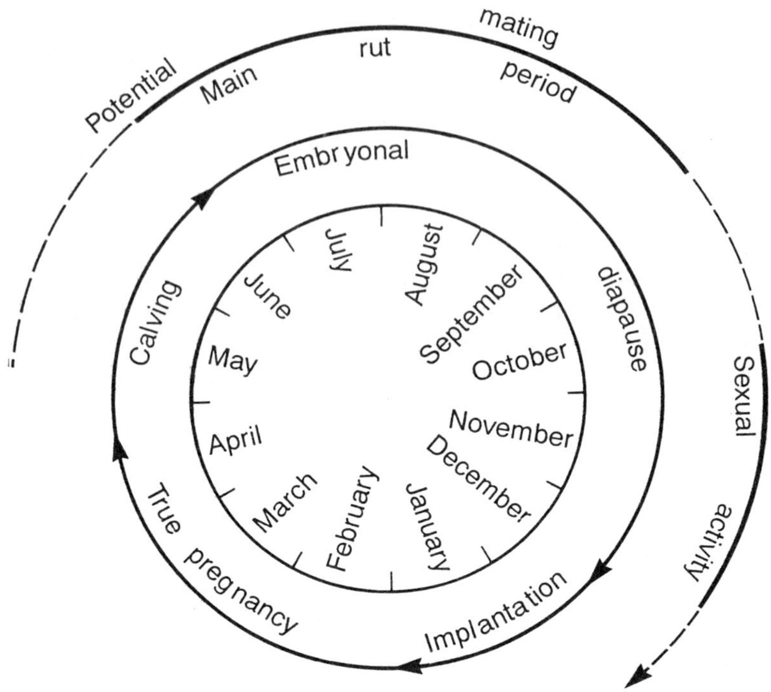

Figure 8.3 Reproductive cycle of *Capreolus*.

Kazakhstan are seasonally polyoestrous with a single long mating period. In January, the gonads contain only follicles in the process of atresia, but by February the follicular zone revives, primary and secondary follicles develop and some vesicular follicles may become as large as 5 mm. During March, the histostructure remains essentially the same, but large secondary follicles become prevalent as early as May. In August, numerous vesicular follicles are present whose emergence and atresia precede the maturation of ovulatory follicles, and it is possible that biostimulators secreted by the former allow the latter to complete growth and maturation. By September or October, doe ovaries contain not only the developing corpora lutea but also one or two large normal follicles which undergo atresia in November and are at an advanced stage of this process as early as December. During the early stages of formation, the corpora lutea seem to promote the development of the next generation of vesicular follicles, thus indicating that roe may indeed be polyoestrous.

A further important fact should be noted: female ovaries may be at different functional stages simultaneously (Popovici *et al.*, 1978). Some follicles contain two ova, each of which can contain two nuclei (Tsaplyuk, 1977). Thus, the possible distribution of corpora lutea in the two ovaries upon ovulation are: 0 + 1, 0 + 2, 0 + 3, 0 + 4, 1 + 1, 1 + 2, 1 + 3, 2 + 2 (Borg, 1970; Prusaite *et al.*, 1977; Blant, 1987). This also indirectly indicates that non-synchronous maturation of ova is possible, which may occasionally result in a second oestrus. Ellenberg (1978) observed such a phenomenon in various female roe after 9, 14, 17, 27, 32 and 84 days. Strandgaard (1972a) also recorded repeated manifestation of sexual behaviour after 20 and 22 days.

These different-length intervals between repeat oestrous cycles add support to the possibility of multiple oestrous cycles; nevertheless, most does do become pregnant after a single short rut (Chapter 7). Thus, fertilization of female roe can, in principle, take place from May (or possibly April) to December (Figure 8.3), but at the beginning of this period (in May and June) successful coupling for most does is practically impossible owing to their continuing pregnancy and subsequent intensive lactation. Towards the end of the mating period (from October to December), fertilization becomes increasingly difficult due to the cessation of spermatogenesis among bucks.

The surge of roe deer sexual activity sometimes observed during the autumn is probably largely due to the manifestation of sexual behaviour among early maturing kids and a corresponding reaction to this behaviour among adults. Subsequently, the implantation of the fertilized blastocyst in December and January causes marked changes in the sexual functioning of pregnant does, probably accompanied by an intensification of glandular function. As a result, bucks sometimes become sexually excited at this time and even attempt copulation; however, this does not result in fertilization as the doe is already pregnant (Schumacher, 1939; Bottmann, 1961). Overall, any manifestation of sexual behaviour during autumn or winter is therefore of little significance in a reproductive sense. Ellenberg (1978) clearly demonstrated this by placing European roe doe adults in enclosures with bucks from October to the following summer and showing that no pregnancies resulted. In *Capreolus*, autumnal sexual activity may reflect the rutting patterns of the ancestral roe deer species.

8.3 PREGNANCY

The roe deer is the only ungulate, indeed the only species of artiodactyl, which has a latent period of pregnancy (delayed implantation) and thus the roe's reproductive cycle (Figure 8.3) differs markedly from the cycles of even closely related species. At the morula stage, the fertilized ovum penetrates the uterus where it rapidly cleaves to form a blastocyst measuring 0.1 mm in diameter. Subsequently, there is a 4.5–5-month period when mitotic activity is minimal and a process of long, slow development increases the blastula diameter to only 5 mm. In this pre-implantation phase, from July or August through to December or January, the blastula remains free within the uterus and is sustained by 'uterine milk' (Short and Hay, 1966; Aitken, 1974, 1981). During this period, blastocysts may migrate from one uterine horn to the other, but in the end (for does carrying two embryos) each horn invariably contains a single embryo whether the ova originated from the same or from different ovaries (Strandgaard, 1972b). Identical monovular twins are very rare (Hamilton *et al.*, 1960; Borg, 1970).

Unlike the delayed implantation observed in weasels (Mustelidae), for example, this phenomenon in roe deer is probably not photoperiodically controlled. Experimental shortening and subsequent prolongation of day length, accelerating the photoperiodic cycle to a point 2 months in advance of the natural seasonal pattern, failed to produce any changes in the length of embryonic diapause, though the deer responded markedly in other ways – by moulting 2 months ahead of the normal date (Lincoln and Guinness, 1972). Moreover, unlike other mammals which exhibit delayed implantation, roe deer demonstrate very high ovarian activity (Aitken *et al.*, 1973; Tsaplyuk, 1977). During the diapause stage, the progesterone level among pregnant does is comparable to that of non-pregnant deer but is much higher than the level among kids, strongly indicating high secretory activity of the corpora lutea. Following implantation, progesterone production rises sharply, whereas among non-pregnant adults it drops to a very low level equivalent to that observed among kids, probably due to the degeneration of the corpora lutea (Hoffmann *et al.*, 1978; Sempere, 1982).

In *Capreolus*, it seems most likely that embryonic diapause is caused by a shortage of substances which trigger and sustain embryonic development. From the beginning of January these substances are present in the secretions produced by the uterine glands which contain specific proteins, blood plasma proteins, about 20 free amino acids, glucoproteins, glucose and free ketose. The sudden elongation of the blastocyst is caused by an increase in oestrogen concentration which seems to stimulate the secretory activity of uterine glands. However, ovarian function does not increase during this period; therefore it seems most likely that the higher oestrogen level is a consequence of embryo growth rather than its cause (Aitken *et al.*, 1973; Aitken, 1974, 1981). In the light of this research, the term 'delayed implantation' does not accurately reflect the process since it is the development of the blastocyst itself that is hindered, not the formation of an anatomical link attaching the blastocyst to the uterus. 'Embryonic diapause' may be a more suitable description.

In about 50% of the does of central Europe, this pre-pregnancy stage ends in the middle of January, i.e. on average embryonic diapause lasts about 168 days (Wandeler, 1975). In areas where the climate is more severe, this stage ends earlier (Hubner, 1938). The few does which were mated in late autumn practically have no pre-pregnancy stage and gestation effectively begins at once (Figure 8.3). In January, placental attachment (implantation) of the embryo takes place while it weighs less than 1 or 2 g (Aitken, 1974; Wandeler, 1975). However, it has also been observed that at the end of the pre-pregnancy period the germinal vesicle may enter the uterus many days before development commences (Short and Hay, 1966).

The gestation period is characterized by rapid growth of the embryo and intensive secretory activity of the uterine horn glands, which increases the level of carbohydrates, especially of fructose, in the amniotic fluid by about 15–30 times, compared with the level during diapause (Aitken, 1976). In primiparous individuals, the embryos develop more slowly and the does usually give birth a week later than mature adults (Ellenberg, 1978). In more northern and alpine areas, gestation is markedly longer: in Sweden, for example, embryos take nearly a month longer to develop than in Germany. On the western coast of Jutland, gestation lasts 12–14 days longer than on the eastern coast where the climate is milder (Strandgaard, 1972b), and similar reports have been provided by other researchers (Rieck, 1955; Sagesser and Kurt, 1966; Kurt, 1968b; Wandeler, 1975). Thus, there seems to be a noticeable climatic influence on the term of roe deer pregnancy.

When air temperatures are higher in spring, the onset of the growing season is earlier and vegetation grows more rapidly, providing a more abundant and better food source for mothers, thus accelerating embryonic development (Sagesser, 1968). Therefore, even within a single geographical area, the length of pregnancy terms can differ from year to year, accompanied by a shift in the corresponding date of kidding (Table 8.2). For example, when spring was late in Germany in 1973, the gestation period for roe deer was on average 10 days longer (Ellenberg, 1978). Similarly, in the spring of 1987 in the Belgorod region of Russia, vegetation growth was delayed by a month resulting in a birth period concentrated into July, 10–15 days later than expected, with only the earliest kids born at the end of June. This occurred despite the fact that the rut took place at the normal time the previous year.

From published reports, the full term of a pregnancy varies from 264 to 318 days for European roe deer but may be as short as 5 months following a fertilization in late autumn. However, pregnancy is commonly much less variable in length, lasting somewhere between 294 and 300 days across various regions of Europe (Prior, 1968; Wandeler, 1975; Ellenberg, 1978). Highly synchronous kidding may be selected for in roe

Table 8.2 Dates-of-birth period across the roe deer's geographic range

Location	Start	Finish	Reference
		European roe deer	
France	May 13	June 5	Sempere, personal communication
France	April 17	June 11	Gaillard *et al.*, 1993c
Switzerland	April 24	June 16	Sagesser and Kurt, 1966
Sweden	May 21	July 3	Espmark, 1969
Germany	May 1	June 3	Ellenberg, 1978
Germany	May 3	June 15	Stubbe and Stubbe, 1985
Denmark	May 5	July 11	Strandgaard, 1972b
Hungary	April 14	May 17	Szederjei and Szederjei, 1971
Poland	May 15	June 20	Kaluzinski, 1982a
Bulgaria	April 24	June 5	Dragoyev, 1965
Byelorussia	May	July	Serzhanin, 1955
Georgia	May	June	Arabuli, 1966
N. W. USSR	April	June	Vereshchagin and Rusakov, 1979
Russia	May 14	July 11	Danilkin's data
		Siberian roe deer	
Moscow (captive)	June 2	June 17	Gromov and Danilkin, 1984
Urals	May 17	June 20	Dvornikov, 1984
Urals	May 23	June	Kiselev, 1979
Urals	May 10	July	Averin, 1949
Urals	May 15	June 23	Danilkin's data
Urals	May 31	–	Filonov, 1974
Kirghizia	May	–	Aizin, 1969
Kazakhstan	May 20	July	Pole, 1973
E. Siberia	May	July	Fetisov, 1953
Trans-Baikal	May	June	Smirnov, 1978
Yakutia	May	June	Yegorov, 1965
Far East	April	June	Bromlei and Kucherenko, 1983
Far East	May 1	July 25	Darman, 1986
Mongolia	May	July 17	Danilkin's data

deer due to the difficulty experienced by females which give birth late in the season in regaining condition for the next summer rut (Gaillard *et al.*, 1993c). In the Siberian roe deer, pregnancy is slightly shorter with a later rut. For example, the length of the pregnancy period averages 287 days in the Urals, varying from 279 to 296 days, and 284 days in the Far East (Darman, 1986). The full term of pregnancy of European and Siberian does kept in the same pen near Moscow also differed, averaging 306 days and 280 days, respectively (Gromov, 1986a).

Embryonic diapause during the period of pregnancy has a great significance in the reproductive cycle of roe deer, concentrating reproductive effort (both the rut and the birth period) into the summer months (Loudon, 1982). Thus, it seems that the arrest of embryonic development is probably adaptative, permitting roe deer to inhabit harsh northern climates with large seasonal variation in environmental conditions. However, it is still unclear why this breeding peculiarity evolved only in roe deer and not in other similar-sized ruminants inhabiting similar climatic zones and habitats. The fact that pregnancy can proceed either with or without a period of embryonic diapause (for example, among kids in their first cycle of sexual activity, aged 4–6 months) indicates that this phenomenon evolved through an adjustment of the female breeding cycle to coincide with the period of highest male sexual activity, which had shifted from autumn to summer. This shift was probably associated with territoriality, and may have been driven by promoting female mate choice at a time when high quality resources are available and male–male competition is most intense (Sandell, 1990). Territoriality, occurring at the onset of the growing season, means that roebucks reach the peak of their hormonal and related sexual activities early, which in turn stimulates sexual responses among the females, promoting successful fertilization in mid-summer. However, due to embryonic diapause, the fertilized ovum does not develop (under some as yet unknown system of control) until the onset of winter, in contrast to the reproductive cycle of other species of Cervidae.

8.4 BIRTH

European roe deer kids are usually born sometime between the end of April and the middle of June, whereas Siberian kids are generally born rather later – from the second half of May through to the middle of July (Table 8.2). The length of the birth period does not usually vary significantly from year to year (Gaillard *et al.*, 1993c), although the peak in number of births may shift slightly due to climatic influences on the length of the pregnancy term. Using data from 1600 marked kids, Rieck (1955) demonstrated that, across Europe, 96% of kids are born in May and June, with a birth peak on 2 June. Over this range, the average parturition date becomes later moving from south-west to north-east, as well as with altitude. Such variation is almost certainly linked to resource availability, often mediated by climatic factors. In Switzerland, the peak of roe deer births occurred on 3 June in 1965, but was more than one week earlier in 1966, on 26 May, due to mild climatic conditions which promoted vegetation growth (Sagesser, 1968). Ellenberg (1978) reported variation in the mean date of birth in Germany from 17 to 29 May over five consecutive years and similar data from France (Gaillard *et al.*, 1993c) found no year-to-year variation in the timing of the birth period, with an overall mean of 15 May. In the Far East, the average birth date is 29 May (Darman, 1986); in the Urals the equivalent date is 9 June. Thus, the birth period is highly synchronous for roe deer, and almost all offspring are produced during the season which provides the most favourable conditions, with mild weather and abundant vegetation for cover and food.

About a month before giving birth, the adult doe separates from its group and takes up a small kidding range which remains the same from year to year, protecting it from other does (Kurt, 1966; Sokolov and Danilkin, 1981). Kids may even be found on the same spot in several consecutive years. The choice of kidding range depends on available cover, feeding conditions and, in arid areas, the availability of free water. Visibility within the kidding range is generally much lower than that of the surrounding habitat. The proximity of human habitation is less important: where roe deer are not extensively hunted, some does may produce their offspring only 100–200 m from the nearest building. However, high levels of disturbance may cause a pregnant doe to emigrate immediately prior to parturition (Sokolov *et al.*, 1986d). The most preferred birth sites are forest edges and flood plain meadows covered with dense, high grass. In forest–steppe habitats, most does drop their offspring in meadows; for example, in Switzerland, 74% of kids are born in this habitat type, while 23% are born in forests and only 3% in open fields. Of those born in meadows, the majority (51%) occurred in grass between 20 and 50 cm high, while only 17% were in grass of less than 20 cm and 32% in grass taller than 50 cm (Blankenhorn, 1975). In mountains and hilly country, southern slopes are the most attractive and competition to obtain kidding ranges on them becomes severe when the population density is high (Sagesser and Kurt, 1966). Some observations show that once a doe kid matures, if the mother has died or emigrated, she may occupy her mother's range and so produce offspring on her own birth range (Aylmer, 1977).

Most births occur during daylight hours. The doe gives birth while in either the lying (Bubenik, 1965; Espmark, 1969) or the standing position (Holler and Prilkofer, 1980). When several kids are born, she may alternate between these two postures (Sokolov and Danilkin, 1981; Gromov and Danilkin, 1984). The process of parturition begins with labour when the doe becomes unsettled, periodically standing up and licking the vagina, and the amnion becomes visible. During contractions, which repeat every 2–9 minutes, the doe attempts to lie on her side and, when straining, turns her head back and gives a sound resembling a constrained cry. About half an hour after the contractions start, the hooves and legs of the kid emerge and are licked clean by the mother, now in an upright position. The first kid is born about 8–40 minutes after the limbs first emerge from the genital tract, delivered forelegs and head first. The second and third kids then appear at 10–40-minute intervals; subsequently the afterbirth separates 12–50 minutes after the birth of the last kid. Generally, delivery takes 1.5–2 hours, but rarely lasts as long as 4–5 hours (Figure 8.4).

8.5 ONTOGENY, NEONATAL AND MATERNAL BEHAVIOUR

The behavioural development of European and Siberian roe kids is closely linked with maternal behaviour; therefore, this section focuses on the mother–offspring relationship during the first few months of life, drawing in particular on the studies of Bubenik (1965), Dathe (1966), Kurt (1968a), Espmark (1969), Sokolov and Danilkin (1981), Gromov and Danilkin (1984) and Sokolov *et al.* (1985c).

8.5.1 Juvenile and maternal behaviour in the post-natal period

As soon as the kid is born, it struggles to free itself from the membraneous caul, usually giving soft squeaks. This is the kid's first attempt at communication to which the mother responds with hisses, which may be muffled. The mother cleans the amnion from her offspring (Figure 8.5a), lies beside it and spends 5 to 20 minutes licking the kid almost incessantly (Figure 8.5b). Most probably, licking dries the skin and normalizes heat exchange; it may also stimulate locomotory activity in the kid. In turn, maternal behaviour seems to be intensified by these tactile and olfactory cues (Blauvelt, 1955; Lent, 1971; Baskin, 1976). When several kids are born, the doe generally pays most attention to the first delivered, and in one recorded case the mother only began to groom her second and third kids 5 minutes and 1.5 hours after birth, respectively (Gromov and Danilkin, 1984). Thus, there seems to be some decrease in the doe's maternal instinct, which may sometimes result in the death of the last kid to be delivered while still in its caul (R. Prior, personal communication). This may partly explain the fact that does with triplets at heel are encountered much less often than expected based on the frequency of observation of three embryos in a uterus. This may be an important factor limiting population growth in highly productive populations.

The mother then immediately eats the caul membranes and surrounding grass from the site where the kids were dropped. Subsequently she also consumes the afterbirth, which is usually expelled 30–40 minutes after the last kid is born. The kid, still wet, crawls towards the mother, pokes its nose into various parts of the mother's body until the teat is located and suckles for 2–3 minutes in several short pulses (Figure 8.5c). Once it has fed sufficiently, the kid rests and the mother may then

(a)

(b)

Figure 8.4 Roe does deliver their kids in either (a) lying or (b) standing position.

(a)

(b)

(c)

Figure 8.5 Juvenile and maternal behaviour of Siberian roe deer during immediate postnatal period. (a) Removing the amnion; (b) licking kid; (c) first suckling.

(d)

(e)

(f)

Figure 8.5 *contd* (d) kid stands; (e) resting; (f) typical suckling posture.

(g)

(h)

Figure 8.5 *contd* (g) leading kid away from birth site; (h) hiding in response to danger.

produce a second or third kid. If the mother does not change her posture after this first feeding, the kid may suckle as many as six or seven times for a total duration as long as 5–9 minutes.

Having rested, the kid attempts to stand, straightening the hind legs and standing on widely splayed front legs (Figure 8.5d) or on the bent phalange of the forelegs, and generally only manages one or two steps before collapsing. In this way, standing and collapsing again, the kid covers the few metres to the nearest patch of high grass, where typically it curls up to rest once more (Figure 8.5e). However, some kids may stand again after only a few minutes and approach the mother or even wander several metres away from her. The second feeding bout occurs from either a lying or a standing position and takes place 1–5 hours after the first one. It is generally initiated by the mother when she approaches her offspring, sniffs and licks it, and lies down beside or stands above the kid. In the latter case, the kid stands while searching for the udder, initially unsuccessfully, before gradually moving towards the rear of the mother's body and assuming a typical suckling posture (Figure 8.5f), while the mother licks clean the ano-genital region.

Once the suckling bout is terminated, the kid may move between 3 and 30 m from the mother to lie down alone in high grass. Occasionally, a newborn kid may attempt to follow the mother, who evidently tries to lead it away from the birth site (Figure 8.5g), but the kid soon becomes tired and lies down in dense grass or in shadow provided by a tree or bush. The maximum recorded distance covered by a free-ranging kid in the southern Urals during its first 5 hours of life was about 170 m away from the birth site.

During the first few hours, kids can be induced to follow any object which is of similar size to their mother (imprinting), and only towards the end of their first week do they begin to discriminate her positively from other roe deer. However, humans may be perceived as a threat as early as 1 or 2 days after birth. The doe probably recognizes her kid by smell and she can usually discriminate her own offspring from other kids as early as the end of the initial post-natal period, i.e. 2–5 hours after birth. From experimental manipulation, however, it has been shown that it is possible to substitute kids with a foster mother up to 3 weeks following birth (Bubenik, 1965; Kurt, 1968a). It seems the doe is unable to identify her own kid from vocal or visual stimuli, and she responds to the calls of any kid without a mother, or even to simulated kid vocalizations. The sequence and the time of occurrence after birth of these post-natal behaviours for Siberian roe deer is summarized in Table 8.3. Subsequently the period of intensive post-natal care, characterized by close spatial proximity of mother and kid, ends.

8.5.2 Early behavioural development

Ungulates have been classified into two main types: as 'hiders' or as 'followers', indicative of the nature of the mother–offspring relationship during the first few weeks or months of life. In the former category, the young occupy a small range, generally isolated and providing good cover; in the latter category, offspring follow the mother during her normal movement patterns (Walther, 1968). Both the European and the Siberian roe deer clearly belong to the former category, and this hiding phase generally lasts for about 2.5–3 months. This period, however, constitutes a time of marked behavioural change, from a situation characterized by a high degree of dissociation between family members and a lack of synchronization of their activities, to one of highly synchronized patterns of behaviour in time and space. Below, we shall consider separately

Table 8.3 First observation (number of minutes after birth) of some post-natal behaviours in newly born Siberian roe deer (Sokolov *et al.*, 1985c)

Activity	Average (n=7)	Range
Raise head	6.5	2–12
Straighten ears	11.3	10–12
Attempt to stand	12.4	2–33
Crawl	17.8	6–34
Squeal	23.5	1–51
Suckle – lying position	27.3	7–102
Stand	29.7	20–47
Walk on toe phalangae	29.3	21–47
Walk on straight legs	40.2	28–51
Suckle – standing position	69.5	45–94
Rest away from birth site	94.0	79–113

the development of the main behavioural patterns during this period (Sokolov *et al.*, 1985c).

(a) Defensive behaviour

Defensive behaviour during the first 2.5–3 months of life consists of minimizing contact between family members and maintaining a high degree of individual dispersion over the range. The mother rests separately from her offspring, which themselves lie up between 5 and 200 m apart. During the first few weeks, the doe suckles the kids alternately and then, following a feeding bout, the kids move to another bedding site, but all their movements are confined to a very small range. When a source of danger is detected, the mother gives an alarm signal by jumping and running away, producing a considerable amount of noise. In response, the kids run towards the nearest cover (high grass, trees, bushes) where they lie down and freeze in a characteristic posture, the legs drawn up to the body and the head pressed to ground (Figure 8.5h). However, during its first 2–5 days, the immobile roe kid does not attempt such an escape when approached and it is even possible to touch animals of this age, although a kid will give a shrill cry when handled. At 3–7 days of age, the kid will run whilst giving the same shrill cry if approached to within 1 or 2 m, but this distance quickly increases to 10 m by 1–3 weeks and subsequently to 15–30 m, at which point the kid no longer cries while in flight. Initially, during the first one or two weeks, the doe responds aggressively to such an approach and her kid's vocal signal, but later her dominant reaction becomes a defensive one, although she remains aggressive towards other does until about 2 months after birth.

Other characteristics of the roe deer which assist in the initial avoidance of detection by predators include the cryptic coloration of the kid's spotted coat and the early stage of development of the skin glands, which do not leave a strong scent. In addition, the doe removes all traces of the act of birth by consuming the membranes of the caul, the afterbirth and the grass from the birth site so as not to attract predators.

Freezing when in danger remains the dominant defensive behavioural reaction of roe deer kids until September, when they begin to show more diverse behavioural responses. This coincides with the autumn moult and the emergence of the large white 'targets' which may encourage a 'follow-the-leader' response, so that the defensive behaviour of family groups becomes synchronized: they generally make an escape as a unit. By this time kids have also learnt by imitation to demonstrate some other defensive reactions such as the 'alarm' posture, re-grouping behaviour, etc. Indeed, kids tend to express such defensive behaviours (particularly running away) more frequently than adult animals who are better able to assess the seriousness of any potential source of danger.

(b) Exploratory behaviour

The first behavioural reactions (visual inspection, sniffing, listening) to various environmental stimuli are displayed by roe kids from the first hour onwards. Once they are 3 days old, a considerable amount of time is spent inspecting various objects, and any motion may trigger an immediate oriented reaction. Any meeting between the kid and another doe involves nasal contact. At 2 weeks of age, kids begin to respond to urine, faeces and the markings of bucks on the ground and on trees, sniffing them for a long time. At this time, the exploratory behaviour of juveniles practically does not differ from that of adults.

(c) Suckling behaviour

During the first two or three weeks the doe suckles her offspring one at a time, following a certain order which is rarely altered. If a kid does not respond, failing to rise from the bedding site as the mother approaches, the doe leaves the area to return some time later, repeating this approach several times per hour until successful. Later, the mother generally feeds her kids together, but the frequency of feeding bouts remains the same.

Initially, the mother initiates any contact between herself and her offspring, possibly stimulated by the filled state of the udder. At this point the doe ceases to graze; her gait becomes uneasy and then becomes short bursts of running. She heads purposefully towards the bedding site, often stopping to survey the area. Once the kid has been located, she sniffs it and licks it clean, whereupon the kid stands and begins to suckle. After several days, the doe's behaviour alters when approaching her offspring: she stops several metres from the bedding site, extending her neck towards the prone kid, and sometimes nodding two or three times. The kid then stands, makes nasal contact and starts to suckle.

During the first 10 days, the kid often gives insistent squeaks to call its mother to suckle, but remains at the bedding site. Subsequently, a kid may actively search for its mother in the near vicinity of this site, giving occasional calls and running towards any other roe deer which pass by. In these circumstances, adult bucks tend to show exploratory rather than aggressive behaviour towards kids. From about 1.5 to 2 months of age onwards, roe kids may begin to take vegetative food and run to their mother only when she is within visual contact range. Generally, does do not suckle kids which are not their own offspring, usually driving them off or running away themselves after sniffing the individual.

Kids learn to find the udder almost immediately during the post-natal period, but initially they often take several attempts to assume the correct feeding posture (Figure 8.5f) in which the rear half of the body is accessible to the mother. During the first few days, suckling begins immediately the kid makes contact with the nipple, but later on it make take several thrusts before the milk begins to flow. During feeding, the kid may squeak impatiently when particularly hungry and often changes nipples. At one feed there are usually several suckling bouts interspersed with periods of rest when the kid may wander some distance from the doe before resuming suckling. During the first 10 days, on average roe kids suckle 4 or 5 times per feed, with a observed maximum of 11, but by 1.5–2 months of age, this average decreases to one bout per feed. Accordingly, the total suckling duration falls from 10 to 12 minutes initially, to one minute. During the first 10–15 days the suckling bout is usually terminated by the kid, but subsequently it is the doe who withdraws, stepping away from the kid and walking or running away, ensuring the kid cannot resume suckling.

The total number of feeds a kid can obtain varies according to the litter size being reared by the mother, but may be as many as 5–9 per day during the first month, decreasing to 2–4 per day in the second month and then dropping down to one or two per day. Despite this decrease in feeding frequency, family members actually spend progressively longer engaged in joint activities, which increase from 30–40 to 70 minutes over the same period. Lactation generally ceases in August and only rarely continues longer (occasionally to October or even November). From this time, kids convert completely to a diet of vegetative food, which they may have begun to consume in small quantities as early as the end of their first 10 days.

(d) Comfort behaviour

Newly born roe kids show almost no activity that can be considered characteristic of comfort behaviour, which usually comprises grooming, scratching, mutual cleaning and preparation of the bedding site. The first signs of head and neck scratching with the hind leg are observed at 2–3 days of age, but coordination of the movements is imperfect and so these activities are not always successful. The first attempts to lick the body are observed after 4–6 days, and by the end of the first 10 days or so the kid is able to reach its ano-genital region. At 3–4 weeks old the kid begins to groom itself, but until the age of 2–2.5 months this function is performed mostly by the mother during feeding. These bouts of grooming are especially long and thorough during the first fortnight, when kids have difficulty excreting. Mutual cleaning is observed from the end of the first 10 days and probably has not only a hygienic but also a social function.

It is the kid itself that chooses the bedding site from the first day onwards, clearly preferring the cover of high grass sward and shrubs and avoiding direct sunlight. By one month of age, this site may be changed less often and at this time kids frequently use the same area for several days consecutively, returning there after feeding. Preparation of the bedding site (scratching the ground with the front legs and removing the top soil) is first observed at 2.5 months of age. In winter, they may remove snow in exactly the same manner until the bare ground is reached, but when the snow is deep, they may stop once they have dug an oval crater in the snow bed.

(e) Play behaviour

Vertical jumps and kicking are observed among kids from the third day. By the end of the first week, play regularly involves fast, unmotivated running around the mother with sudden stops. Play behaviour becomes most intensive during periods of joint feeding with siblings and often involves mutual chasing in which the mother may also take part. At 1.5 months old, chases are often interspersed with threat postures, attacks and butting. Buck kids may start to indulge in sexually related forms of play behaviour from 1.5 to 2 months old, and during this period they also develop activities resembling marking such as head and neck rubbing on trees and shrubs. At 4 months of age their play includes the entire set of postures peculiar to the antagonistic behaviour of adult

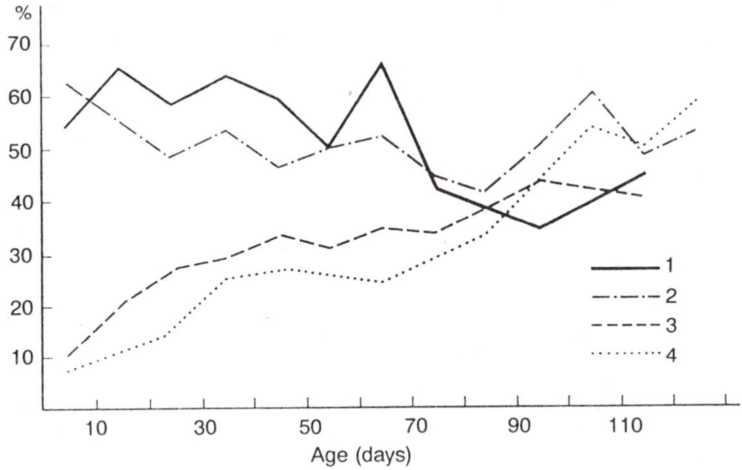

Figure 8.6 Daily activity (percentage of time active) of mothers and kids. (1) European mother; (2) Siberian mother; (3) European kid; (4) Siberian kid. (From Espmark, 1969; Sokolov *et al.*, 1985c)

bucks. Interestingly, this burst of antagonistic behaviour among kids corresponds to the period of growth of the sexual organs and consequently increased hormonal levels.

(f) Activity

During the first few days, the activity of roe deer kids solely concerns feeding, each feed lasting for about half an hour. Kids stand only when their mother approaches and immediately lie down again as soon as she leaves. But as early as their second week, kids may walk for several minutes before and/or after suckling when their mother is no longer present. From one month old, the length of each period of activity increases up to 1 or even 1.5 hours; at 2 months, 1.5–2 hours of continuous activity may be observed. Half of this total activity time is spent alone, with grazing in the proximity of the bedding site constituting 30–40% of activity. Though the frequency of suckling gradually decreases with age, the number of active periods remains approximately the same (about 5–7 per day) but, in contrast, the total time active greatly increases as the kid becomes older. The activity level of the doe remains at a constantly high level throughout the period of lactation.

When the kid is 2.5–3 months old, the activity of family members becomes synchronous, both in space and in time (Figure 8.6). They graze together, rest at the same time and will run as a unit from any perceived source of danger. At this stage, the offspring have learnt all the main patterns of behaviour from their parents; however, the territorial and sexual behaviour of juveniles still differs greatly from that of adults.

In comparative analysis of the behavioural ontogeny, as well as maternal behaviour, we have not found any appreciable differences between the European and the Siberian roe deer. The behaviour patterns of the two species develop in a similar manner and seem identical during the post-natal period. In addition, both show an approximately equal level of overall activity at this time, again highlighting their close taxonomic relationship.

8.6 EUROPEAN AND SIBERIAN ROE DEER HYBRIDIZATION

Reproductive or genetic isolation (lack of hybridization) is one of the most important criteria defining a species; therefore hybridization experiments involving European and Siberian roe are of great interest. The most rigorous experiments were conducted in Germany (Stubbe and Bruchholz, 1979, 1980) when two Siberian roebucks were mated with European does 32 times, of which 13 attempts were unsuccessful. Of 19 births, a caesarean delivery was necessary in nine cases and an overly large fetus had to be extracted manually in another three. Subsequently, two hybrid F1 bucks were crossed with hybrid F1 does a total of 10 times, but no offspring were produced from these attempts. In contrast, backcrosses between hybrid does with pure European and with pure Siberian roebucks produced viable offspring. This series of experiments suggests that hybrid males are sterile and also shows that European does mated by Siberian bucks cannot normally deliver the larger fetus which is produced. Under natural conditions, therefore, such crosses would normally result in the death of the doe.

Similarly, experiments conducted at the Moscow Region Experimental Base of the Russian Academy of Sciences have confirmed that hybridization of the European and Siberian forms is rarely successful (Sokolov and Gromov, 1985b; Gromov, 1986a). Of six mating attempts between two Siberian bucks and three European does, only three resulted in the production of a fetus. Of these cases one doe died before delivery, another during, and only the third actually bore a hybrid kid (buck). Furthermore, all 12 attempts to cross two European bucks with nine Siberian does were unsuccessful; however, the back-crossing of the hybrid F1 buck with both European and Siberian does proved successful, with a total of four kids reared, demonstrating that in fact not all hybrid roebucks are sterile. On the other hand, the hybridization of European bucks with Siberian does seems to be virtually impossible, perhaps because European bucks do not normally respond to the stimulation of Siberian does during the period of rutting activity. It has been suggested that European does may excrete some olfactory component which is also attractive to Siberian roebucks, but Siberian does do not possess such an attractant for European bucks.

Despite the above results, successful hybridization has been reported from crosses of European roebucks and Siberian does in Leningrad Zoo (*International Zoo Yearbook*, 1970, **10**, p. 249; 1971, **11**, p. 259). Experiments here have convincingly shown that the rutting of such pairs is normal, the buck readily responds to the olfactory stimuli of a doe on heat, and the level of fertility is normal. However, it should be noted that the number of stillborn fetuses is relatively high (Danilkin, 1986b). Of eight experimental matings, two produced no offspring; in two further cases the kids were stillborn, while the remaining four does produced twins. No experimental back-crossing with hybrid males was carried out, so it is impossible to judge if these bucks were sterile, but it was noted that hybrid bucks did respond to Siberian does on heat and attempted to mate them. A back-cross between a hybrid doe and a European buck produced no offspring, although rutting and mating did take place normally. The hybrid female showed some abnormal development, producing an antler when one year old; by the following autumn this had become 6–7 cm high, growing upward but also to the sides from its base. A second antler emerged in the April of the doe's fourth year, growing several centimetres, before she was shot the same year. When culled, this hybrid doe weighed 35 kg – substantially more than the weight of a European doe. The weight of the three newborn hybrids which were produced averaged 1.66 kg, a value similar to that of Siberian kids (Gromov and Danilkin, 1984).

Gromov (1986a) also described differences in sexual behaviour of the European and Siberian roe deer, which he considered to be of taxonomic importance. During the rutting period, European roe are characterized by a specific ritual of sexual behaviour where the mating pair run in rings around some natural feature, but Gromov did not observe such behaviour among Siberian deer. However, these observations were in captivity. Under natural conditions we have more than once observed Siberian roe pairs displaying this behaviour typical of sexual activity, running around shrubs, trees or hummocks on paths describing a circle or a figure-of-eight. This feature of sexual behaviour has been described previously (e.g., Pole, 1973, p. 139; Sludski and Baidavletov, 1984, p. 71; Sokolov and Danilkin, 1981, p.95; Darman, 1986).

The above hybridization experiments, summarized in Table 8.4, have produced some conflicting results. In particular, it is not clear whether hybrid males are indeed sterile, or whether European roebucks will respond to Siberian does on heat. The cause of such inconsistency is not obvious but could involve the different origins or genotypes of the parent pairs, differing feeding regimes in the various experiments, or even climatic factors. It is not possible to quantify the influence of these potential

Table 8.4 Results of experimental hybridization of Siberian and European roe deer (pooled data from Stubbe and Brucholz, 1979, 1980; Sokolov and Gromov, 1985b; Danilkin, 1986b)

Cross[1]		Result		Offspring	
Female × male	*n*	Successful mating[2]	Normal delivery[3]	*n*	Mass (kg)
Sib. × Eur.	19	7	4	3	1.7
Eur. × Sib.	38	22	8	25	2.0–2.2
F1 × Eur.	4	3	2	4	1.45
F1 × Sib.	3	3	3	4	2.2
Eur. × F1	11	1	1	2	1.9
Sib. × F1	2	1	1	2	1.45
F1 × F1	10	0	–	–	–
BC1Sib. × Sib.	2	2	1	3	2.1
BC1Eur. × Eur.	2	2	–	–	–

1. Sib. = Siberian roe; Eur. = European roe; F1 = first generation hybrid; BC1Sib. = progeny of F1 doe × Siberian buck cross; BC1Eur. = progeny of F1 doe × European buck cross.
2. Embryos were produced.
3. Unassisted birth of live kids.

factors on hybridization and for the moment we can only speculate. Reproduction of roe deer in captivity may well differ from that observed under natural conditions. Indeed, the almost entirely unsuccessful attempts to mate a European roebuck with females of the same species in Leningrad Zoo over several years indicate either that some males are naturally subfertile or that captivity interferes with normal reproductive behaviour. In some further experiments (Sokolov and Gromov, 1985b), some does were unable to mate successfully with a particular male but produced hybrid offspring following copulation with a second individual. Similarly, only one of two does (both originating from Lithuania) was able to reproduce successfully with one particular buck.

Further experiments are necessary, carried out under various environmental conditions with larger sample sizes, before this question can be resolved in more detail. From the results presented here, it seems that the hybridization of European and Siberian roe deer can be relatively successful in both directions, with Siberian buck–European doe pairs as well as with European buck–Siberian doe pairs, and hybrids may be fertile. However, at the same time, crossing of these two species is not simple (Table 8.4), as in only about 20% of all cases do does succeed in producing live kids normally without some form of assistance. This provides direct evidence of the existence of some reproductive barriers between these species, and this is also supported indirectly by the fact that the many introductions of Siberian roe deer into the range of the European roe over the last century have generally been unsuccessful. As has been noted previously (Turcek, 1951), the smaller sized European does which are mated by Siberian roebucks often die giving birth to the large fetus. It seems justified to consider the European and Siberian roe deer as different species, although hybridization and the production of fertile offspring may be possible.

Thus, sexual activity is first displayed by roe deer at the age of 4–6 months. Roebucks come into puberty at the beginning of their second year, but does may do so as early as their first year, although most females actually become pregnant for the first time at 14 or 15 months, producing their first offspring when two years old.

Does are seasonally polyoestrous with one long mating period. European roe deer mate from June to August, and Siberian roe slightly later, from July to September, but these mating periods do overlap. A few female kids may come into oestrus somewhat later.

European roe deer kids are usually born from the end of April until the middle of June, while the equivalent dates for Siberian roe are from the second half of May to the middle of July. The length of the full term of pregnancy normally varies from 264 to 318 days in European roe does, but may be as short as 5 months following a fertilization in late autumn; it is slightly shorter in the Siberian roe deer, varying from 279 to 296 days, due to a somewhat later oestrus in this species.

The roe deer is the only ungulate which exhibits a period of latency during pregnancy (embryonic diapause). It seems likely that this arrest of embryonic development is adaptative, and it is probably associated with territoriality. The facts that pregnancy can proceed either with or without this 5-month diapause and that the first cycle of sexual activity among kids (aged 4–6 months) occurs during autumn suggest that it is the breeding cycle of females which has adjusted in line with the period of highest sexual activity among bucks. This adaptation ensures that all reproductive activity (rutting and birth) is concentrated into the more hospitable summer months and may promote female choice at a time when males are in peak condition. The fertilized blastocyst does not commence development until the onset of winter and so kids are born (as for most other species of Cervidae) during the following spring.

Hybridization of the European and the Siberian roe deer is possible but many hybrid males are sterile and European does mated by Siberian bucks experience severe complications when delivering the larger than normal fetus, generally resulting in death under natural conditions.

9

Population structure and dynamics

9.1 SEX–AGE STRUCTURE

9.1.1 Birth sex ratios

In populations of both European and Siberian roe deer, the birth sex ratio may differ substantially from parity (Table 9.1) and this is probably related to environmental variation. Ellenberg (1978) quite clearly demonstrated that does on a high nutritional plane during the summer which were consequently in good condition at the onset of the rutting period produced three times as many female offspring, while under adverse conditions the birth sex ratio was 1 : 3 in favour of males. Similarly, Hewison (in press c) discovered an inverse relationship of the population birth sex ratio with both average maternal body weight and average female fecundity across 14 British roe deer populations. These results support the hypothesis that does in poor condition (low body weight and low fecundity) tend to produce male offspring while those in good condition produce more females. This relationship between maternal condition and progeny sex was also observed within each of these 14 populations, with lighter individual mothers tending to give birth to buck kids and heavier mothers producing does.

Hewison (in press c) proposed that local resource competition may explain these variations in birth sex ratio for roe deer. Under this hypothesis, females in poorer than average condition should invest in that sex of offspring more likely to disperse from the natal area. In roe deer, dispersal is more common and occurs at an earlier age among male juveniles, particularly at high densities when resources may be limiting. Thus, where environmental conditions are limiting, roe does tend to produce male biased litters in order to avoid potential future competition posed by female offspring. As might be predicted under such a hypothesis, both young and

Table 9.1 Population sex ratio of embryos, kids and adults across the roe deer's geographic range

Sex Ratio: Embryos/Kids		Sex Ratio: Subadults/Adults		Reference
n	Male : female	*n*	Male : female	
		European roe deer		
1595	1 : 0.40–2.00	–	–	Hewison, 1993, in press
679	1 : 0.79	–	–	Kurt, 1968b
238	1 : 0.75	–	–	Wandeler, 1975
32	1 : 1.0	–	–	Stubbe and Passarge, 1979
55	1 : 0.9	–	–	Prior, 1968
475	1 : 0.9	–	–	Borg, 1970
186	1 : 0.73	655	1 : 1.34	Nikolandic, 1968
487	1 : 0.89	327	1 : 1.9	Strandgaard, 1972b
191	1 : 0.85	172	1 : 1.87	Ellenberg, 1978
37	1 : 0.6	–	–	Espmark, 1969
763	1 : 1.0	–	1 : 1.03–1.65	Engl, 1982
1722	1 : 0.9	3202	1 : 1.38	Pikula *et al.*, 1985a
–	–	–	1 : 1.5	Cotta *et al.*, 1961
111	1 : 1.25	9737	1 : 1.2–2.5	Fruzinski and Labudski, 1982a,b
138	1 : 1.3	949	1 : 1.03–1.65	Kaluzinski, 1982a
42	1 : 1.0	104	1 : 1.06–1.83	Pielowski and Bresinski, 1982
–	–	3415	1 : 2.52	Zejda, 1978
–	–	–	1 : 1.1–1.5	Stangl and Wenter, 1977
16	1 : 0.8	129	1 : 1.1	Arabuli, 1966
389	1 : 1.19	686	1 : 1.31–1.56	Bluzma, 1975
–	–	–	1 : 1.3–1.4	Prieditis, 1975
179	1 : 0.97	–	1 : 1.82	Randweer, 1985, 1989
116	1 : 0.93	496	1 : 1.31	Danilkin's data (Belgorod and Kursk)
		Siberian roe deer		
–	–	899	1 : 1.2	Averin, 1949
46	1 : 0.9	–	–	Pole, 1973
44	1 : 1.2	156	1 : 1.19	Smirnov, 1978
116	1 : 1.0	288	1 : 1.0	Lavov, 1971
37	1 : 0.95	754	1 : 1.1	Darman, 1986
16	1 : 1.3	–	–	Bromlei and Kucherenko, 1983
217	1 : 0.97	1382	1 : 1.38	Danilkin's data (Urals)
681	1 : 1.20	2645	1 : 1.28	Danilkin's data (Amursk)

old roe doe mothers (generally the age classes most susceptible to the adverse effects of competition) more commonly produce male kids.

9.1.2 Population age–sex structure

Among the population as a whole, despite a tendency for doe kids to experience slightly higher levels of juvenile mortality than buck kids (Gaillard *et al.*, 1993d), the sex ratio generally shifts towards a female bias (Table 9.1), probably due to a variety of causes which may be important in population regulation. Because of the high level of competition for territories and the aggression of adult territorial bucks, most juvenile males are forced to emigrate from their birth range, whereas most juvenile does are philopatric, staying on or near their mothers' ranges. This reduces the total population number of male roe, which are then exposed to much higher risks of mortality during dispersal across unfamiliar country and may often die before they reach 2 years of age. In populations at high density, even 2-year-old males may not be able to occupy a permanent territory and may therefore be forced to disperse.

Other factors which may cause a female bias in the adult sex ratio of a roe deer population include a naturally lower survival rate among male adults (Gaillard *et al.*, 1993d), trophy hunting in many parts of Europe, and the highly aggressive nature of male–male combats during the territorial period which can result in injury and even death in some cases.

There is no clear relationship between sex ratio and population density, even when densities differ by a factor of 40 or 50 (Table 9.2). This is not particularly surprising in view of the complexity of factors which may influence the adult sex ratio in a population. Furthermore, a simple measure of density does not account for variation in habitat quality across study sites and therefore any such comparison is by nature oversimplistic.

From published studies, it seems that sex–age structure does not differ greatly between populations of European and Siberian roe deer. Plotting the available data on age composition for populations from various regions (obtained mainly from large-scale culling operations) (Andersen, 1953; Strandgaard, 1972a; Pielowski and Bresinski, 1982; Pikula *et al.*, 1985a; Lavov, 1971; Smirnov, 1978), some regularities become apparent (Figure 9.1). The plot shows that, with increasing age, a decreasing proportion of the population is represented within a given age class and only a few deer live to more than 7–10 years old. Gaillard *et al.* (1993d) showed that roe deer survival decreased after 7 years of age in two populations in France. The maximum age among free-ranging roe deer is generally 10 ±2 years but some marked individuals have lived 15–17

Table 9.2 Variation in the adult sex ratio at different levels of population density

Population density (Deer/100 ha)	Sex ratio (Male : female)
1–10	1 : 1.0–1.8
10–20	1 : 1.0–2.5
20–30	1 : 1.2–2.2
30–40	1 : 1.2–2.1
40–50	1 : 1.8–2.1

References: Lavov, 1971; Prusaite *et al.*, 1973; Smirnov, 1978; Bluzma, 1975; Sokolov and Danilkin, 1981; Randweer, 1989; Eiberle, 1962; Kurt, 1968a; Strandgaard, 1972a; Fruzinski and Labudski, 1982a; Pielowski and Bresinski, 1982.

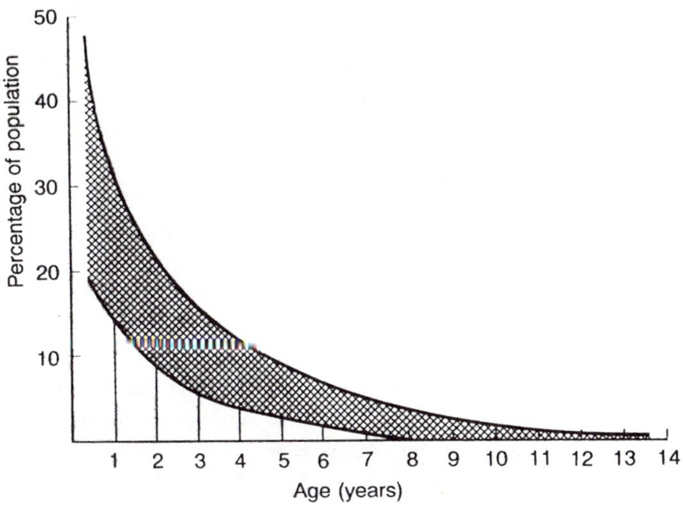

Figure 9.1 Age structure of roe deer populations (percentage representation of each age class), based on data from large-scale culling. Shaded region incorporates variation in age composition across populations.

years and, in captivity, some deer have survived for as long as 19–25 years (Strandgaard, 1972a; Stubbe and Passarge, 1979).

The average age of a population is very variable, almost certainly due to the differing levels of hunting pressure experienced. For example, average age has been estimated as 2.85 years in Czechoslovakia (bucks 2.7 years, does 2.98 years) (Pikula *et al.*, 1985a), 2.49 years in Denmark (Strandgaard, 1972a), 2.72 years in Germany (Ellenberg, 1978), from 2.4 to 3.5 years over various parts of Poland (Bobek *et al.*, 1974; Kaluzinski, 1982a; Pielowski and Bresinski, 1982; Fruzinski and Labudzki, 1982a), from 2.25 to 3.16 years in the Trans-Baikal area (Lavov, 1971; Smirnov, 1978) and 3.8 years for males and 4 years for females in Lithuania (Padaiga, 1975b).

The population average age varies over the year and will be lowest immediately following the birth season; for example, it fell from 2.4 years prior to kidding to 1.8 years immediately afterwards in one study in Germany (Stubbe and Stubbe, 1985). In fact, the percentage of the population made up of kids can be as high as 50% immediately after the birth period but high levels of juvenile mortality may reduce this proportion each month, depending on the harshness of ecological conditions experienced by the population. By autumn, in European roe deer, kids have been reported to constitute 18–47% of the population in studies in Denmark (Strandgaard, 1978), 27.1–36.1% in Lithuania (Prusaite *et al.*, 1974; Bluzma, 1975; Padaiga, 1975b), 18.9–39.1% in various parts of Estonia (Randweer, 1985), and 28.7–48.5% in Poland (Fruzinski and Labudzki, 1982a; Pielowski and Bresinski, 1982). Equivalent figures for Siberian roe deer populations in autumn are approximately similar, ranging from 20 to 42.4% in the Trans-Baikal area (Lavov, 1966, 1971; Smirnov, 1978), 30.6–41.6% in the Far East (Darman, 1986) and 35.1–48.1% in the southern Urals (Kiselev, 1979; Sokolov and Danilkin, 1981 and unpublished data; Dvornikov, 1984).

This marked variation within single populations indicates the potentially dramatic effects on juvenile mortality of year-to-year variation in environmental conditions (Gaillard *et al.*, unpublished data). The representation of kids within the population may fall even further towards spring, especially when winter conditions are particularly severe (Gaillard *et al.*, 1993d), and has been reported as 14.6–34.1% in the Far East and 18.3–36.2% in the southern Urals. Similar figures have been presented for field roe populations in Europe (Kaluzinski, 1982a).

In recent years we have obtained reliable data on the structure of a migratory Siberian roe population living in the Amursk Region (Table 9.3). In September, kids make up the largest proportion of the population – between 35.1 and 40.6%. Females older than one year, on average, constitute 35.5% of the population, with males averaging 27.2% (18.0% adults and 9.2% yearling bucks) (Table 9.3). Considering animals aged over one year only, the male : female sex ratios of the population from 1987 to 1993 were: 1 : 1.41, 1 : 1.71, 1 : 1.12, 1 : 1.22, 1 : 1.37, 1 : 1.33 and 1 : 1.37, an overall average of 1 male to 1.30 females. The sex ratio among

Table 9.3 The autumn sex–age structure of a migratory population of Siberian roe deer in the Amursk region (1987–1993) (Danilkin *et al.*, 1995)

	Males						Females				
	Adults		Yearlings		Total		> 1 year		Kids		Total
Year	*n*	%	*n*	%	*n*	%	*n*	%	*n*	%	*n*
1987	–	–	–	–	164	25.3	231	35.7	253	39.0	648
1988	–	–	–	–	69	22.4	118	38.3	121	39.8	308
1989	218	18.7	108	9.3	326	28.0	365	31.4	472	40.6	1163
1990	197	17.5	129	11.4	326	28.9	398	35.2	405	35.9	1129
1991	186	18.8	75	7.5	261	26.3	357	36.0	374	37.7	992
1992	201	17.8	113	10.0	314	27.8	419	37.1	397	35.1	1130
1993	175	17.4	100	10.0	275	27.4	376	37.4	353	35.2	1004
TOTAL	977	18.0	525	9.7	1735	27.2	2264	35.5	2375	37.3	6374

kids in autumn remained approximately similar over five years, averaging 1 : 1.30 in 1989, 1 : 1.24 in 1990, 1 : 1.09 in 1991, 1 : 1.12 in 1992 and 1 : 1.42 in 1993, a combined average ratio of 1 male to 1.22 females.

The average age composition in autumn of non-exploited European and Siberian roe deer populations were approximately the same (Figure 9.2a) but the age composition of exploited populations may also vary according to the distribution of hunting pressure on each sex–age class. In practice, even where selective shooting is implemented, often a disproportionate number of medium aged deer are shot, generally the most fecund age class within the population (Fruzinski and Labudzki, 1982b; Pielowski and Bresinski, 1982). The hunting methods practised in many parts of the former USSR during autumn and winter often involve the culling of the largest mature roe deer (up to 77% of the total take) – mostly does (Figure 9.2b) – causing a severe impact on this age class and leading to an abrupt decrease in population productivity and growth.

9.2 FACTORS DETERMINING POPULATION SIZE

Roe deer population size is determined by the ecological capacity of the habitat and abiotic factors which may limit population density through their effects on reproduction, mortality and migration. If birth and mortality rates balance each other, the population may be relatively stable for some time, provided that ecological conditions do not change dramatical-

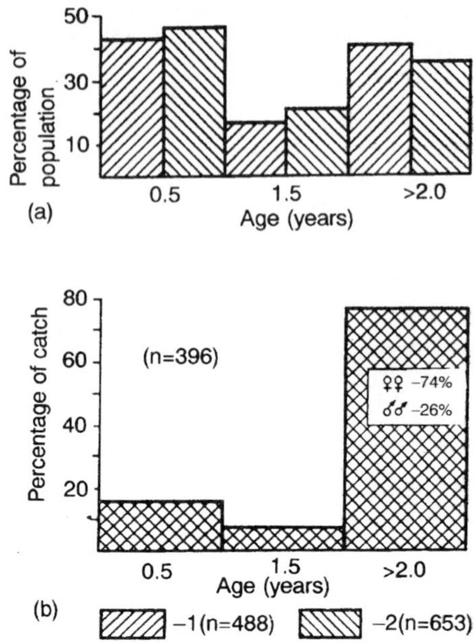

Figure 9.2 (a) Age structure of two non-exploited populations of roe deer: (1) Siberian roe (Chelyabinsk); (2) European roe (Belgorod). (b) Distribution of culled animals according to age class in former USSR.

ly. In practice, this situation is rare and population size usually varies widely according to the level of hunting pressure, predation and climatic and other environmental factors. Factors influencing roe deer population dynamics (natural and anthropogenic, intrinsic and extrinsic, biotic and abiotic) are interrelated and often, if not always, interact. Nevertheless, it is often methodologically simpler to examine the effects of a particular factor on a particular component of the dynamics of a population in isolation.

9.2.1 Reproduction

The rate of reproduction is one major component of population dynamics and is determined by the length of the period to puberty, the breeding proportion of the female population and the average litter size. Roe deer are distinguished from other cervids by their high levels of production reflected in an early sexual maturity (does may exceptionally give birth at one year of age), a high level of fecundity among adult does and, unlike most other cervid species, production of litters of two, three or even four offspring (Stubbe and Danilkin, 1992).

(a) Breeding proportion of female population

Generally, the majority of studies have reported very high values for the proportion of adult roe does breeding. For example, in Lithuania, of 206 females examined from all age classes, only seven (3.4%) were not pregnant (Prusaite *et al.*, 1977), although in some years as many as 9% of females did not breed (Prusaite and Baleishis, 1975). Similarly, only 6.5% (Kaluzinski, 1982a) and 14% (Fruzinski and Labudzki, 1982b) of females were not pregnant in two studies in Poland, while in Denmark this proportion varied from 0 to 5% according to population density (Strandgaard, 1972a). In a single population in France, Gaillard *et al.* (1992) found more than 98% of adult does pregnant, despite a high population density. Furthermore, although often based on rather small sample sizes, almost all studies of Siberian roe have reported 100% pregnancy among adult does – in various parts of Kazakhstan (Pole, 1973), in the Trans-Baikal area (Lavov, 1971) and in our own studies in the Urals. Only Darman (1986) reported observations of non-pregnant Siberian does, finding three out of 18 females examined without a fetus in the Far East; however, this material largely consisted of animals found dead.

Thus, roe does generally do not seem to be forced to breed only every other year, as is the case for other species of cervid such as red deer, as a result of the costs associated with gestation and lactation. However, in a study of 15 roe populations from across Britain, Hewison (1993, in press a) provided values for percentage pregnancy among adult does considerably lower than any previously reported for this species. Although the percentage of adult does pregnant exceeded 80% in nine these populations, in three poor performance study sites, the equivalent figures fell between 67 and 70%.

(b) Litter size

Most European and Siberian does produce twins, more rarely singletons or triplets, but quadruplets are very much an exception, although four embryos are not infrequently found in the uterus at post mortem (Table 9.4). Occasional reports describing observations of five (Averin, 1949; Sablina, 1955; Serzhanin, 1955; Yegorov, 1965; Myrberget and Milton, 1971; Vereshchagin and Rusakov, 1979) and even six (Cherkasov, 1962) embryos exist but are exceptional.

From published data, over the entire range of the species the potential litter size varies widely as indicated by the average number of

Table 9.4 Fecundity of roe does across their geographical range

Location	Corpora lutea/doe		Embryos/doe		Potential litter size (%)				Reference
	n	Av.	*n*	Av.	1	2	3	4	
					European roe deer				
Switzerland	–	–	573	1.69	–	–	–	–	Kurt, 1968a
Switzerland	114	2.12	438	1.90	17	76	7	1	Wandeler, 1975
Switzerland	99	2.00	–	–	10	77	10	1	Blant, 1987
Sweden	–	–	47	2.30	15	47	32	6	Essen, 1966
Sweden	179	2.51	362	2.19	16	50	33	1	Borg, 1970
Denmark	46	1.98	–	–	9	85	6	0	Andersen, 1953
Denmark	41	1.9	–	–	7	93	0	0	Strandgaard, 1972b
Denmark	8	2.4	–	–	0	63	37	0	Strandgaard, 1972b
Denmark	–	–	24	1.9	–	–	–	–	Strandgaard, 1972b
Great Britain	32	1.77	26	1.8	15	85	0	0	Hamilton *et al.*, 1960
Great Britain	–	–	29	1.9	17	79	4	0	Chaplin *et al.*, 1966
Great Britain	–	–	38	1.9	13	82	5	0	Prior, 1968
Great Britain	35	1.80	14	1.8	14	86	0	0	Short and Hay, 1966
Great Britain	17	1.71	54	1.9	17	78	4	2	Chapman and Chapman, 1971
Germany	–	–	56	1.91	–	–	–	–	Ellenberg, 1978
Germany	–	–	149	1.85	21	72	7	0	Stubbe *et al.*, 1982

Poland	–	–	71	1.56	–	–	–	–	Fruzinski and Labudski, 1982b
Poland	–	–	138	1.88	12	68	12	1	Kaluzinski, 1982a
Lithuania	152	2.16	–	–	13	62	22	3	Prusaite *et al.*, 1977
Latvia	–	–	–	1.7	–	–	–	–	Prieditis, 1985
				Siberian roe deer					
Kazakhstan	–	–	25	1.9	28	56	16	0	Sludski *et al.*, 1984
Kazakhstan	–	–	21	2.3	25	42	33	0	Sludski *et al.*, 1984
Urals	12	2.21	28	2.1	14	64	18	4	Danilkin's data
Ante-Baikal	20	2.15	–	–	–	–	–	–	Ustinov and Lobanov, 1983
Trans-Baikal	31	1.7	–	–	29	71	0	0	Lavov, 1971
Far East	–	–	13	1.92	15	77	8	2	Darman, 1986
Yakutia	–	–	8	2.25	0	75	25	0	Tavrovski *et al.*, 1971

corpora lutea per doe, which ranges from 1.28 to 2.5, and the average number of embryos per doe (1.13–2.3) (Table 9.4). Litter size is most commonly around two per female, but the lower end of the above ranges again concern populations with particularly poor performance from the British study of Hewison (1993, in press a).

The difference between the number of corpora lutea and the number of embryos observed is due to failure to implant the fertilized blastocyst in mid-winter and/or resorption of the embryo during pregnancy. While in some populations, failure to implant may be in the region of 20–30% or higher (Hewison, 1993), in healthy adults losses due to resorption are minimal, generally in the order of 1–4% (Short and Hay, 1966; Chaplin *et al.*, 1966; Strandgaard, 1972b; Wandeler, 1975; Sludski and Baidavletov, 1984; Darman, 1986), although they may be as high as 9–22% among sick individuals (Borg, 1970; Wandeler, 1975).

(c) Age at first breeding

Both the proportion of breeding does and the average litter size depend to some extent on age and are lower among the juvenile age classes. A significant proportion of doe kids may ovulate and be fertilized in the first few months of life in some populations, but such precocious reproductive activity rarely results in successful implantation of a fetus (Hewison, 1993, in press a). More commonly, age at first breeding is 15 months in roe deer, with most does producing their first kid at 2 years of age. This may differ between populations (Hewison, 1993, in press a), although fecundity among this yearling age class is generally lower than that of more mature adults (Borg, 1970; Gaillard *et al.*, 1992). Similarly, potential litter size is generally lower among yearlings (Borg, 1970; Hewison, 1993, in press a) (Table 9.5).

Among deer older than 2 years of age, Gaillard *et al.* (1992) found that the proportion of does breeding did not vary with age and Hewison (1993, in press a) did not observe any change in this same parameter, nor in potential litter size, with increasing age among adults, although embryo mortality may be higher in older does. Other studies have reported some age-related pattern in fecundity among adult roe. Borg (1970) found the average number of embryos per female was 2.23 for 2-year-olds, 2.35 for 3–5-year-olds, 2.14 for 6–7-year-olds and 1.88 for does aged 8 years and older. In a study in Germany, the average number of kids per female increased with age from 1.76 to a peak of 1.94 at 4 years old, declining to 1.73 after the age of 7 years (Stubbe *et al.*, 1982). Approximately the same pattern has been reported in Poland (Kaluzinski, 1982a; Fruzinski and Labudski, 1982b) and Czechoslovakia (Pikula *et al.*, 1985b). This suggests that fecundity of female roe deer may

Table 9.5 Variation in female fecundity with age in the European roe deer

Average no. corpora lutea/doe			Average no. embryos/doe			
1–2 Yrs	3–5 Yrs	> 6 Yrs	1–2 Yrs	3–6 Yrs	> 7 Yrs	Reference
1.94	2.17	2.54	–	–	–	Prusaite *et al.*, 1977
2.38	2.54	2.54	2.23	2.35	1.88–2.14	Borg, 1970
1.56–1.93	1.92–2.00	2.20	–	–	–	Georgii, 1974
1.94	2.08–2.22	2.20	–	–	–	Wandeler, 1975
1.57	2.08	2.08	–	–	–	Strandgaard, 1972b
1.82–1.87	2.00–2.36	2.00	–	–	–	Blant, 1987
–	–	–	2.00	2.00–2.54	1.60–1.78	Kaluzinski, 1982a
–	–	–	1.40	1.67–2.00	2.00	Fruzinski and Labudski, 1982b

reach a peak some time around 4 or 5 years of age and therefore these animals may be the most productive in the population.

As is generally the case in mammals, sexual maturity of roe deer seems to depend on body weight, with puberty dependent on some critical threshold body mass (Gaillard *et al.*, 1992). Furthermore, subsequent pregnancy success may also depend to a large extent on maternal body weight in roe deer (Loudon, 1987; Hewison, 1993, in press a). Blant (1987) showed that the average maternal weight increased with potential litter size from 16.2 kg for does producing none or one corpus luteum, 18.2 kg for those with two corpora lutea and 21.3 kg for those with three corpora lutea. This was also supported by Ellenberg's (1978) data (Table 9.6). Roe deer body weight, and hence fecundity, is probably largely a function of the interaction between population density and habitat quality, but other variables such as climate and social factors may also be important.

9.2.2 Mortality

The annual probability of survival is highest among adult roe deer between the ages of 2 and 7 years, estimated as over 0.8 for both sexes, but after 7 years old survival probability falls, possibly due to the effects of tooth wear (Gaillard *et al.*, 1993d). Survival among prime-age adults may be relatively insensitive to external influences with significant mortality only in extremely severe conditions (see below). Generally, female adults survive better than males (see also Strandgaard, 1972a), whereas among kids it is males which tend to exhibit lower levels of mortality

Table 9.6 Variation in female fecundity in relation to age and body weight in European roe deer (from Ellenberg, 1978)

n	Age	Average body weight (kg) in December	Breeding result
7	Kid	12.4	Died during winter
8	Kid	13.1	Kids died
9	Kid	15.2	Reared 1 kid each
22	Kid	15.9–17.6	Reared 2 kids each
20	Yearling	19.7	Barren or kids died
11	Yearling	22.7	Reared 2 kids each
10	Adult	22.7	Reared 'poor condition' kids
30	Adult	24.8–25.1	Reared 'good condition' kids

(Gaillard *et al.*, 1993d). This may be due to the formation of matriarchal groups in high density populations and differential mother–young relationships according to the sex of the offspring.

Juveniles may experience higher rates of mortality than older deer and are particularly sensitive to environmental conditions (Gaillard *et al.*, 1993d). Mortality is particularly high during the first summer of life – often greater than 50% (Strandgaard, 1972a; Prusaite *et al.*, 1974; Ratcliffe and Rowe, 1985; Blant, 1987). It does not differ between the sexes but varies greatly between years, producing a marked cohort effect in roe deer populations (Gaillard *et al.*, unpublished data). It may be particularly pronounced among young mothers giving birth for the first time. In Germany, 10% of kids died within the first week alone (Stubbe and Stubbe, 1985). Total mortality in Poland from May to November averaged 37.6% (Fruzinski and Labudzki, 1982b). Mortality to autumn among Siberian roe kids was reported as 16% in Kazakhstan (Pole, 1973) and between 13.7 and 32.3% in the Far East (Darman, 1986). From our observations in a reserve in the Urals, kid mortality varied between 11.5 and 19.3%, while in the Ilmen reserve, it was higher, at 25%. This generally poor early survival rate may be due to the high level of energy invested by a roe doe mother at each breeding attempt, making further investment during times of resource restriction unlikely (Gaillard *et al.*, unpublished data). Mortality during the act of birth may be significant in some instances, reported as 4.6% of pregnant does in captivity by Ellenberg (1978).

Juvenile mortality during winter is probably linked to the rate of maturity and body weight at 8 months of age (Gaillard *et al.*, 1993d). Mortality at this stage may also be substantial (Sablina, 1955; Petrov *et al.*, 1968; Lavov, 1971; Padaiga, 1971; Pole, 1973; Filonov, 1974; Prusaite *et al.*, 1973, 1974; Bobek *et al.*, 1974; Randveer, 1985; Prieditis, 1985; Stubbe and Stubbe, 1985) and can have a pronounced impact on roe deer population dynamics. More commonly, one of two fawns survives winter (Pole, 1973; Eiberle, 1976; Sokolov and Danilkin, 1981; Fruzinski and Labudzki, 1982b; Kaluzinski, 1982a; Darman, 1986; Blant, 1987) but family effects may depend on the success of the cohort (Gaillard and Delorme, in press).

9.2.3 Population density

From longitudinal studies of populations over a number of years, a negative relationship between population density and productivity (number of kids per doe) of female roe deer has been established (Boutin *et al.*, 1987; Vincent *et al.*, 1995). In the same way, Ueckermann (1964) found in one area that an increase in deer density from 17 to 40 head/100 ha was concomitant with a decrease in annual population growth from 129% to 74% of the total number of females. In a second area, a doubling of the population from 150 to 324 led to the rate of growth falling from 68% to 57%. In northern Lithuania, the growth of a non-exploited population stopped as soon as the population density reached a level of 19–20 animals/100 ha with a winter doe : kid ratio of between 0.54 to 0.76, although no impoverishment of the habitat was noted (Prusaite *et al.*, 1973). Hein (1966) reported an increase in the rate of population growth from 67% to 93% associated with a drop in population density from 21 to 10 animals/100 ha. This density-dependent relationship was also noted by Kurt (1968a) in Switzerland and was attributed to a delay in puberty, an increase in the rate of embryo resorption and high neonatal mortality among kids during the first weeks following birth at high density. From data such as winter doe : kid ratios, it is difficult to ascertain whether density-dependent responses within a population are due to changes in female fecundity or juvenile mortality, although Vincent *et al.* (1995) suggested that juvenile mortality may be more sensitive to density variation (Gaillard *et al.*, unpublished data).

In across-population analysis, the reproductive rate of low density populations in Sweden is higher than that in central Europe (Essen, 1966; Borg, 1970). In Denmark, average corpora lutea production was found to be 1.6 per doe in a high density population but between 1.9 and 2.4 for a population at a more moderate density, typical for that area (Strandgaard, 1972a). Such comparisons between populations are more difficult, as density is only a relative term and the extent of increasing competition at higher density will depend on the level of available resources. Even where densities are the same in two populations, the reproductive rate may vary between them due to a difference in the quality of the resources the deer must share. Bobek (1977) highlighted the lack of statistically significant differences in the breeding performance of roe deer in populations at different densities but this analysis did not account for possible habitat quality variation.

9.2.4 Habitat quality

Based on experimental manipulation of roe deer in captivity, Ellenberg (1974b, 1978) suggested that the effects of population density on reproductive rate and juvenile survival were far less important than food availability, as productivity remained near a maximum (150–212%) even when density was extremely high but resources were abundant. This again illustrates the fact that density and resource availability are factors which are almost always interdependent. However, Vincent *et al.* (1995) did report density-dependent effects on population productivity, despite an apparent lack of resource limitation, and attributed this to the social problems of overcrowding at very high density.

Where the quality of resources is high and/or competition for those resources is low, kids may grow and develop faster and achieve higher body weights, leading to high rates of juvenile survival (Gaillard *et al.*, 1993d, in press). In contrast, juveniles may develop more slowly where resources are limiting and therefore suffer higher levels of mortality. Dominant does may be able to monopolize the best ranges and thus produce larger kids with better over-winter survival than subordinate females (Gaillard *et al.*, in press). This retarded growth may mean that these offspring are unable to achieve the threshold body weight necessary for successful reproduction as yearlings (Gaillard *et al.*, 1992) and so puberty is delayed a year and population productivity falls.

Inadequate nutrition prior to the rutting period may also adversely affect the body weight and therefore reproductive performance of adult does which, although they will still generally breed, may produce a smaller litter (Blant, 1987; Loudon, 1987; Hewison, 1993, in press a). Under natural conditions, Prieditis (1985) suggested that reproductive productivity of a roe deer population decreases when, during summer, 30–40% of all vegetation shoots show signs of damage, and that resource limitation in this season cannot be easily compensated for. Bobek (1977) proposed that summer food availability was the major factor limiting roe deer population size.

As outlined above, the birth sex ratio may vary according to environmental conditions; thus, roe deer populations in rich habitats may produce a high proportion of female offspring. As population growth is largely a function of the number of does available to breed, such changes in the primary sex ratio will in turn mean that growth will accelerate even faster when the habitat is of good quality. Further, when nutritional conditions are poor during pregnancy or immediately following birth, not only do females tend to produce more male offspring but there is also high post-natal mortality (up to 75%), particularly among female kids (Ellenberg, 1978).

Mortality among adults from starvation is rare and only occurs under very extreme conditions, particularly in poor habitats and severe winters with a significant accumulation of snow. Roe deer have a considerably lower metabolic rate and consequently lower food requirement during winter (Weiner, 1977). Normally, food is in short supply only when population density is high; for example, in Lithuania the winter pastures in deciduous forest habitat were completely degraded when density rose to between 15 and 35 deer/100 ha. In poor quality habitats such as mixed spruce–deciduous forests and pine forests, the resource base may become over-exploited at densities as low as 6–8 head/100 ha and mass die-offs can occur (Padaiga, 1965), usually at the very end of winter and the beginning of spring. At this time of year, the deer are in poor condition and (probably due to longer day length) metabolic rate increases, but this increased energy demand cannot be satisfied because of the shortage of high quality foods (Ellenberg, 1978).

The quality of the available food is equally important and it has been reported that daily protein consumption should be not less than 13% for wild ungulates (Smith *et al.*, 1956; Dasmann, 1966; Ullrey *et al.*, 1967). At a critical level of 7 or 8% the resources are so poor that starvation ensues, possibly killing as much as 40% of the population. The protein content of most winter foods exploited by the roe deer falls below this critical level, and in addition digestibility is generally 1.75 times poorer than for summer foods (Drozdz, 1979). However, according to Eisfeld (1974a,b, 1976), protein content is not the most important factor for roe deer: it is the energy value of the food which is decisive. When the energy value of available food is low or energy demands are high, an animal may be unable to meet its energy requirements from the contents of a full rumen, but may also be unable to feed due to the limited rumen capacity of roe. This situation is likely to occur at high density, when all energy-rich foods have been consumed selectively and only poor quality browse remains. Thus, resource limitation of roe deer during winter is essentially determined not only by the quantity of available foods but also by their nutritional value and digestibility.

Resource restriction may cause an increase in roe deer juvenile mortality, delay the onset of puberty, inhibit the rate of reproduction and induce large-scale emigration among juvenile bucks. When feeding conditions are favourable, population productivity is high due to increased rates of juvenile survival, precocious breeding among juvenile does, high levels of doe fecundity and a bias towards the production of doe kids.

9.2.5 Behavioural regulation

Several authors have previously suggested that roe deer populations may be essentially regulated by behavioural mechanisms arising from territoriality, through the dispersion of juveniles (Kurt, 1968a; Bramley, 1970; Strandgaard, 1972a; Bobek, 1977). Under this hypothesis, the size of territories adjusts according to the availability of resources and increase in population density is prevented by the aggression of territorial adults towards juveniles and/or non-territorial animals forcing emigration (Bobek, 1977). In this way, territoriality ensures that the deer are distributed in space, regulating population size immediately prior to the birth of a new generation (Strandgaard, 1972a) and creating an area of reduced competition within the range of a territorial buck for pregnant or lactating females he may have mated (Ellenberg, 1978). From this point of view, territorial behaviour is the basis of a homeostatic mechanism which maintains the population's size in a stable equilibrium with the environment that it utilizes (Panov, 1983). However, Ellenberg (1978) suggested that this would be insufficient to regulate roe deer numbers and this suggestion is supported by recent reports of density-dependent responses in this species (Boutin *et al.*, 1987; Vincent *et al.*, 1995; Gaillard et al., in press).

Behavioural regulation is often considered important in high density roe deer populations, but even at low density competition may be strong for the most favourable ranges. Furthermore, if such behavioural factors do play a role in the regulation of roe deer numbers, this will occur as soon as the bucks seize their territories in spring, long before the feeding grounds are at saturation density and food resources are depleted. Under this hypothesis, population density reaches a certain level where it stabilizes and this 'threshold' is probably determined by the maximum size of buck territories and doe home ranges according to their energy requirements at a given body weight. Studies of the spatial structure of European and Siberian roe deer populations (Sokolov and Danilkin, 1981; Danilkin and Minayev, 1988) indicate that populations may exist in some regions at low density (3–5 animals/1000 ha of suitable feeding grounds), despite the fact that the summertime 'threshold' population density in the best habitats is much higher (40–50 animals/1000 ha). This may be due to juvenile emigration maintaining population density at a stable low level – a social carrying capacity (Bobek, 1977) – probably until adjacent habitats are fully saturated (Strandgaard, 1972a). Only then does increasing competition lead to a decrease in range size and density-dependent responses as the population density rises to a new higher level. This concept of threshold densities may explain why unexploited populations inhabiting the best habitats stop growing at a certain point and stabilize at a level which is much lower than the carrying capacity of that habitat.

This mechanism of regulation (through juvenile dispersal) would be important mainly at the intra-population group level and would be effective only when population density is relatively low. It would not limit overall population growth; rather, it would encourage dispersal and colonization of new habitats. At higher densities, emigration would be balanced by immigration of juveniles from neighbouring ranges and there would be no net effect on population size. This situation is mirrored by observations in captivity where juveniles are unable to migrate and density-dependent responses are observed. In such cases the response to an increase in density is a gradual reduction in the size of home ranges within the population (Vincent *et al.*, 1995). As feeding conditions deteriorate, density-dependence is also observed in body weight, fecundity and juvenile mortality (Vincent *et al.*, 1995; Gaillard *et al.*, unpublished data). Indeed, these factors are of much greater importance than behavioural responses in regulating roe deer population size. Any impact of a behavioural mechanism would occur with some sort of time lag and would generally not be effective in maintaining the population at a level within the carrying capacity of the habitat. This has important implications for the management of roe deer populations, particularly where no large predators are present.

9.2.6 Climate

Climatic factors not only determine overall roe deer distribution but may also have a very significant effect on population dynamics, causing marked fluctuation in population size. Mortality may be especially high during severe winters (Cederlund, 1982; Kaluzinski, 1982a; Gaillard *et al.*, 1993d). This is illustrated by the situation during the particularly harsh winter of 1972/3 in the Far East when snow fell as early as the end of October to a depth of between 1 and 1.5 m, cutting the deer off from their winter ranges. In addition to herds of 20 or 30 dying from starvation or exposure, many animals drowned when attempting to cross the rivers along the migration route and stressed individuals became particularly prone to predation and poaching. By the spring, population numbers had fallen to one third or one quarter (Shvets, 1975; Kucherenko and Shvets, 1977).

In the north of the European roe deer range, the main natural cause of mortality in such conditions is starvation, accounting for 32% of deaths in the north-west of the former USSR (Vereshchagin and Rusakov, 1979) and 58.8% in Sweden (Borg, 1970), where the mortality in the severe winter of 1976/77 was higher than that characteristic of normal winters by nearly an order of magnitude, particularly among kids (Cederlund and Lindstrom, 1983). During the winters of 1969/70, 1976/77 and 1978/79, losses in Poland were as high as a third of the autumn population, increasing with density, and were also especially heavy among young (77–90%) and old individuals, as well as in field populations (up to 51%) (Fruzinski and Labudzki, 1982b; Kaluzinski, 1982a). Similarly, unusually cold winters drastically reduced roe deer numbers in Germany when mortality was recorded as high as 50% in many field populations. Again, mortality was not random with respect to age and sex; it affected the young and old age classes, in particular females, more severely (Stubbe and Passarge, 1979; Briedermann and Ahrens, 1980; Stubbe, 1987).

Mass roe deer die-offs in the Kaliningrad region were recorded in the harsh winter of 1979 due to starvation, intestinal inflammation and pneumonia (Romanov and Romashin, 1982). Deep snow during the winter of 1969/70 reduced the roe population of the Bialowieza Primeval Forest by a factor of 3 (Smoktunovich, 1980), and high levels of mortality were observed in the Bryansk region in 1964 and 1967 when snow accumulated to 70 cm depth and extended periods of extremely low temperatures (–30°C) exacerbated the problems. Older deer and juveniles were the most susceptible, and were often killed by stray dogs and wolves when searching for food in more open areas (Vatolin, 1975).

It seems that severe winter weather often affects roe deer populations in interaction with predation. Deep snow, by restricting mobility and increasing the energetic demands of movement (Borg, 1970), may increase susceptibility to predation (Cederlund and Lindstrom, 1983). In the Urals, a larger wolf population and increased levels of poaching have been responsible for depressing roe deer numbers over a long period (Kirikov, 1952; Ushkov, 1954; Filonov, 1974). Deep snow and an increase in wolf predation nearly eradicated roe deer from northern Kazakhstan, and this situation was repeated across the republic in the winters of 1945/46 and 1959/60 and in the mountains of the southeast during 1965/66 and 1975/76 (Sludski and Baidavletov, 1984). In Estonia (Randweer, 1985), an abrupt fall-off in roe deer numbers, mainly in kids (56–71%), caused by hard winters was accentuated by high hunting and predation pressure (Figure 9.3); for example, the combination led to a decrease of 34–40% in 1939/40 and 15–20% in 1981/82. During the severe winters of 1969–71, the large Lithuanian roe deer population was reduced, on average, by 34.4%; 17 000 animals died in the 1960s alone (Figure 9.4), half of which were juveniles, while 34% were old and 16% prime aged animals (Bluzma, 1975). In Latvia, mass die-offs have accounted for between 12 and 20% of the entire population (Prieditis, 1975).

In the Baltic Republics, winter severity seems to be the most important factor influencing population dynamics, with extended periods of low temperatures and deep snow particularly difficult for roe deer to cope with. The timing of the onset and end of winter and the degree of midwinter severity are critical. In severe winters the most important variables are the total number of days with snow cover and the depth of snow; in more moderate years average temperatures may be more influential; in mild conditions the number of days of snow cover and snow fall in April are the crucial variables. In spring, the ambient temperature is critical during the kidding period: rainy, cool weather in the first few weeks of life may lead to high levels of post-natal mortality (Kiili, 1965). In contrast, Gaillard *et al.* (unpublished data) found a positive relationship between kid survival over the summer and rainfall in May and June in two roe populations in France, probably due to the promotion of vegetation growth and because droughts are less likely in years with wetter springs. However, Padaiga (1971) suggested that malnutrition and disease, directly related to population density, are the main factors responsible for roe deer mortality and that harsh snow conditions only

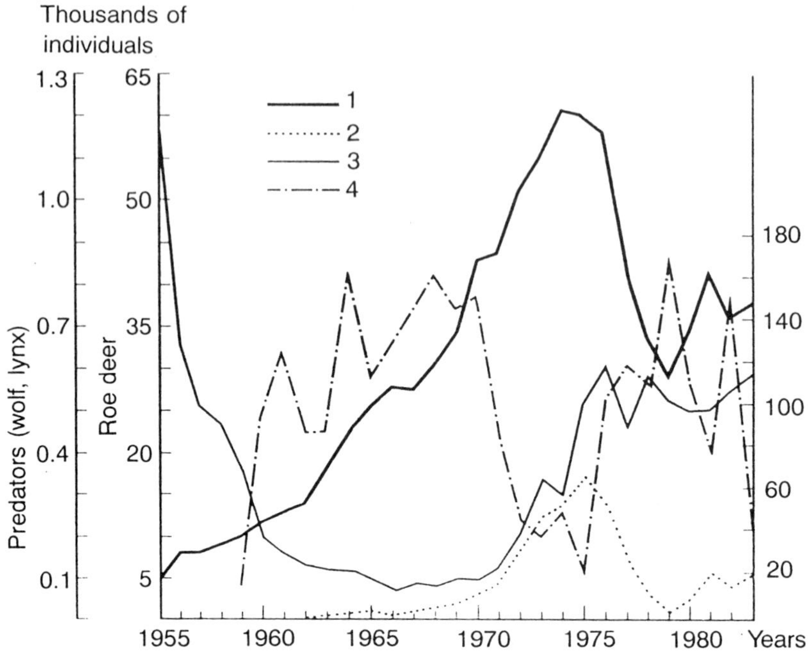

Figure 9.3 European roe deer population dynamics in Estonia (adapted from Randweer, 1985). (1) Roe deer numbers; (2) roe deer cull; (3) predator (wolf and lynx) numbers; (4) winter severity index (A+B)*0.5, where A = mean temperature of coldest three months (expressed as percentage of mean temperature for that period) and B = mean maximum snow depth for given decade (expressed as percentage of mean snow depth for that period).

accentuate this relationship as an increase in population density boosts winter mortality to a higher level than that observed when snow depth increases. This view was supported by Smoktunovich (1980) from studies of the population dynamics of ungulates in the Bialowieza Primeval Forest, highlighting how these factors interact in determining roe deer population size.

The impact of harsh climatic conditions is also reflected in a lower reproductive productivity. Hewison (1993) found that although the success of ovulation and fertilization at the rut seemed to depend on body weight thresholds (see also Hewison, in press a), climatic conditions had a strong, independent influence on the proportion of roe deer that subsequently became pregnant and the number of fetuses implanted (Putman *et al.*, in press). The critical factors were found to be the amount of winter rainfall and the average mid-winter temperature, which are probably responsible for setting the costs of thermoregulation and main-

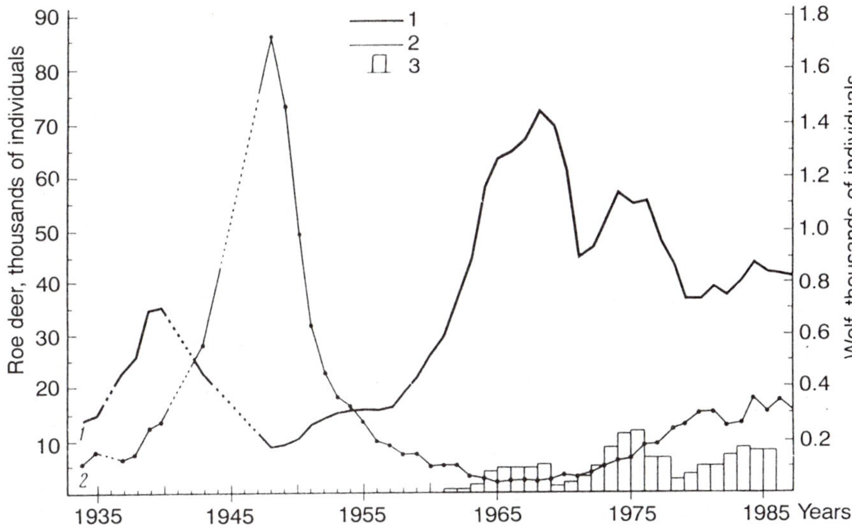

Figure 9.4 European roe deer population dynamics in Lithuania (from Bluzma, 1975, with additions). (1) Roe deer numbers; (2) wolf numbers; (3) roe deer cull.

tenance activities. This suggests a dual mechanism influencing reproductive productivity in roe deer, with ovulation in the summer rut dependent on body weight thresholds and therefore habitat quality and density, but with subsequent implantation in mid-winter dependent on climatic variables. This may have arisen due to the separation of ovulation and fertilization from implantation by 5 months of embryonic diapause and may allow roe deer populations to respond to large seasonal fluctuations in environmental conditions.

9.2.7 Disease and parasitism

Roe deer are susceptible to infectious and parasitic diseases, most of which occur in many species of wild and domesticated ungulate. In roe deer, this has been studied in some detail (Necas, 1963; Stubbe and Passarge, 1979) but the role of disease in population dynamics is still not clear. Infection is often triggered by inadequate nutrition resulting from overstocking or overgrazing (Goreglyad, 1970) and the degree of infestation of some parasites is generally proportional to roe deer population density (Padaiga and Marma, 1970; Padaiga, 1971). The rate of infection also depends on the habitat and soil type as well as the wetness of the habitat (Pujauskas, 1970). However, it is rare that a parasitic infection is directly responsible for the death of roe deer: the organisms generally work indirectly, weakening the host and only occasionally causing mass mortality within the population, especially during severe winters (Rikovski, 1970; Prieditis, 1970). The proportion of animals which have died of parasitic infection is especially high in central Europe (where population density is generally the highest overall), affecting as much as 40% of all dead animals (Siefke, 1966). In Sweden, roe deer deaths were attributed to disease and parasites in 17.7% of cases (Borg, 1970); in Czechoslovakia, 46.3% of 1285 roe deer carcasses investigated during the period 1972–1981 were deaths due to parasitic infection, mainly involving pulmonary helminthiasis (about 70%). The majority of such mortality was observed during autumn and winter (Sterba and Zamek, 1985).

Pulmonary nematodes and hepatic trematodes were common among roe deer in Britain (McDiarmid, 1974); in Bulgaria, deaths due to encephalomyelitis infection have been recorded (Gabrashanski, 1970). Rabies has also caused some mortality in some parts of Europe (Stubbe and Passarge, 1979), while anthrax (2.2%), blood parasites (6.6%), internal diseases (5.4%), pulmonary helminth infection (1.6%) and intestinal helminths (0.6%) were all cited as causes of mortality in the Crimea (Rukhlyadev, 1941). During the winter of 1961/62, about 25% of the population died in a game reserve in the Volyn region, presumably because of a listeria infection (Adamovich and Oitsos, 1963), and in the Stavropol region roe deer were frequently lost due to tick-borne toxicosis (Nikolski and Pozov, 1973). Despite the numerous and widespread forms of disease and parasitism that may affect the roe deer, many researchers suggest that they do not have a significant impact on population dynamics over most of the roe deer's range, where population density is comparatively low (Table 9.7).

9.2.8 Competition

Differences in diet selection allow roe deer and other ungulates to coexist in a single habitat but avoid severe competition for food (Schroder and Schroder, 1982; Hearney and Jennings, 1983). Closer examination makes it clear that, in some seasons and in particular habitats, the trophic niche of roe deer overlaps with some closely related species such as red deer, fallow deer and moose, while chamois and hares can also pose serious competition for food (Vladishevski and Yelski, 1974; Kaluzinski and Bresinski, 1976; Long, 1980; Smoktunovich, 1983; Picard *et al.*, 1985).

Roe deer browse selectively, taking the most nutritious and easily digested foods, whereas other ruminants often are more bulk feeders, taking low quality inferior vegetation. Competition for quality food sources may be particularly serious for the roe deer and this may be aggravated by their comparatively small body size which makes the

Table 9.7 Main causes of mortality (%) of roe deer in different regions of its overall geographical range

Location	*n*	Human impact[1]				Predation					Disease[2]	Misc. etc.	N/K[3]	Reference
		P'ch	RTA	Agr'	TOTAL	Wolf	Lynx	Dog	Fox	TOTAL				
Switz'nd	12950	–	13.5	5.5	82.5	–	–	0.9	–	0.9	6.0	2.1	8.5	Lehman and Sagasser, 1986
Poland	81	6.2	7.4	1.2	35.8	–	–	9.9	–	9.9	19.7	3.7	30.9	Pielowski, 1984
Kal'grad	78	37.0	6.8	12.3	57.5	–	–	6.8	–	19.6	4.1	1.4	–	Romanov and Romashin, 1982
Belgorod	42	45.2	–	7.1	52.3	4.7	–	16.7	4.7	26.1	–	7.1	14.2	Danilkin's data
Ukraine	64	12.5	–	–	12.5	10.9	–	10.9	20.3	42.1	18.7	–	26.7	Karpenko, 1966
NW USSR	82	–	1.2	–	2.4	23.2	14.6	4.9	2.4	45.1	31.7	4.9	15.9	Vereshchagin and Rusakov, 1979
Urals	1868	32.6	–	–	32.6	50.4	7.0	0.4	–	57.8	3.6	1.7	3.3	Filonov, 1974
Urals	122	7.8	3.5	–	11.3	–	78.9	1.8	–	80.7	–	1.8	4.4	Dvornikov, 1982, 1984
Urals	164	29.3	3.0	1.8	34.1	48.2	8.5	1.8	–	58.5	–	–	7.3	Danilkin's data
Alma-Ata	36	–	–	–	–	30.5	11.1	–	–	55.6	5.6	8.2	27.8	Zhiryakov, 1979
T-Baikal	46	–	–	–	52.2	8.7	36.9	–	–	45.6	2.2	–	–	Smirnov, 1978
Khingan	69	24.6	–	–	24.6	33.3	15.9	2.9	–	53.6	–	2.8	18.8	Darman, 1982a

1. P'ch = Poaching; RTA = Road traffic accident; Agr' = Agricultural machinery.
2. Dis. = Disease, including emaciation.
3. N/K = Cause of death not known.

browse above a certain height inaccessible (Ellenberg, 1978). Competition increases with increasing snow depth since woody browse is predominant in the diet of many ungulates in the winter season (Dulkeit, 1964; Roucher *et al.*, 1985; Gordiyuk, 1986). Red, sika and fallow deer may actually displace roe deer from the best feeding grounds and may also prevent them from feeding at sites where winter supplementary food is provided by driving them away from feeding troughs and racks.

Although populations of wild ungulate may increase in number independently until their nutritive base is exhausted, competition may limit this increase. In some regions, competition from the elk is particularly intense (Kiselev, 1976). In others, competition with sika deer causes roe deer population growth to slow at a sika density of about 20 animals/1000 ha, virtually ceasing at 30 sika/1000 ha, with further increase in sika density accompanied by an abrupt fall-off in roe numbers (Vladishevski, 1968; Skriba, 1975; Shostak, 1978; Smoktunovich, 1980, 1983). A further striking example of the effects of competition was recorded in the Bogdo-Ula reserve in Mongolia between Siberian roe deer and Altai red deer (*Cervus elaphus sibiricus*). At the beginning of the 1930s, roe deer were widespread in the forests of this reserve (Kozlov, 1924), and until the 1950s they were common despite a low density of Altai red deer. During the 1970s and 1980s, when the red deer population density increased up to 55–90 animals/1000 ha, roe deer disappeared from the reserve (Danilkin and Dulamtseren, 1981). It seems that domestic cattle may be an even more serious competitor for roe deer, practically obliterating their food base in some regions (Lavov, 1978; Nikolayev, 1982; Pereira, 1985). The transition to housing cattle indoors during winter in western Europe has greatly reduced the potential for competition, encouraging expansion of roe deer populations.

Coexistence with some species of ungulate may have beneficial effects. For example, wild boar and domestic ungulates that dig the snow in winter (over the Asian section of the roe deer's range) may clear sites of snow that roe deer will readily exploit for feeding. In addition, in areas of Russia where snow may accumulate to some depth, the survival of roe deer in winter strongly depends on other species of deer, moose and especially boar which tread paths in the snow, facilitating the movement of roe.

9.2.9 Predation

Where large predators exist within the range of roe deer, mainly in eastern Europe and over nearly all of Siberia, predation has a significant effect on roe numbers, with the wolf and the lynx the most important species (Table 9.7). In the north of the Trans-Baikal area, **wolves** kill

28.5–32.4% of the autumn population, taking animals at random with respect to sex and age. After predators had been removed from an experimental plot of 4500 km^2, the roe deer population density rose from 18 to 27 head/1000 ha over two years (Lavov, 1974, 1982). Wolf predation increases by severalfold when the winter is snowy and the formation of a frozen snow-crust makes movement difficult.

In the Far East, two or three wolves may kill 12–14 roe deer of a population per day, and in areas of winter herding each predator may take as many as 20–30 animals during the period of snow cover and up to 35–40 roe deer per year. It has been estimated that predation accounts for 2000 roe deer annually in the Primorski Territory and up to 1500 roe deer in the Khabarovsk Territory. In various forest populations, 20–25% of the population may suffer predation (Kucherenko, 1976, 1979; Kucherenko and Zubkov, 1980; Kucherenko and Shvets, 1977; Bromlei and Kucherenko, 1983).

Roe deer are also the main prey species in eastern Siberia for predators which, as in the Far East, follow roe herds during the period of migration (Fetisov, 1953). Predation pressure was especially heavy in Yakutia when snow lie was deep – for example, in the winter of 1957/58, when the total ungulate population in the Tokko and Chara river basins fell to less than one tenth of its previous size (Tavrovski *et al.*, 1971). This was also the case in the north of Kazakhstan, where deep frozen snow occurs frequently and roe deer may constitute between 2.9 and 98.2% of the wolf's diet in areas lying to the south of Balkhash (Sludski, 1953; Sludski and Baidavletov, 1984) and 15% in the Trans-Ili Alatau (Zhiryakov, 1980). During the period from 1936 to 1946, wolves annually accounted for 10–29% of the Ilmen reserve roe population in the Urals and roe deer remains were found in the excreta of predators in 95.9% of all cases (Averin, 1949; Filonov, 1974). In a single night in March 1977 in the Zhiguli reserve, wolves killed a total of six roe deer (one adult buck and five does) (Belyanin, 1979).

It is clear that wolves have also played a role in the dynamics of European roe deer populations in the Baltic Republics, particularly in the post-war years (Figures 9.3, 9.4). In Latvia, the roe population did not begin to grow until the eighteenth century when bounties were paid for culled predators, which consequently became much rarer (Garozs, 1987). In the Pripyat reserve in Byelorussia, roe deer account for 11.1% of the wolf's diet (Litvinov *et al.*, 1979) and in the Odessa region a wolf pack of four or five was reported to kill no fewer than 60 roe deer each winter (Gurski and Nazarenko, 1969).

The **lynx** is also a major predator of roe and may have a greater impact on populations than the wolf in some regions. In the Amursk region alone, the lynx may kill as many as 10 000 roe deer annually (Dimin, 1975b; Kucherenko and Shvets, 1977; Bromlei and Kucherenko, 1983).

When prey is abundant and easily taken, this predator may kill more than is necessary to satisfy feeding requirements, killing almost daily and sometimes even taking two or three deer per day (Dimin and Yudakov, 1967). Roe deer are the main component of the lynx diet in the Trans-Baikal area (Smirnov, 1978) and in the Altai (Dulkeit, 1953). In the winter of 1969/70, over an area of 50 km^2 in the south of Baikal region, five lynx reduced roe deer numbers by 80% (Lavov, 1970). In the Angara region, the lynx follows migrating roe herds to the wintering ranges where roe deer remains have been found in all lynx faeces sampled (Yelski and Shishikin, 1979). Lynx predation is also significant in the Urals (Kiselev, 1974, 1979) and in Georgia (Baratashvili, 1982), where this predator takes up to 45% of the population. Radiotracking of lynx in the Swiss Alps has established that, on average, an adult lynx kills one ungulate every 6.6 days and a female with two cubs kills once every 2.7 days, so that the annual requirement of one predator is about 60 roe deer (Brietenmoser and Haller, 1987).

Where predation includes the combined pressure of the wolf and lynx, roe deer populations may be unable to recover and this is evident in data from Latvia (Garozs, 1979, 1987) and Estonia (Randweer, 1985). In such situations, predation may account for as much as 37.8% of all roe deer deaths in the north-west of the former USSR (Rusakov, 1978) and 67.5% of the mortality in the Carpathians during the harsh winter of 1975/76, one third involving wolves and two thirds due to lynx (Kerechun, 1979). In the Far East, predators of the roe deer include the wolf, the lynx, the tiger and the yellow-throated marten (*Martes flavigula*) which respectively accounted for 64.5%, 28.7%, 3% and 2.8% of a total of 625 recorded instances of predation (Rakov, 1979).

However, predation by the wolf and lynx tends to be in some kind of balance (Bromlei and Kucherenko, 1983; Matyushkin, 1985): when wolf numbers fall and consequently the pressure on the prey population slackens, predation by the lynx tends to increase. For example, in prewar years in the Ilmen reserve, the large wolf herd accounted for 67.5% of the total number of roe deer deaths, while the lynx was responsible only for 3%. Subsequently, wolf numbers decreased abruptly after the war and mortality due to wolf predation was practically nil, whereas that due to lynx predation increased up to 84% (Filonov, 1974, 1975). A similar situation was recorded in the Khingan reserve in the Amursk region (Darman, 1982a,b).

Stray **dogs** can also have some impact on roe deer populations, occasionally comparable with that inflicted by the wolf. For example, in the Ukraine, wild dogs and wolves each accounted for about 10% of all reported roe deer deaths (Karpenko, 1966). Dogs killed 22% (and wolves 18.5%) of roe deer in the Zhiguli reserve (Belyanin, 1980b) and 16.7% in the Les na Vorksle reserve and its surroundings in the Belgorod region (Table 9.7). Wild dogs sometimes form hunting packs in Kazakhstan (Pleshak, 1980; Sludski and Baidavletov, 1984). Several packs of wolf–dog hybrids have also been reported in one district of the Chelyabinsk region where they nearly eradicated a roe deer population that initially consisted of about 500 animals (Sokolov and Danilkin, 1981; Danilkin, 1982).

The **fox**, too, may have a noticeable effect on roe deer numbers, reducing the rate of population growth appreciably (Raesfeld, 1956; Borg, 1962; Pielowski, 1970; Strandgaard, 1972a; Wandeler, 1975; Stubbe and Passarge, 1979; Cederlund and Lindstrom, 1983). In the Bialowieza Primeval Forest, periods of roe population increase coincide with years of intensive fox control (Vladishevski, 1968). Foxes may account for more than 50% of roe deer deaths in the Crimea (Rukhlyadev, 1948) and up to 20% in the Ukraine (Karpenko, 1966). In Kazakhstan, roe kid remains are often found near fox dens and in their faeces. Observations of foxes chasing kids and even adult roe deer during periods of deep snow have been noted (Sludski and Baidavletov, 1984). Similar reports have been recorded in Yakutia (Tavrovski *et al.*, 1971) and in the Angara area (Yelski and Shishikin, 1979).

The impact of predation by **jackals** is probably comparable to that of the fox but much less mortality is inflicted by other predators such as the wolverine, the yellow-throated marten, the tiger and the brown bear. In addition, newborn kids may be killed occasionally by badgers, raccoon dogs, martens, golden and steppe eagles, the European wild cat (*Felis sylvestris*) and snow leopards (Fetisov, 1953; Arabuli, 1966; Aizin, 1969; Bromlei and Kucherenko, 1983; Sludski and Baidavletov, 1984; Zhiryakov and Djansinayev, 1986). In regions of high density, wild boar may also have a significant impact on roe kids (Stubbe and Passarge, 1979).

There is no consensus as to whether predation on roe deer is selective (Malafeyev and Kryazhimski, 1984). According to Bubenik (1966b), mortality due to the wolf and lynx is nearly twice as high among does as among kids and bucks. This was supported by data from Czechoslovakia, where predation was again about twice as high on does as on either bucks or kids (Hell, 1979). However, Filonov (1977) reported that bucks made up nearly twice as much of the wolf's diet relative to does (26%) and kids (21%), whereas the lynx took adult does in 90% of cases and bucks only 10% of the time. In contrast, Smirnov (1978) reported that kids fell prey to lynx more often (50%) than adult and subadult does (35.7%) or bucks (14.3%) in the west of the Trans-Baikal area. According to our observations (n = 14), kids fall prey to the lynx more rarely (21%) than adult does (64%). Half of these deer were killed by the predator and then left intact, and post-mortem examination revealed that all the prey were in good condition and showed no sign of previous injury. Of those roe deer taken by wolves or wolf–dog hybrids (n = 79), 25.3% were bucks, 39.3% were does and 35.4% were kids, and marked abnormality was only noted in five individual animals. It seems that wild dogs usually kill the weaker juveniles which tend to run at the rear of their group, while foxes, jackals and other small predators take kids, regardless of sex or condition.

9.2.10 Hunting and traffic accidents

In many areas, the major cause of roe deer mortality is hunting by humans to control population growth and prevent damage to the habitat. Over-hunting can result in an abrupt drop in roe deer numbers, such as occurred in Russia during the late nineteenth and early twentieth centuries. According to Turkin and Satunin (1902), in the southern Urals thousands of pitfall traps were set for roe deer in the Kasli Forest alone with most hunters having more than 100 each. In many areas, systems of pitfall traps and capture pens were set up over enormous distances connected by pole fences (Cherkasov, 1962). In regions to the south of Krasnoyarsk, some fences were 40 km long (Kornilov, 1856), and in the Trans-Baikal area the total length of fences was greater than 2000 km and more than 20 000 pitfall traps were in operation (Bioncourt, 1901).

Trapping of roe deer was also a widespread practice throughout the year. For instance, in the Urals, reports exist of individual hunters taking more than 500 over a period of 9 years, or even 1000 roe deer (Turkin and Satunin, 1902). Other methods of capture included the use of leg clamp traps, noose traps, bows and crossbows, drive capture (which often involved hundreds of horsemen) (Fetisov, 1953) and capture with nets on routes of migration in which as many as 200 animals were caught at a time (Deyev, 1911). During the migration season, some hunters covered 200–300 km to traditional sites of river crossings (Nikolayev, 1890).

The most widespread and destructive method was drive-catches across the frozen snow in winter. These often resulted in the near complete extermination of a herd due to the difficulties roe deer experience in moving over deep snow. The scope and consequences of this style of hunting are well illustrated by Sabaneev (1875) who reported that, in the spring of 1851 in the southern Urals, more than a thousand roe deer were taken in the Kasli region alone when snow persisted into March. After such large-scale culls, the total number of roe deer taken the following year was generally not even half as high. Similarly, in 1882 in the Altai mountains, inhabitants of the village of Uimon culled as many as 300 roe deer after driving them into the open steppe where the snow could not support their weight (Nikolski, 1883). This destructive method of hunting was equally widespread in eastern Siberia (Silantiev, 1898), where at the end of 1770s peasants of the village of Balaganka took 30

Table 9.8 Estimated total population size and annual cull during the 1980s in various countries (some figures are very approximate)

Country	Population size (1000)	Annual cull (1000)
	European roe deer	
Former USSR	350–400	10–40
Poland	500	115–128
Romania	290	30–35
Czechoslovakia	290	90
Yugoslavia	350	60
Hungary	220	41–80
Bulgaria	120	7
W. Germany	1600–1700	550–687
E. Germany	310	100–188
Austria	1000	130–237
Sweden	400	80
Norway	150	11
Finland	5	2
France	270–300	62
Denmark	140	39–59
Italy	100	–
Switzerland	120	45
Netherlands	25	5.0–6.5
Belgium	25	8–9
Spain	20	2
Luxembourg	13	5
Great Britain	400	200
	Siberian roe deer	
Former USSR	300–350	5–10
Mongolia	100–150	30
China	200–300	50
Korea	10	3

roe deer per farm, killing as many as 1500 head in several days during heavy snows. In 1875, inhabitants of Motskoye village in the Irkutsk region killed 500 roe deer in a week, most of which were pregnant does. This provides some examples of the consequences of over-hunting, depressing population growth to such an extent that it takes years for the population to recover to its former size. Roe deer have also often been exploited by the army (Sperk, 1885) and at the beginning of the twentith century the availability of long-range weapons following several wars led to the extermination of roe deer in many regions.

Roe deer are hunted in practically all countries across their range but the total cull differs from country to country (Table 9.8). They are intensively exploited in western Europe, where approximately 20–40% of the population are shot annually. These figures are highest in Germany (30–60%), Denmark and Switzerland (about 40%) and are lowest (less than 10%) in the former USSR, Norway, Spain, Bulgaria and a few other Asian countries.

These official statistics do not reflect the true situation in regions where **poaching** is widespread. The extent of roe deer poaching in the former USSR is illustrated by the situation in the Sverdlovsk region where poachers take up to 10% of the local population (Kiselev, 1979). In the Amur Ussuri region, amateur hunters are granted a cull of up to 2000 deer per season but the actual number of roe taken was generally three times as great (Kucherenko, 1976). In the Amursk region, the number of poached deer is estimated to exceed the official take by more than five or six times (Morozov, 1975). The effects of such over-exploitation are evident in central Yakutia where a population of 10 000 roe deer in 1963 was reduced to 800 animals by 1970 due to poaching (Popov *et al.*, 1980). In the Tien Shan, poachers may kill up to 58.3% of the population (Belousov, 1973). Smirnov (1974) referred to well-armed poachers using salt licks to attract deer, noose traps and drive hunting with cars resulting in an illegal roe cull in the west of the Trans-Baikal area more than double the official one. Similar year-round poaching is reported elsewhere (Shargayev, 1972; Nikolayev, 1980).

Roe deer poaching is equally widespread in Mongolia (Danilkin and Dulamtseren, 1981; Sokolov *et al.*, 1982a). Unfortunately, in the former USSR, the level of poaching is high even in reserves and other protected areas (Table 9.7). In a radiotracking study, eight of the 40 roe deer (20%) fitted with radio transmitters in the Les na Vorskle reserve in the Belgorod region were killed or wounded by poachers (Sokolov *et al.*, 1986d).

Recently, poachers have been operating under cover of licence, when they are granted a licence for one animal but take many more. An anonymous poll of hunters in the Far East revealed that of 66 men who had bought a licence, only one individual restricted hunting to the single animal, while the others took between 2 and 20 roe deer (Danilkin and Darman, 1987).

Migrating populations suffer the heaviest pressure during traditional meets along migration routes, including poachers in vehicles and sometimes armed with submachine guns. For example, in 1983–1985, several cases were reported of poachers using the headlights of armoured personnel carriers, tractors and other vehicles to dazzle roe deer before shooting them in the Amursk region, and a total of more than 1000 men were detained for illegal roe deer hunting during the period 1984–1986. Thus, it is almost impossible to estimate the actual total roe deer cull in the former USSR, but it is certainly much greater than the official legal hunt (Metelski, 1976; Lavov, 1978). Tentative estimates which include adjustments for estimated poaching indicate that no less than 25 to 30% (150 000 to 200 000 head) of the total population are taken annually (Danilkin, 1982, 1985a).

A further important cause of roe deer mortality concerns **road traffic accidents**. In some parts of Switzerland, vehicles account for between 8 and 27% of roe deer deaths (Eiberle, 1972). In Germany and Austria, 50 000 to 110 000 head (approximately 3–7% of the total population) are killed every year on the autobahns and perhaps as many again die in accidents with high speed agricultural machinery. Losses as high as 1–5 roe deer/100 ha have been reported in central Europe (Stubbe and Passarge, 1979; Beklova *et al.*, 1982; Engl, 1982; Kaluzinski, 1982b; Reimoser and Zandl, 1986). In agricultural habitats, there have been a few cases of roe deer dying due to chemical poisoning, particularly when fertilizers or pesticides are improperly stored or used. However, the indirect effects of human activities such as the destruction of habitats and disturbance are far more detrimental to roe deer

9.2.11 Habitat modification and migration

Roe deer require both food and cover, and any modification of the habitat may affect the capacity of the habitat to meet these particular needs and so have an impact on the dynamics of the population. This includes both natural agents, such as fires, and human-mediated transformations such as felling and ploughing operations, which may radically affect the carrying capacity of a particular habitat. Processes such as the ploughing of steppe habitats and land improvement schemes are usually to the detriment of roe deer, particularly in areas of heavy snow where the deer are unable to colonize agricultural fields. For example, in the 1960s and 1970s, large tracts of virgin land were ploughed in Siberia and the Far East, leading to an appreciable decrease in the available area of quality habitat and thus reducing overall population numbers. The previously large-scale migrations of roe deer in these areas became less marked and ceased entirely in some places (Baikalov and Semenov, 1971; Shvets, 1975). In contrast, burning and clearing small areas (which then regenerate) and tree planting improve the quality of habitat exploitable by the roe deer and may lead to population growth. However, many roe deer die during forest fires (Stubbe and Passarge, 1979; Darman, 1982a), especially during spring in Kazakhstan when locals seek to improve the hay lands. Such fires may last for several days or even weeks, destroying all wildlife in the area (Sludski and Baidavletov, 1984). Intensive forestry and agriculture are generally beneficial for the roe deer, and this fact is indicated by their widespread distribution in some of the most modified environments of western and central Europe. These activities have an overall negative effect on roe deer population dynamics only in certain parts of the roe deer's range (Chapter 4).

The building of the Krasnoyarsk Hydroelectric Station Basin in Siberia has destroyed traditional routes of roe deer migration as a 100 km stretch of the Yenisei downstream of the dam does not freeze, thus creating a serious barrier for the deer. Attempted crossing of the river more often than not results in death (Ziryanov, 1975) as the width of the river and the nature of the ice make the crossing more hazardous than before, when the ice did not hamper movement (Lapsin and Sokolov, 1971). Previously, as many as 1000 roe deer crossed the Yenisei's frozen surface, but now only 150–200 individuals migrate to the right bank of the river, and most of the population is forced to remain on the eastern bank where wintering conditions are less favourable. When the Zeya Water Reservoir was created in the Far East, roe deer routes of migration were blocked by wide bays, resulting in an abrupt fall off in population size (Shchetinin, 1975; Bromlei, 1984; Iliashenko, 1984).

Seasonal migrations of populations or individuals are a response to adverse conditions or habitat alterations and do not play a direct role in the regulation of numbers. However, mortality is high among migrating deer, partly due to risks such as crossing rivers but also because of an increased susceptibility to predation. This mortality generally involves the weakest individuals (juveniles, old or sick deer and lactating does), which may have an impact on the sex–age composition of the population and hence the reproductive productivity. In addition, occasional migrants may remain on a new range, reducing the size of the parent population. In fact, increased mortality during seasonal migration occurs over most of the roe deer's range and may have a significant impact on population size.

Thus, the growth of a roe deer population is not constant but varies considerably, depending on environmental factors. From the information presented in this chapter, it seems that the reproductive rate of roe deer is often relatively stable in suitable conditions and at moderate density. However, fecundity does vary widely, particularly at high density (perhaps over 20–30 deer/100 ha) or in poor habitats. In many populations, the rate of growth depends on mortality factors, particularly among juveniles, and although the factors influencing roe deer population dynamics are somewhat compensatory, they do generally work in sets.

Towards the west of the total range, hunting and losses caused by traffic and agricultural accidents have a very significant effect on roe deer populations. Parasitic diseases may also have an impact, especially when density is high, but predation is minimal in this region, although the fox may take a considerable number of kids in some areas. Severe winters may increase the level of mortality, particularly among young, old or weak individuals, and may compensate for the lack of predation in this region. Hunting may replace natural predation in some areas, controlling population growth.

In the north of the roe deer's range, natural mortality through starvation may be higher and predation is much heavier, particularly in winters of deep snow, but disease is comparatively rare. In the east of the European roe deer range and all over Siberia, predation and poaching are important factors determining roe deer numbers. Although licensed shooting is minimal, the combined total of the official cull with the poached estimates are equal to or greater than population losses from other causes. In low density populations, roe deer mortality from disease and parasites is negligible. The wolf and the lynx are the main predators of roe deer, and where they are both present the extent of predation by one depends on the other. Heavy predation pressure all but eliminates mortality due to disease. However, healthy individuals are also taken by these predators and high wolf or lynx density causes roe deer population growth to slow.

Finally, it should be stressed that the European and Siberian roe deer are species with high potential for exploitation. In the eastern part of Europe and over the whole range of the Siberian roe deer, it would be possible to increase the number of deer significantly. In Russia alone, there is potential habitat for an estimated 4 million roe deer. It is hoped that this book will provide information both for those who study and for those who manage the roe deer.

References

Abramov, A.V., Naumenko, B.N., Danilkin, A.A. and Struchkov, A.Yu. (1989) Using cluster analysis in roe deer craniometry. *Ecology, Morphology, Exploitation and Conservation of Wild Ungulates*, Moscow, Part 1, pp. 144—145. (In Russian.)

Adamovich, V. and Oitsos, I. (1963) Causes of roe mortality in the Volyn Region. *Okhota i Okhotnich'e Khozyaistvo* **4** 27–28. (In Russian.)

Adolf, T.A. (1954) Some data on Astrakhan steppe saiga biology. *Learned Transactions of the Potemkin Moscow Pedagogical Institute* **2** (2) 247–256. (In Russian.)

Aitken, R.J. (1974) Delayed implantation in roe deer (*Capreolus capreolus*). *Journal of Reproduction and Fertility* **39** (1) 225–233.

Aitken, R.J. (1976) Uterine secretion of fructose in the roe deer. *Journal of Reproduction and Fertility* **46** (2) 439–440.

Aitken, R.J. (1981) Aspects of delayed implantation in the roe deer (*Capreolus capreolus*). *Journal of Reproduction and Fertility* **29** (1) 83–95.

Aitken, R.J., Burton, J., Hawkins, J. *et al.* (1973) Histological and ultrastructural changes in the blastocyst and reproductive tract of the roe deer (*Capreolus capreolus*) during delayed implantation. *Journal of Reproduction and Fertility* **34** (3) 481–493.

Aizin, B.M. (1969) Ungulates. *Game Mammals of Kirghizia*. Frunze. pp. 78–111. (In Russian.)

Alekseyeva, E.V. (1980) *Mammals of the Pleistocene of the south-east of West Siberia*. Moscow: Nauka Publishers. 137 pp. (In Russian.)

Alekseyeva, L.I. (1977) Mammalian fauna of the Early Anthropogene of Eastern Europe. *Proceedings of the Geological Institute*. Moscow. Iss. 300. 216 pp. (In Russian.)

Allen, G.M. (1930) Pigs and deer from the Asiatic Expedition. *Amerivan Museum Novitates* **430** 1–19.

Amrud, J. and Nes, N. (1966) The chromosomes of the roe (*Capreolus capreolus*). *Hereditas* **56** (2) 217–220

Andersen, J. (1953) Analysis of a Danish roe-deer population based upon the extermination of the total stock. *Danish Review of Game Biology* **2** 127–155.

Angibault, J.M., Bideau, E., Vincent, J.P. *et al.* (1993) Détermination de l'âge chez le Chevreuil (*Capreolus capreolus* L.). Test de critères morphologiques ... partir d'animaux d'âge connu. *Mammalia* **57**, 579–587.

Antipin, V.M. (1941) *Mammals of Kazakhstan: Ungulates.* Alma-Ata: Kazgosizdat Publishers. Vol. 3. 107 pp. (In Russian.)

Arabuli, A.B. (1963a) On feeding of the European roe deer (*Capreolus capreolus* L.) on the Tsiv-Gombor Ridge. *Reports of the Georgian Academy of Sciences* **XXX** (4) 467–474. (In Russian.)

Arabuli, A.B. (1963b) Migrations of the European roe deer (*Capreolus capreolus* L.) in Kakhetia and their causes. *Zoologicheskii Zhurnal* **42** (7) 1113–1115. (In Russian.)

Arabuli, A.B. (1966) Roe deer in East Georgia. Author's abstract of the Dissert. for the degree of Cand. Sc. (Biology). Tbilisi. 22 pp. (In Russian.)

Arabuli, A.B. (1967) On studying ecology of the roe (*Capreolus capreolus* L.) on the Cahetian Caucasioni. *Proceedings of the Georgian Academy of Sciences' Institute of Zoology (Materials on the Georgian fauna)* **2** 166–169 (In Russian.)

Aulak, W. and Babinska-Werka, J. (1990a) Preference of different habitats and age classes of forest by roe deer. *Acta Theriologica* **35** 289–298.

Aulak, W. and Babinska-Werka, J. (1990b) Use of agricultural habitats by roe deer inhabiting a small forest area. *Acta Theriologica* **35** 121–127.

Aulak, W. and Babinska-Werka, J. (1990c) Estimation of roe deer density based on the abundance and rate of disappearance of their faeces from the forest. *Acta Theriologica* **35** 111–120.

Averin, Yu.V. (1949) Ecology of the roe (*Capreolus capreolus pygargus* Pall.) of the Ilmen Reserve. *Proceedings of the Ilmen Reserve* **4** 9–62. (In Russian.)

Averin, Yu.V. (1960a) Main changes in the specific composition of the mammalian and avian fauna of Moldavia in the 17th to 19th centuries. *MNS Bull. Biol. Dept.* **65** (2) 5–12. (In Russian.)

Averin, Yu.V. (1960b) Some changes in game fauna in recent centuries. *Nature Conservation in Moldavia* **1** 125–132. (In Russian.)

Aylmer, F. (1977) What do the does do? *Deer* **4** (4) 206 –207.

Baccus, R., Ryman, N., Smith, M.H. *et al.* (1983) Genetic variability and differentiation of large grazing mammals. *Journal of Mammalogy* **64** (1) 109–120.

Bader, O.N. (1950) Some problems of the paleography of the Urals and north-east Europe in the light of archeological data. *Materials on the Quaternary period of the USSR*. Moscow-Leningrad: USSR Academy of Sciences' Publishing House. Iss. 2. pp. 140–149. (In Russian.)

Baikalov, A.F. and Semenov, G.K. (1971) Aerial census of the roe in Khakassia. *Problems of Game Management in the Krasnoyarsk Territory*. Krasnoyarsk. pp. 104–108. (In Russian.)

Baleishis, R.M. and Prusaite, Ya.A. (1980) European roe feeding in a small deciduous forest of North Lithuania in 1975–1976. *Proceedings of the Lithuanian Academy of Sciences*. Ser. B (V). **1** (89) 85–91. (In Russian.)

Bannikov, A.G. (1954) *Mammals of the People's Republic of Mongolia*. Moscow: USSR Academy of Sciences' Publishing House. 669 pp. (In Russian.)

Barabash-Nikiforov, I.I. (1957) *Mammals of the south-east of the Chermozem Centre.* Voronezh. 370 pp. (In Russian.)

Barabash-Nikiforov, I.I. and Formozov, A.N. (1963) *Theriology*. Moscow: Vysshaja Shkola Publishers. Vol. 1. 394 pp. (In Russian.)

Barancheyev, L.M. (1962) Mass non-periodic migrations of roe deer in the Amursk Region. *Migrations of Animals*. Moscow. Iss. 3. pp. 26–36. (In Russian.)

Baratashvili, T. (1982) The lynx in the Borzhom Preserve. *Okhota i Okhotnich'e Khozyaistvo* **2** 12–13. (In Russian.)

Barclay, E.N. (1933) Notes on the roe-deer. *Annals and Magazine of Natural History (London)* **10** (12) 66–80.

Barishnikov, G.F., Garutt, V.Ye., Gromov, I.M. *et al.* (1981) *The Catalogue of Mammals of the USSR*. Leningrad: Nauka Publishers. 455 pp. (In Russian.)

Barth, D., Gimenez, T., Hoffmann, B. and Karg, H. (1976) Testosteronkonzentrationen im peripheren Blut beim Rehbock (*Capreolus capreolus*). Seasonale und rhythmische Veranderungen sowie Beziehungen zu einigen Verhaltenweisen und anderen saisonabhangigen Parametern. *Zeitschrift fur Jagdwissenschaft* **22** (3) 134–148.

Bartos, L., Hyanek, J. and Zirovnicky, J. (1981) Hybridization between red and sika deer. *Zoologischer Anzeiger* **207** (5–6) 260–270.

Baskin, L.M. (1976) *Behaviour of Animals*. Moscow: Nauka Publishers. 295 pp. (In Russian.)

Bayern, A. and Bayern, J. (1975) *Uber Rehe in einem steirischen Gebirgsrevier*. Hamburg: J. Bauer, 207 pp.

Becker-Dillingen, J. (1945) *Die Ernahrung des Waldes in der freien Wildbahn*. Donauworth. 251 pp.

Beklova, M., Koubek, P., Pikula, J. and Zejda, J. (1982) Game losses during the harvest of perennial fodder plants. *Folia Zoologica* **31** (1) 37–54.

Belik, V.I. (1953) Composition and distribution of game mammalian fauna of Yakutia. *Yakutian Game Fauna and Wildlife Management*. Yakutsk. Iss. 1. pp. 5–19. (In Russian.)

Belishev, B.F. (1934) Distribution of roe deer in West Siberia. *Okhotnik Sibiri* **11–12** 27.

Belousov, N.I. (1973) Some aspects of alpine roe deer ecology. *Problems of Siberian Vertebrate Ecology*. Kemerovo. pp. 164–171. (In Russian.)

Belov, S.N., Nikolayev, V.V., Unzhakov, V.V. and Yudin, B.S. (1980) Ungulate population dynamics in the Novosibirsk Region and its determining factors. *Ungulates of the USSR*. Moscow: Nauka Publishers. pp. 67–69. (In Russian.)

Belyanin, V.N. (1975) Distribution and numbers of roe deer in the Zhiguli Preserve. *Ungulates of the USSR*. Moscow: Nauka Publishers. p. 63. (In Russian.)

Belyanin, V.N. (1979) Wolves in the Zhiguli Preserve. *Ecological Principles of Predatory Mammal Management*. Moscow: Nauka Publishers. pp. 86–87. (In Russian.)

Belyanin, V.N. (1980a) Distribution and numbers of roe deer (*Capreolus capreolus pygargus* Pall.) in the Zhiguli Preserve. *Physiological and Population Ecology of Animals*. Saratov: Saratov University Press. Iss. 6(8). pp. 186–189. (In Russian.)

Belyanin, V.N. (1980b) Materials on ungulate mortality in the Zhiguli Preserve. *Ungulates of the USSR*. Moscow: Nauka Publishers. pp. 122–123. (In Russian.)

Berg, F.C. (1978) Zum Raum-Zeit-System des Rehes. *Allgemeine Forst. 1978*. **33** (3) 48–50.

Bettmann, H. (1961) Winterbrunft beim Rehwild. *Zeitschrift fur Jagdwissenschaft* **7** (4) 169.

Bideau, E., Vincent, J.P., Quéré, J.P. and Angibault, J.M. (1983a) Occupation de l'espace chez le chevreuil (*Capreolus capreolus* L., 1758). 1. Cas des males. *Acta Oecologica-Oecologica Applicata* **4** (2) 163–184.

Bideau, E., Vincent, J.P., Quéré, J.P. and Angibault, J.M. (1983b) Note sur l'evolution de l'association mère-jeune chez le chevreuil (*Capreolus capreolus* L., 1758) étudies par la technique du radio-tracking. *Mammalia* **47** (4) 477 –482.

Bideau, E., Vincent, J.P., Maublanc, M.L. and Gonzalez (1987) Dispersion chez le jeune Chevreuil (*Capreolus capreolus*): étude sur une population en milieu forestier. *Acta Oecologica-Oecologica Applicata* **8** 135–148.

Bideau, E., Gerard, J.F., Vincent, J.P. and Maublanc, M.L. (1993) Effects of age and sex on space occupation by European roe deer. *Journal of Mammalogy* **74** 745–751.

Bioncourt, A.A. (1901) On Russian hunting. *Priroda i okhota*. Moscow: Book VIII. pp. 10–16. (In Russian.)

Bischoff, T.L.W. (1854) *Entwicklungsgeschichte des Rehes*. Gieben, Ricker.

Blackler. (1916) *Annals and Magazine of Natural History*, Ser. 8, **18** 78.

Blankenhorn, H.J. (1975) Kitz-Markierungs-Aktion 1970–1971. *Jager* **9** 42–45.

Blant, M. (1987) *Dinamique de population, condition et constitution du chevreuil (*Capreolus capreolus *L., 1758) dans les cantons de Neuchatel et Vaud (ouest de la Suisse)*. Univ. Neuchatel, Switzerland, 160 pp.

Blant, M., Chappuis, F., Maire, J.A. *et al.* (1982) *La biologie du chevreuil* Capreolus capreolus *(L.) dans le canton de Neuchatel (Suisse)*. Univ. Neuchatel, Switzerland, 105 pp.

Blauvelt, H. (1955) Dynamics of the mother–newborn relationship in goats. *Group. Proc.* **1** 211–258.

Bluzma, P.P. (1975) The ecology and morphology of Lithuanian roe deer. Author's abstract of the dissert. for the degree of Cand.Sc.(Biology). Moscow. 30 pp. (In Russian.)

Bobek, B. (1977) Summer food as the factor limiting roe deer population size. *Nature* **268** (5615) 47–49.

Bobek, B., Drozdz, A., Grodzinski, W. and Weiner, J. (1974) Studies on productivity of the roe deer population in Poland. *XI International Congress of Game Biologists, Stockholm*. pp. 115–123.

Boessneck, J. (1956) Zur Grosse des mitteleuropaischen Rehes (*Capreolus capreolus* L.) in alluvial-vorgeschichlicher und fruher historischer. *Zeitschrift fur Saugetierkunde* **21** (3–4) 121–131.

Boisaubert, B. (1980) Aussetzung von Rehwild – Vorlaufige Ergebnisse. *Bulletin Mensuel* **41** 11–18.

Borg, K. (1962) Predation on roe deer in Sweden. *Journal of Wildlife Management* **26** (2) 133–136.

Borg, K. (1970) On mortality and reproduction of roe deer in Sweden during the period 1948–1969. *Viltrevy* **7** (2) 121–149.

Borowski, S. and Kossak, S. (1975) The food habits of deer in the Bialowieza Primeval Forest. *Acta Theriologica* **20** (32) 463–506.

Boutin, J.M., Gaillard, J.M., Delorme, D. and Van Laere, G. (1987) Suivi de l'evolution de la fecondite chez le chevreuil (*Capreolus capreolus*) par l'observation des groupes familiaux. *Gibier Faune Sauvage* **4** 255–265.

Bramley, P.S. (1970) Territoriality and reproductive behaviour of roe deer. *Journal of Reproduction and Fertility*, Suppl. No. 11. pp. 43–71.

Bramley, P.S. (1972) Management of roe deer (*Capreolus capreolus*) in small deciduous forests in England. *Forestry* **45** (2) 211–221.

Brauner, A.A. (1915) Which species do roe deer of Southern Russia and the Crimea belong to? *Transactions of the Crimean Naturalists' Society*. Simferopol. Vol. 5. pp. 112–114. (In Russian.)

Brauner, A.A. (1923) *Agricultural Zoology*. Kiev: Ukrizdat Publishers. 436 pp. (In Russian.)

Bresinski, W. (1982) Grouping tendencies in roe deer under agrocenosis conditions. *Acta Theriologica* **27** (29) 427–447.

Briedermann, L. (1974) Wass ast unser Rehwild? *Unsere Jagd.* **24** (4) 110–111.

Briedermann, L. and Ahrens, M. (1980) Bestande und Verbreitung des Schalenwildes in der DDR. *Unsere Jagd*. **30** (1) 4–6.

Brietenmoser, U. and Haller, H. (1987) Zur Nahrungsokologie des Luchses *Lynx lynx* in den scweizerischen Nordalpen. *Zeitschrift fur Saugetierkunde* **52** (3) 268–291.

Bromlei, G.F. (1984) The order Ungulata. *Mammals of the Zeya Reserve.* Vladivostok. pp. 107–117. (In Russsian.)

Bromlei, G.F. and Kucherenko, S.P. (1983) *Ungulates of the Southern Far East.* Moscow: Nauka Publishers. 305 pp. (In Russian.)

Broom, D.M. and Johnson, E. (1980) Responsiveness of hand-reared roe deer to odours from skin glands. *Natural History* **14** (1) 41–47.

Bubenik, A.B. (1959) *Grundlagen der Wildernahrung.* Berlin: Bauernverlag. 229 pp.

Bubenik, A.B. (1960) Le rythme nycthemeral et le régime journalier des ongules sauvages: problèmes théoriques, rythme d'activité du chevreuil. *Mammalia* **24** 1–66.

Bubenik, A.B. (1962) Das Verhalten des Rehes am Tage und in der Nacht in Laufe des Jahres. *Die Pirsch* **24–25** 944–947, 988–1000.

Bubenik, A.B. (1965) Beitrag zur Geburtskunde und zu den Mutter-Kind-Beziehungen des Rehes (*Capreolus capreolus* L.) und Rotwildes (*Cervus elaphus* L.). *Zeitschrift fur Saugetierkunde* **30** (2) 65–128.

Bubenik, A.B. (1966a) Das Geweih. *Entwicklung, Aufbau und Ausformung der Geweihe und Gehorne und ihre Bedeutung fur das Wild un fur die Jagd.* Hamburg–Berlin: Paul Parey. 214 pp.

Bubenik, A.B. (1966b) Vliv rysa (*Lynx lynx* L.) a vlka (*Canis lupus* L.) na strukturu populaci srnci (*Capreolus capreolus* L.) a jeleni zvere (*Cervus elaphus* L.). *Lynx* **6** 7–10.

Bubenik, A.B. (1968) The significance of the antlers in the social life of the *Cervidae. Deer.* No. 1. pp. 208–214.

Bubenik, A.B. (1970) Rehwildhege und Rehwildbiologie. Ernahrung und Beifutterung. *Der Deutsche Jager.* **38** (15) 389–394.

Bump, G. (1970) Acclimatization of game animals in the United States. *IX International Congress of Game Biologists.* Moscow. pp. 136–141.

Burt, W.H. (1943) Territoriality and home range concepts as applied to mammals. *Journal of Mammalogy* **24** 346–352.

Buttner, K. (1980) Untersuchuingen zum Einfluss von Grosse und Zusammensetzung des Wintersprungs bein Rehwild auf die Zeitdauer verschiedener Verhaltenweisen. *Zeitschrift fur Jagdwissenschaft* **26** (4) 181–194.

Cadariu, M., Popovici, N. and Cismas, M.A. (1977) Schimbari ciclice sezoniere in hipofiza capriorului (*Capreolus capreolus* L.) mascul. *Stud. Univ. Babes. Bolyai Biol.* **22** (2) 48–55.

Casanova, P. (1981) Il caprielo (*Capreolus capreolus* L.): un tipice abitatere della macchia. *Italia Forestale e Montana* **36** (5) 251–256.

Cederlund, G. (1981) Daily and seasonal activity pattern of roe deer in a boreal habitat. *Viltrevy* **11** (1) 1–41.

Cederlund, G. (1982) Mobility response of roe deer (*Capreolus capreolus*) to snow depth in a boreal habitat. *Viltrevy* **12** (2) 39–68.

Cederlund, G. (1983) Home range dynamics and habitat selection by roe deer in a boreal area in Central Sweden. *Acta Theriologica* **28** (30) 443–460.

Cederlund, G. (1989) Activity patterns in moose and roe deer in a north boreal forest. *Holarctic Ecology* **12**, 39–45.

Cederlund, G. and Lindstrom, E. (1983) Effects of severe winters and fox predation on roe deer mortality. *Acta Theriologica* **28** (7) 129–145.

Cederlund, G., Ljungquist, H., Markgren, G. and Stalfelt, F. (1980) Foods of moose and roe deer at Grimso in central Sweden. Results of rumen content analyses. *Viltrevy* **11** (4) 169–247.

Chaplin, R.E., Chapman, D.S. and Prior, R. (1966) An examination of the uterus and ovaries of some female roe deer (*Capreolus capreolus* L.) from Wiltshire and Dorset, England. *Journal of Zoology, London* **48** 570–574.

Chapman, D.I. (1975) Antlers – bones of contention. *Mammal Review* **5** (5) 121–172.

Chapman, D.I. and Chapman, N.G. (1971) Further observations on the incidence of twins in roe deer, *Capreolus capreolus. Journal of Zoology, London* **165** 505–509.

Chapman, N.G., Claydon, K., Claydon, M. *et al.* (1993) Sympatric populations of muntjac (*Muntiacus reevesi*) and roe deer (*Capreolus capreolus*): a comparative analysis of their ranging behaviour, social organisation and activity. *Journal of Zoology London* **229** 623–640.

Charlemagne, M. (1937) *Zoogeography of the Ukrainian Workers' Socialist Republic.* Kiev: UWSR Publishing House. 234 pp. (In Russian.)

Chegorka, P.T. (1986) Phenotypic criteria for the diagnosis of red and sika deer and their hybrids in mixed populations. *Gene Pool Conservation and Ecosystem Management in Forest Preserves (All-Union Conference, the Berezina Preserve).* Moscow. Pt. 2. pp. 222–225. (In Russian.)

Cheltsov-Bebutov, A.M. (1976) More on the zoogeography of sand semi-deserts of the Turgai Region. *Theoretical and applied aspects of nature conservation and wildlife management (Collected Papers of the Moscow Veterinary Academy)*, Vol. 84. Moscow. pp. 44–53. (In Russian.)

Cherkasov, A.A. (1962) *Notes of a Naturalist and Hunter.* Moscow: USSR Academy of Sciences Publishing House. 504 pp. (In Russian.)

Corbet, G.B. (1978) *The Mammals of the Palaearctic Region: A Taxonomic Review.* British Museum (Natural History), London. 314 pp.

Cotta, V. (1969) Asupra capriorului sibesian (*Capreolus capreolus pygargus* Pallas) in Romania. *Revista Padurilor*. **84** (7) 345–348.

Cotta, V., Almasan, H.A., Babulia, T. and Scutaru, V. (1961) Raspindirea si densitatea de effectiv a capriorului in R.P.R. Caile pentru obtineria unui efectiv rational. Studii si dercetri. *Inst. Cercetari Forest*. **22** 5–25.

Cronin, M.A. (1991a) Mitochondrial-DNA phylogeny of deer (*Cervidae*). *Journal of Mammalogy* **72** 533–566.

Cronin, M.A. (1991b) Mitochondrial and nuclear genetic relationships of deer (*Odocoileus spp.*) in western North America. *Canadian Journal of Zoology* **69** 1270–1279.

Dahl, S.K. (1930) Materials on the systematics and biology of the Crimean roe deer. *Transactions of the Crimean Naturalists' Society* **12** 64–127. (In Russian.)

Danilkin, A.A. (1977) Vocal communication in the Siberian roe deer. *Behaviour of Mammals.* Moscow: Nauka Publishers. pp. 239–243. (In Russian.)

Danilkin, A.A. (1982) Roe deer resources in the USSR and their effective exploitation. *Game Theriology*. Moscow: Nauka Publishers. pp. 108–115. (In Russian.)

Danilkin, A.A. (1985a) Management of roe deer in Russia. *Okhota i Okhotnich'e Khozyaistvo* **3** 16–18. (In Russian.)

Danilkin, A.A. (1985b) Karyotypes of Eurasian roe deer (*Capreolus* Gray): a speciation hypothesis. *Reports of the USSR Academy of Sciences* **285** (6) 1513–1516. (In Russian.)

Danilkin, A.A. (1985c) Present day ranges of the European (*Capreolus capreolus* L.) and Siberian (*C. pygargus* Pall.) roe deer. *Reports of the USSR Academy of Sciences* **283** (3) 749–752. (In Russian.)

Danilkin, A.A. (1986a) The European and Siberian roe deer: subspecies or species. *Okhota i Okhotnich'e Khozyaistvo* **7** 16–18. (In Russian.)

Danilkin, A.A. (1986b) Hybridization between the European and Siberian roe deer. *Okhota i Okhotnich'e Khozyaistvo* **9** 16–18. (In Russian.)

Danilkin, A.A. (1986c) Modern ideas about roe deer (*Capreolus* Gray) systematics and distribution. *IV Congress of the All-Union Theriological Society: Theses of reports*. Moscow. Vol. 1. pp. 53–54. (In Russian.)

Danilkin, A.A. (1989) Ecology and systematics of roe deer of Eurasia (Biological fundamentals of rational management and conservation). Author's abstract of the dissert. for the degree of DSc (Biology). Moscow, 47 pp. (In Russian.)

Danilkin, A.A. (1990) Embryonal diapause in roe deer (*Capreolus, Cervidae*) and an ethological hypothesis concerning its development. *Zoologicheskii Zhurnal* **69** (3) 120–124. (In Russian.)

Danilkin, A.A. and Baskevich, M.N. (1987) Karyotypes of the Far Eastern roe deer (*Capreolus pygargus*). *Zoologicheskii Zhurnal* **66** (2) 314–317. (In Russian.)

Danilkin, A. and Darman, Yu. (1987) Time presses. *Okhota i Okhotnich'e Khozyaistvo* **4** 6–7. (In Russian.)

Danilkin, A.A. and Dulamtseren, S. (1981) The roe deer in Mongolia. *Okhota i Okhotnich'e Khozyaistvo* **3** 44–45. (In Russian.)

Danilkin, A.A. and Markov, G.G. (1985) On the systematical status of Caucasian roe deer (*Capreolus* Gray). *Reports of the USSR Academy of Sciences* **283** (1) 231–235. (In Russian.)

Danilkin, A.A. and Markov, G.G. (1987) More on the taxonomy of Far Eastern, East Siberian and Tien Shan roe deer (*Capreolus pygargus* Pall.). *Transactions of the USSR Academy of Sciences, Biol. Ser.* **2** 315–318. (In Russian.)

Danilkin, A.A. and Minayev, A.N. (1988) Spatial and ethological structure of a European roe (*Capreolus capreolus*) population inhabiting the Les na Vorksle Preserve. *Urgent Problems of the Morphology and Ecology of Higher Vertebrates*. Moscow. Pt 2. pp. 509–538. (In Russian.)

Danilkin, A.A., Chistova, T.Yu., Yatsenko, V.N. and Chudonovskaya, G.A. (1983) On the chromosomal polymorphism of roe deer. *Population Variation of the Species and Problems of Mammalian Gene Pool Conservation*. Moscow. pp. 37. (In Russian.)

Danilkin, A.A., Markov, G.G., Gerasimov, S. and Nikolov, C. (1985) A craniometric study of the Siberian roe deer (*Capreolus pygargus* Pall.). *Reports of the USSR Academy of Sciences* **281** (2) 503–506. (In Russian.)

Danilkin, A.A., Struchkov, A.Yu., Mazkov, G.G. *et al.* (1989) Findings of a craniometric analysis of Eurasian roe deer. *Ecology, Morphology and Management of Wild Ungulates*. Moscow. Pt 1. pp. 148–149. (In Russian.)

Danilkin, A.A., Minayev, A.N. and Darman, Yu.A. (1991) Migration of a Siberian roe deer population. *Ongules/Ungulates 91, Proceedings of the International Symposium, Toulouse, France*, (eds F. Spitz *et al.*), INRA. p. 21.

Danilkin, A.A., Markov, G.G., Stubbe, C. and Struchkov, A.Yu. (1992a) Morphometric analysis. *European and Siberian Roe Deer*. Moscow: Nauka Publishers, pp. 25–43 (In Russian.)

Danilkin, A.A., Darman, Yu.A. and Minaev, A.N. (1992b) The seasonal migrations of a Siberian roe deer population. *Revue d'Ecologie (La Terre et la Vie)* **47** (3) 231–243.

Danilkin, A.A., Darman, Yu.A. and Minaev, A.N. (1993) Homing in migrating Siberian roe deer. *Reports of the Academy of Sciences* **332** (5) 664–666. (In Russian.)

Danilkin, A.A., Darman, Yu.A., Minayev, A.N. and Sempere, A. (1995) Social structure, behaviour and ecological parameters of a migratory population of the Siberian roe deer *(Capreolus pygargus* Pall.*) Transactions of the Academy of Sciences Biol. Ser.* **1** 48–61.

Danilov, N.I. (1974) Immigration of the wild boar and roe deer to Karelia. *Problems of Animal Ecology*. Petrozavodsk. pp. 158–160. (In Russian.)

Darman, Yu.A. (1982a) An analysis of roe deer mortality in the Khingan Preserve. *Mammals of the USSR. III Congress of the All-Union Theriological Society: Theses of reports*. Moscow. Vol. 1. pp. 182–183.

Darman, Yu.A. (1982b) Big predators in the Khingan Preserve. *Protection of Far Eastern Predatory Mammals*. Vladivostok. pp. 4–7. (In Russian.)

Darman, Yu.A. (1983) Biotopic distribution of roe deer in the Khingan Preserve. *Biology and Exploitation of Game Animals and Birds of Siberia and the Far East*. Irkutsk. pp. 43–51. (In Russian.)

Darman, Yu.A. (1986) The biology of roe deer of the Khingan Preserve. Author's abstract of the dissert. for the degree of Cand. Sc.(Biology). Moscow. 20 pp. (In Russian.)

Darwin, C. (1859) *The Origin of Species*. Moscow: Selkhozgiz Publishers. 631 pp. (In Russian.)

Dasmann, B. (1966) *Wildlife Biology*. New York: Wiley. pp. 61–65.

Dathe, H. (1966) Zum Mutter-Kind-Verhaltnis bei Cerviden. *Beitr. Jagd- und Wildforsch* **5** 83–88.

Dementiev, G.N. (1933) *Roe Deer*. Moscow–Leningrad: KOIZ Publishers. 52 pp. (In Russian.)

Demidov, G. (1940) Chronicle. *Sov. Okhotnik* **6** 49. (In Russian.)

Denisov, G.V. (1978) Results of ungulate introduction in the RSFSR in the 9th Five-Year Period. *Proceedings of the AUCAI*. Moscow. Iss. 149. pp. 24–49. (In Russian.)

Deyev, P. (1911) Natural historical, geographical and economicoecological description of forest plots of the Usk-Kamenogorsk, Bukhtarma Forestries of the Semipalatinsk Region. *From state forests of the Akmolinsk and Semipalatinsk Regions*. Omsk. Iss. 1. pp. 1–23. (In Russian.)

Dimin, V.A. (1975a) Seasonal migrations of roe deer in the Upper Amur Area. In: *Ungulates of the USSR*. Moscow: Nauka Publishers. pp. 92–93. (In Russian.)

Dimin, V.A. (1975b) The effect of predation on wild ungulates of the Upper Amur Area. *Ungulates of the USSR*. Moscow: Nauka Publishers. pp. 194–195. (In Russian.)

Dimin, V.A. and Yudakov, A.G. (1967) The lynx effect on the game fauna of the Upper Amur Area. *Management of Natural Resources of the Upper Amur Area*. Khabarovsk. pp. 164–166.

Dinnik, N.Ya. (1910) Mammals of the Caucasus. Pt. 1. Cetaceans and Ungulates. *Transactions of the Caucasian Branch of the Imperial Russian Geographical Society* **27** (1), 247 pp. (In Russian.)

Dmitriev, V.V. (1938) Ungulates of the Altai Preserve and its surroundings (the Eastern Altai and Wesern Sayani). *Proceedings of the Altai Preserve*. Moscow. Iss. 1. pp. 226–229. (In Russian.)

Dmitriev, V.V., Zolotovski, M.V., Folitarek, S.S. *et al.* (1937) *The Altai State Preserve*. Moscow. 103 pp. (In Russian.)

Doppilmair, G G. (1926) *Sable hunting on the north-eastern coast of Baikal*. Publishing House of the Buryat ASSR State Planning Committee. 272 pp. (In Russian.)

Dragoyev, P. (1965) Year II. Studies of roe deer (*Capreolus capreolus* L.) biology. *Groskostop. nauka*. **6** 501–513. (In Bulgarian.)

Drescher-Kaden, U. (1974) Untersuchungen uber Verdauungstrakt im Pansen von Gamswild post mortem. *Tagungsber. 1. Int. Gamswild-Treffen Oberammergan.* pp. 26–33.

Drozdz, A. (1979) Seasonal intake and digestibility of natural foods by roe deer. *Acta Theriologica* **24** (13) 137–170.

Drozdz, A. and Osiecki, A. (1973) Intake and digestibility of natural foods in roe deer. *Acta Theriologica* **18** (3) 81–91.

Dulitski, A.I., Kormilitsin, A.A. and Kormilitsina, V.V. (1975) Crimean ungulate studies. *Ungulates of the USSR.* Moscow: Nauka Publishers. pp. 88. (In Russian.)

Dulkeit, G.D. (1953) The predatory importance of the lynx and wolverine in the Altai taiga bniocenosis. *Changes in the Vertebrate Fauna of our Country.* Moscow: MNS Press. pp. 147–152. (In Russian.)

Dulkeit, G.D. (1964) Game fauna, problems and methods of assessing productivity of hunting grounds of the Altai-Sayan mountain taiga. *Proceedings of the Stolbi State Preserve.* Iss. 4. 353 pp. (In Russian.)

Durov, V.V. (1980) Roe deer population dynamics and seasonal distribution in the Caucasian Preserve. *Ungulates of the USSR.* Moscow: Nauka Publishers. pp. 147–148. (In Russian.)

Dvornikov, M.G. (1982) Factors determining mortality of roe deer in the Ilmen Preserve. *Problems of Wildlife Management.* Perm. pp. 18–22. (In Russian.)

Dvornikov, M.G. (1984) Ecology and biocenotic role of ungulates in the V. I. Lenin Ilmen State Preserve. Author's abstract of the diss. for the degree of Cand. Sc. (Biology). Sverdlovsk. 25 pp. (In Russian.)

Eberle, J. (1986) Alterstruktur des Rehwildes in der Blattzeit bestimmt. *Unsere Jagd.* **36** (10) 300.

Eiberle, K. (1962) Beobachtungen uber das Verhalten des Rehwildes. *Schweizerische Zeitschrift fur Forstwesen* **113** (11) 660–668.

Eiberle, K. (1967) Wildverteilungsstudien am Rehwild. *Schweizerische Zeitschrift fur Forstwesen* **118** (2) 45–56.

Eiberle, K. (1972) Rehwildhege und Strassenverkehr. *Schweizerische Zeitschrift fur Forstwesen* **123** (4) 201–210.

Eiberle, K. (1976) Rechnerische Moglichkeiten der Bestandeskontrolle beim Rehwild. *Schweizerische Zeitschrift fur Forstwesen* **127** (5) 312–326.

Eisfeld, D. (1974a) Der Eiweis- und Energiebedarg des Rehes, diskutiert anhand von Laborversuchen. *Verhandlungen der Gesellschaft fur Okologie.* Erlangen. pp. 129–139.

Eisfeld, D. (1974b) Haltung von Rehen (*Capreolus capreolus* L.) zu Versuchzwecken. *Zeitschrift fur Saugetierkunde* **39** (3) 190–199.

Eisfeld, D. (1976) Ernahrungsphysiologie als Basis fur der okologische Beurteilung von Rehpopulationen. *Revue Suisse de Zoologie* **83** 914–928.

Ellenberg, H. (1974a) Wilddichte, Ernahrung und Vermehrung beim Reh. *Verhandlungen der Gesellschaft fur Okologie.* Erlangen. pp. 59–76.

Ellenberg, H. (1974b) Die Korpergrosse des Rehes als Bioindikator. *Verhandlungen der Gesellschaft fur Okologie.* Erlangen. pp. 141–154.

Ellenberg, H. (1978) Zur Populationsokologie des Rehes (*Capreolus capreolus* L., *Cervidae*) in Mitteleuropa. *Spixiana, Zeitscrift fur Zoologie,* Suppl. 2. pp. 1–211.

Ellerman, J.R. and Morrison-Scott, T.C.S. (1951) *Check-list of Palaearctic and Indian mammals 1758 to 1946.* London: British Museum, 810 pp.

Engl, D. (1982) Rehwildmarkierung. *Osterr. Weidwerk.* **5** 210–212.

Espmark, Y. (1969) Mother–young relations and development of behaviour in roe deer (*Capreolus capreolus* L.). *Viltrevy* **6** (6) 460–540.

Espmark, Y. (1974) Social behaviour of roe deer at winter feeding stations. *Applied Animal Ethology* **1** (1) 35–47.

Essen, L. (1966) Das Rehwild in Schweden. *Beitr. Jagd- und Wildforsch.* **5** 143–148.

Esser, W. (1958) Beitrag sur Untersuching der Asung der Rehwildes. *Zeitschrift fur Jagdwissenschaft* **4** (1) 1–41.

Fadeyev, Ye.V. (1969a) Results of introduction of some wild ungulate species in game preserves of Russia. *Moscow N.S. Bull., Biol. Dept.* **74** (1) 37–46. (In Russian.)

Fadeyev, Ye.V. (1969b) On releases of roe deer in hunting reserves of Russia. *Moscow N.S. Bull., Biol. Dept.* **74** (1) 37–46. (In Russian.)

Fadeyev, Ye.V. (1981) Roe deer of the European Territory of the USSR. *Okhota i Okhotnich'e Khozyaistvo* **5** 18–20. (In Russian.)

Feng, Z., Cai, C. and Zheng, C. (1983) A checklist of the mammals of Xiang (Tibet). *Acta Theriologica Sinica* **4** (4) 341–358.

Fetisov, A.S. (1947) *Seasonal Foods of the Siberian Roe Deer in the Irkutsk Region.* Irkutsk: Regional Publishing House. 15 pp. (In Russian.)

Fetisov, A.S. (1953) *Roe Deer in East Siberia.* Irkutsk: Regional Publishing House. 73 pp. (In Russian.)

Filonov, K.P. (1974) Peculiarities of the South Urals Siberian roe deer population. *Okhotovedenie.* Moscow: Lesnaya Promishlennost Publishers. pp. 26–40. (In Russian.)

Filonov, K.P. (1975) Ecological replaceability of mortality factors in ungulate populations. *Ungulates of the USSR.* Moscow: Nauka Publishers. pp. 206–207. (In Russian.)

Filonov, K.P. (1977) Ungulate population dynamics in preserves. *Okhotovedenie.* Moscow: Lesnaya Promishlennost Publishers. 229 pp. (In Russian.)

Flerov, K.K. (1928) On diagnostic characters of the genus *Capreolus* Frisch (*Cervidae*). *Reports of the USSR Academy of Sciences*, Ser. A, **23** 479–484. (In Russian.)

Flerov, K.K. (1929) Some new data on roe deer (*Capreolus* Frisch.) of East Asia. *Reports of the USSR Academy of Sciences*, Ser. A, **18** 429–432. (In Russian.)

Flerov, K.K. (1935) *Ungulates (Ungulata) of Tajikistan: (The Tajik Base Proceedings, Vol. 1).* Moscow–Leningrad: USSR Academy of Sciences Publishing House. pp. 93–130. (In Russian.)

Flerov, K.K. (1952) The genera *Moschus* and *Cervus. Fauna of the USSR. Mammals.* Moscow–Leningrad: USSR Academy of Sciences Publishing House. Vol. 1. Iss. 2. 256 pp. (In Russian.)

Formozov, A.N. (1946) *Snow Cover in the Life of Mammals and Birds of the USSR.* Moscow: MNS Press. 141 pp. (In Russian.)

Fruzinski, B. and Labudzki, L. (1982a) Demographic processes in a forest roe deer population. *Acta Theriologica* **27** (25) 365–375.

Fruzinski, B. and Labudzki, L. (1982b) Sex and age structure of a forest roe deer population under hunting pressure. *Acta Theriologica* **27** (26) 377–384.

Fruzinski, B., Labudzki, L. and Wlazelko, M. (1983) Habitat, density and spatial structure of the forest roe deer population. *Acta Theriologica* **28** (16) 243–258.

Gabrashanski, N. (1970) Infectious encephalomyelitis in roe deer. *Proceedings of the IXth International Congress of Game Biologists.* Moscow. pp. 640–642. (In Russian.)

Gaillard, J.M. and Delorme, D. (in press) Family effects in a roe deer population. *The Biology of Deer. Proceedings of the III International Cogress on the Biology of Deer.*

Gaillard, J.M., Sempere, A.J., Boutin, J.M. *et al.* (1992) Effects of age and body weight on the proportion of females breeding in a population of roe deer (*Capreolus capreolus*). *Canadian Journal of Zoology* **70** 1541–1545.

Gaillard, J.M., Delorme, D. and Jullien, J.M. (1993a) Effects of cohort, sex, and birth date on body development of roe deer (*Capreolus capreolus*) fawns. *Oecologia* **94** 57–61.

Gaillard, J.M., Delorme, D. and Jullien, J.M. (1993b) Early growth and body weight at the onset of the winter in roe deer (*Capreolus capreolus*). *Mammalia* **57** 359–366.

Gaillard, J.M., Delorme, D., Jullien, J.M. and Tatin, D. (1993c) Timing and synchrony of births in roe deer. *Journal of Mammalogy* **74** 738–744.

Gaillard, J.M., Delorme, D., Boutin, J.M. *et al.* (1993d) Roe deer survival patterns: a comparative analysis of contrasting populations. *Journal of Animal Ecology* **62** 778–791.

Gaillard, J.M., Delorme, D., Boutin, J.M. and Van Laere, G. (in press) Causes of variability in winter body weight of roe deer fawns: a long term study of two contrasting populations. *Journal of Wildlife Management*.

Gaillard, J.M., Boutin, J.M. Delorme, D. *et al.* (unpublished data) Summer survival patterns in roe deer fawns: a comparative study of the proximate and ultimate causes of variability.

Garozs, V. (1987) Roe deer: why are they not numerous in our forests? *Nauka i Zhizn'*. **1** 14–15. (In Russian.)

Garozs, V.Ya. (1979) The lynx and wolf effect on the Latvian roe deer population. *Ecological Principles of Predatory Mammal Management*. Moscow: Nauka Publishers. pp. 257–261. (In Russian.)

Gatikh, V.S. (1975) The distribution and numbers of wild ungulates in the Pripyat Preserve. *Ungulates of the USSR*. Moscow: Nauka Publishers. pp. 73–74. (In Russian.)

Gebczynska, Z. (1980) Food of the roe deer and red deer in the Bialowieza Primeval Forest. *Acta Theriologica* **25** (40) 487–500.

Geiger, G. and Kramer, A. (1974a) Rank-order of roe deer at artificial winter feeding sites in a Swiss hunting district. *XI International Congress of Game Biologists, Stockholm*. pp. 107–113.

Geiger, G. and Kramer, A. (1974b) Randordnung von Rehwild (*Capreolus capreolus* L.) und der Winterfutterung in einem Schweizerischen Jagdrevier. *Zeitschrift fur Jagdwissenschaft* **20** (1) 53–56.

Geist, V. (1968) Horn-like structures as rank symbols, guards and weapons. *Nature* **220** (5169) 813–814.

Gelder, R.G. (1977) Mammalian hybrids and genetic limits. *American Museum Novitates* **2635** 1–25.

Georgii, B. (1974) Corpora lutea and adrenal weight in roe deer. *XI International Congress of Game Biologists, Stockholm*. pp. 125–132.

Gill, R. (1990) Monitoring the status of European and North American Cervids. *Global Environment Monitoring System*, Gems Inform. Ser. 8. 277 pp.

Gladkov, N.A. and Nikolski, G.V. (1935) Materials on the mammalian fauna of the Middle and Lower Amu Darya. *Proceedings of the MSU State Zoological Museum* **2** 3–16. (In Russian.)

Gonzalez, R. and Angibault, J.M. (1987) Etude du rythme d'activité du chevreuil pàr biotélémetrie: aspect methodologique; Premiers resultats en milieu forestier. *Bull. Sfeca*. **2** (1) 159–164.

Gorbachev, S.N. (1915) Mammals of Orel Province in the light of new data obtained in 1910–1914. *Materials on the Nature of Orel Province* **21** 1–13. (In Russian.)

Gordiyuk, N.M. (1986) Relationships of ungulates in the Bashkiria Preserve. *The Role of Large Predators in Biocenoses of Preserves*. Moscow. pp. 70–81. (In Russian.)

Goreglyad, K.S. (1970) Fundamentals of wild ungulate disease control. *Proceedings of the IXth International Congress of Game Biologists*. Moscow. pp. 575–579. (In Russian.)

Gossow, H. (1977) Waldstrukturen und Wildstandsentwicklung. *Wald und Wild Seminar*. pp. 1–27.

Graczyk, R. (1978) Character powiazan ekolohicznych populacji sarn (*Capreolus capreolus* Linnaeus, 1758) z ekosystemami lesnym i polnym. *Rocz. AR Pozn.* **100** 23–29.

Graczyk, R. and Bereszynski, A. (1978) Observacje nad tworjeniem sie, organizacja i liczebnoscia gesienno-zimowych skupien sarn (*Capreolus capreolus* L.) w ekosystemie polnym. *RAR Pozn.* **100** 31–38.

Grafodatski, A.S., Sharshov, A.A. and Shutov, V.V. (1990) Karyotypic relationships between *Cervidae*. *Zoologicheskii Zhurnal* **69** (4) 101–114. (In Russian.)

Gray, J. (1821) On the natural arrangement of vertebrate animals. *London Med. Reposit.* **15** (1) 296–320.

Grayevskaya, B.M., Zolotareva, N.N., Danilkin, A.A. and Yelfimova, S.S. (1980) A comparative study of metabolism in the European (*Capreolus capreolus* L.) and Siberian (*Capreolus pygargus* Pall.) roe deer. *Ungulates of the USSR*. Moscow: Nauka Publishers. pp. 332–333. (In Russian.)

Grayevskaya, B.M., Zolotareva, N.N. and Yelfimova, S.S. (1988) The biochemical characterization of the European and Siberian roe deer as species. *Reports of the USSR Academy of Sciences* **303** (5) 1259–1261. (In Russian.)

Grechushkin, V.S., Kharchenko, B.I., Molodan, G.N. and Mglinets, S.M. (1975) The present day state of ungulate populations in the Donetsk Region. *Ungulates of the USSR*. Moscow: Nauka Publishers. pp. 78–79. (In Russian.)

Grigorov, G.R. (1976) Year XIII. Studies of wintertime roe deer feeding in the Gabrovo District. *Gorskostop. Nauka*. **6** 67–75. (In Bulgarian.)

Gromov, V.I. (1948) The paleontological and archeological basis of the stratigraphy of Quaternary deposits in the USSR (mammals, the Paleolith). *Proceedings of the Institute of Geological Sciences*. Iss. 64, Geol. Ser. No. 17. 521 pp. (In Russian.)

Gromov, V.S. (1986a) The morphological variability, behaviour and systematics of the roe deer. Author's abstract of the dissert. for the degree of Cand. Sc. (Biology). 27 pp. (In Russian.)

Gromov, V.S. (1986b) On differences in body colour between the European (*Capreolus capreolus*) and Siberian (*C. pygargus*) roe deer. *Zoologicheskii Zhurnal* **65** (11) 1693–1703. (In Russian.)

Gromov, V.S. (1988a) Age-related and seasonal food and energy reguirements of the European (*Capreolus capreolus*) and Siberian (*C. pygargus*) roe deer. *Urgent Problems of the Higher Vertebrate Morphology and Ecology*. Moscow. Pt. 11. pp. 495–508. (In Russian.)

Gromov, V.S. (1988b) Growth regularities of the European (*Capreolus capreolus*) and Siberian (*C.pygargus*) roe deer. *Zoologicheskii Zhurnal* **67** (9) 1381–1392. (In Russian.)

Gromov, V.S. and Danilkin, A.A. (1984) Doe–fawn relationships in the Siberian roe deer. *The Signalling and Ecology of Mammals and Birds*. Moscow: Nauka Publishers. pp. 93–107. (In Russian.)

Gromova, V.I. (1948) Mammal remains of Early Slavic sites near the city of Voronezh. *The Archeology of the USSR: Materials and studies.* No. 8. (In Russian.)

Groves, C.P and Grubb, P. (1987) Relationships of living deer. *Biology and Management of the Cervidae* (ed. C.M. Wemmer), Washington and London: Smithsonian Institution Press, pp. 21–59.

Grubb, P. (1993) Order Artiodactyla. *Mammal Species of the World: a Taxonomic and Geographic Reference.* Washington and London: Smithsonian Institution Press, pp. 337–414.

Gunther, G. (1978) Ein entschlossener Versuch. Sichere Ermittlung vorhandener Wildbestande. *Die Pirsch* **30** (23) 1563–1567.

Gurski, I.G. (1975) The wild boar, roe deer and red deer in steppes lying to the north-west of the Black Sea. *Ungulates of the USSR.* Moscow: Nauka Publishers. pp. 79–80. (In Russian.)

Gurski, I.G. and Nazarenko, L.F. (1969) The wolf effect on roe deer numbers in steppes near the Black Sea. *The Capability and Productivity of Hunting Grounds of the USSR.* Kirov. Pt. 2. pp. 168–169. (In Russian.)

Gusev, A.A. (1984) The role of wild ungulates in the function of biogeocenoses of the Central Chermozem Preserve. Author's abstract of the dissert. for the degree of Cand. Sc. (Biology). Moscow. 19 pp. (In Russian.)

Gustavsson, I. (1965) Chromosome studies in live species of deer representing the four genera *Alces, Capreolus, Cervus* and *Dama. Mammal. Chromosomes Newsl.* **18** 149.

Gustavsson, I. and Sundt, C.O. (1968) Karyotypes in five species of deer (*Alces alces* L., *Capreolus capreolus* L., *Cervus elaphus* L., *Cervus nippon nippon* Temm. and *Dama dama* L.). *Hereditas* **60** (1–2) 233–248.

Haffer, J. (1986) Superspecies and species limits in vertebrates. *Zeitschrift fur Zoologische Systematik und Evolutionsforschung* **24** (3) 169–190.

Haltenorth, T. (1963) Klassifikation der lebenden Saugetiere: Artiodactyla. *Handbuch der Zoologie.* Berlin. p. 167.

Hamilton, W.J., Harrison, R.J. and Young, B.A. (1960) Aspects of placentation in certain *Cervidae. Journal of Anatomy* **94** 1–33.

Hanus, V. and Fiser, Z. (1981) Telesny vyroi srnci zvere. *Prace vyzk. Ustavi Lesn.* **58** 191–212.

Harrington, R. (1985a) Evolution and distribution of the *Cervidae. Biology of Deer Production* (eds P. Fennessy and K. Drew). The Royal Society of New Zealand Bulletin, Vol. 22. pp. 3–11.

Harrington, R. (1985b) Hybridization in deer – its detection and uses. *Biology of Deer Production* (eds P. Fennessy and K. Drew). The Royal Society of New Zealand Bulletin, Vol. 22. p. 62.

Harrison, D.L. (1968) *The Mammals of Arabia.* London: Ernest Benn. Vol. 2. 381 pp.

Hartl, G.B., Reimoser, E., Willing, R. and Koller, J. (1991) Genetic variability and differentiation in roe deer (*Capreolus capreolus* L.) of Central Europe. *Genetics, Selection, Evolution* **23** 281–299.

Hearney, A.W. and Jennings, T.J. (1983) Annual foods of the red deer (*Cervus elaphus*) and the roe deer (*Capreolus capreolus*) in the east of England. *Journal of Zoology, London* **201** (4) 565–570.

Hein, J. (1966) Untersuchungen uber Wege zur Verbesserung der Qualität eines Rehwildbestandes. Diss. Gottingen.

Hell, P. (1979) *Srncia zver.* Bratislava: Priroda. 310 pp.

Hell, P. and Herz, J. (1970) Rozsirenie, hustota osidlenia a produkcia srnca horneho europskeho (*Capreolus c. capreolus* L., 1758) na Slovensku. *Acta Zootechn. Nitra.* **20** 227 –241.

Helle, P. (1980) Food composition and feeding habits of the roe deer in central Finland. *Acta Theriologica* **25** (3) 395–402.

Hennig, R. (1962) Uber des Revierverhalten des Rehbocke. *Zeitschrift fur Jagdwissenschaft* **8** (2) 61–81.

Hennig, R. (1962) Uber einige Verhaltenswesen des Rehwildes (*Capreolus capreolus*) in freier Wildbahn. *Zeitscrift fur Tierpsychologie* **19** (2) 223–229.

Henry, B.A.M. (1978a) A comparison of the winter diet of roe deer and sheep. *Journal of Zoology, London* **185** (2) 270–273.

Henry, B.A.M. (1978b) Diet of the roe deer in an English conifer forest. *Journal of Wildlife Management* **42** (4) 937–940.

Henry, B.A.M. (1981) Distribution patterns of roe deer (*Capreolus capreolus*) related to the availability of food and cover. *Journal of Zoology, London* **194** 271–275.

Heptner, V.G., Nasimovich, A.A. and Bannikov, A.G. (1961) *Mammals of the Soviet Union. Artiodactyles and Perissodactyles.* Moscow: Vysshaja Shkola Publishers. Vol. 1. 776 pp. (In Russian.)

Herz, J. (1971) Zhodnotenie polovnickeho obhospodarovania zveri srbcej v okrese Trnava. *Folia Venatoria* **1** 201–213.

Hewison, A.J.M. (1993) The reproductive performance of roe deer in relation to environmental and genetic factors. Ph.D. thesis, University of Southampton, Southampton, UK.

Hewison, A.J.M. (1995) Isozyme variation in roe deer in relation to their population history in Britain. *Journal of Zoology, London* **235** 279–288.

Hewison, A.J.M. (in press a) Variation in fecundity of roe deer: effects of age and body weight. *Acta Theriologica.*

Hewison, A.J.M. (in press b) Cranial variation of British roe deer and correlated differences in fecundity. The Biology of Deer. *Proceedings of the III International Congress on the Biology of Deer.*

Hewison, A.J.M. (in press c) Birth sex-ratios and local resource competition in roe deer. *Journal of Mammalogy.*

Hewison, A.J.M., Angibault, J.M., Boutin, J. *et al.* (in press a) Annual variation in body composition of roe deer. *Canadian Journal* of *Zoology.*

Hewison, A.J.M, Vincent, J.P., Bideau, E. *et al.* (in press b) Density-dependent variation in cohort mandible size as an index of roe deer population trend. *Journal of Zoology, London*

Hinde, R.A. (1975) *Animal behaviour, A synthesis of ethology and comparative psychology*. Moscow: the Miy Publishers. pp. 454 (In Russian.)

Hoffman, B., Barth, D. and Karg, H. (1978) Progesterone and oestrogen levels in peripheral plasma of pregnant and non-pregnant roe deer (*Capreolus capreolus*). *Biology of Reproduction* **l9** 931–936.

Hofmann, R.R. (1978a) Die Ernahrung des Rehwildes im Jahresablauf nach Modell Weichselboden. *Wildbiolog. Inform. Jager. Stuttgart*. pp. 121–136.

Hofmann, R.R. (1978b) Die Verdauungsorgane des Rehes und ihre Anpassung an die besondere Ernahrungsweise. *Wildbiolog. Inform. Jager. Stuttgart*. pp. 103–112.

Hofmann, R.R., Geiger, G. and Konig, R. (1976) Vergleichend-anatomische Untersuchungen an der Vormagenschleimhaut von Rehwild (*Capreolus capreolus*) und Rotwild (*Cervus elaphus*). *Zeitschrift fur Saugetierkunde* **3** 167–193.

Holand, O. (1992) Seasonal variation in body composition of European roe deer. *Canadian Journal of Zoology* **70** 502–504.

Holand, O. (1994) Seasonal dynamics of digestion in relation to diet quality and intake in European roe deer (*Capreolus capreolus*). *Oecologia* **98** 274–279.

Holisova, V., Obrtel, R. and Kozena, I. (1982) The winter diet of roe deer (*Capreolus capreolus*) in the Southern Moravian agricultural landscape. *Folia Zoologica* **3** (3) 209–225.

Holisova, V., Kozena, I. and Obrtel, R. (1984) The summer diet of field roe bucks (*Capreolus capreolus*) in Southern Moravia. *Folia Zoologica* **33** (3) 193–208.

Holisova, V., Obrtel, R. and Kozena, I. (1986a) Rumen content vs. faecal analysis to estimate roe deer diets. *Folia Zoologica* **35** (1) 21–32.

Holisova, V., Obrtel, R. and Kozena, I. (1986b) Seasonal variation in the diet of field roe deer (*Capreolus capreolus*) in Southern Moravia. *Folia Zoologica* **35** (2) 97–115.

Holisova, V., Obrtel, R., Kozena, I. and Danilkin, A. (1992) Feeding. *European and Siberian Roe Deer*. Moscow: Nauka Publishers, pp. 124–139. (In Russian.)

Holler, W. and Prilkofer, K. (1980) Graviditat und Geburt beim Reh (*Capreolus capreolus* L.). *Physiol. Patholog. Therapiemoglichk.* **12** 368–374.

Holmes, F. (1973) *Following the Roe. A Natural History of the Roe Deer*. Edinburgh. 110 pp.

Honacki, J.H., Kinman, K.E. and Koeppl, J.W. (1982) *Mammal Species of the World*. Lawrence: Kansas. 694 pp.

Honak, B. (1970) *Mesi Kamerou a Puskou*. Brno: Nakladatelstvi Blok. 79 pp.

Hromas, J. (1983) Relations between the trophy quality and living conditions of roe deer in the CSSR. *Lesnictvi*. **29** (7) 591–610.

Hubner, F. (1938) Das Rehwild. *Weidwerk der Welt*. pp 197–200.

Iliashenko, V.Yu. (1984) The effect of the Zeya Water Storage Basin on terrestrial vertebrates of mountain taiga ecosystems. Author's abstract of the dissert. for the degree of Cand. Sc. (Biology). Moscow. 18 pp. (In Russian.)

Ishunin, G.I. (1961) Mammals (Predators and Ungulates). *Fauna of the Uzbek SSR*. Tashkent: Uzbekistan Academy of Sciences' Publishing House. Vol. 3. 231 pp. (In Russian.)

Jackson, J.E. (1974) Feeding habits of roe deer. *Mammal Review* **4** 93–101.

Jackson, J.E. (1980) The annual diet of the roe deer (*Capreolus capreolus*) in the New Forest, Hampshire, as determined by rumen content analysis. *Journal of Zoology, London* **192** (1) 71–84.

Jacobi, J. (1932) Kraniometrische Untersuchungen und Stammesgeschichtliche Ableitung der Rehe (*Capreolus* H. Smith). *Jena. Z. Naturwissenshaft* **67** 326–345.

Jaczewski, Z. (1976) The induction of antler growth in female red deer. *Bull. Acad. Pol. Sci. Ser. Sci. Biol.* **24** (1) 61–65.

Jaczewski, Z., Doboszynska, T. and Krzywinski, A. (1976) The induction of antler growth by amputation of the pedicle in red deer (*Cervus elaphus* L.) males castrated before puberty. *Folia Biologica (PRL)* **24** (3) 299–307.

Janeau, G., Quere, J.P., Spitz, F. and Vincent, J.P. (1981) Etude par radiotracking des variations saisonieres de l'etendue du domaine vital de cinq chevreuils (*Capreolus capreolus* L.). *Biological Behaviour* **6** (4) 291–304.

Janson, J. (1975) Morphology of a local roe deer (*Capreolus capreolus capreolus* L.) population. *Proceedings of the Latvian Agricultural Academy*. Iss. 83. pp. 52–55. (In Russian.)

Janson, J. and Padaiga, V. (1975) Die naturliche Asung des Renwilldes in der Schneefreich zelt des Jahres. Jagid Wissenshaft bei intensiver Forstwirt Schaft. Wiss. Konf. Kaunas. p. 98–100.

Jeppesen, J.L. (1989) Activity patterns of free-ranging roe deer (*Capreolus capreolus*) at Kalo. *Danish Review of Game Biology* **13** 1–30.

Jewell, P.A. (1966) The concept of home range in mammals. *Symposium of the Zoological Society of London* **18** 85–109.

Juon, P. (1963) Uber neuere Erkenntnisse zur Frage der Rehwildernahrung. *Schweizerische Zeitschrift fur Forstwesen* **114** (3) 98–117.

Jurgenson, P.B. (1968) *Game Mammals and Birds*. Moscow: Lesnaya Promishlennost Publishers. 308 pp. (In Russian.)

Jurgenson, P.B. (1969) State and prospects of game resources. *Proceedings of the Zavidovo Hunting Reserve*. Iss. 1. pp. 79–208. (In Russian.)

Kahlke, G.D. (1971) The family *Cervidae* Gray, 1821. *The Pleistocene of Tiraspol*. Kishinev: Shtiintsa Publishers. pp. 137–156. (In Russian.)

Kahlke, H.D. (1955) *Grossaugetiere im Eiszeitalter*. Leipzig-Jena: Urania-Verlag. 88 pp.

Kahlke, H.D. (1956) *Die Cervidenreste aus den altpleistozanen Ilmkiesen von Sussenborn bei Weimar*. Tell 1. Akademie Verlag, Berlin, 62 pp.

Kahlke, H.D. (1958a) Die Cervidenreste aus den altpleistozanen Tonen von Voigtstedt bei Sangerhausen. Tell I. Die Schadel, Geweihe und Gehorne. *Abh. Dtsch. Akad. Wiss. Berlin. Kl. Chemie, Geol. Biol. Jahrg. 1956*. Berlin, No. 9. pp. 1–51.

Kahlke, H.D. (1958b) Die Cervidenreste aus den Altpleistozanen sanden von Mosbach (Biebrich-Wiesbaden). Tell. 1. *Abh. Dtsch. Akad. Wiss. Berlin. Kl. Chemie, Geol. Biol., Jahrg. 1956*, No. 7. pp. 1–75.

Kahlke, H.D. (1969) Die Rhinocerotiden-Reste aus den Kiesen von Sussenborn bei Weimar. Das Pleistozan von Sussenborn. *Pal. Abh. Palaozoologie* **3** (3–4) 667–709.

Kalninsh, A.I. (1950) *Hunting and Wildlife Management in the Latvian SSR*. Riga: Latgosizdat Publishers. 540 pp. (In Russian.)

Kaluzinski, J. (1974) The occurrence and distribution of field ecotype of roe deer in Poland. *Acta Theriologica* **19** 291–300.

Kaluzinski, J. (1978) Badania biometryczne i obserwacje biologiszne sarny (*Capreolus capreolus* L.) populacji polnej. *Rocz. AR Pozn.* **100** 73–81.

Kaluzinski, J. (1982a) Dynamics and structure of a field roe deer population. *Acta Theriologica* **27** (27) 385–408.

Kaluzinski, J. (1982b) Roe deer mortality due to mechanization of work in agrocenoses. *Acta Theriologica* **27** (30) 449–455.

Kaluzinski, J. (1982c) Composition of the food of the roe deer living in fields and the effects of their feeding on plant production. *Acta Theriologica* **27** (30) 457–470.

Kaluzinski, J. and Bresinski, W. (1976) The effect of the European hare and roe deer populations on the yields of cultivated plants. *Ecology and Management of the European Hare Population*. Warsaw. pp. 247–253.

Kampmann, H. (1983) Untersuchungen uber die Auswirkungen des Seitentriebverbisses durch Rehwild auf das Hohenwachstum von Fichtenpflanzen. *Zeitschrift fur Jagdwissenschaft* **29** (4) 235–243.

Kapitonov, V.I. (1980) The effect of the summer drought of 1974 and 1975 on ungulates of the Kazakh Upland. *Ungulates of the USSR*. Moscow: Nauka Publishers. pp. 163–165. (In Russian.)

Karpenko, A.V. (1966) The bioecological basis of a program aimed at protecting Left Bank Ukraine forests against their destruction by the roe deer. Author's abstract of the dissert. for the degree of Cand. Sc. (Biology). Kharkov. 21 pp. (In Russian.)

Kashkarov, D.N. (1931) *Animals of Turkestan*. Tashkent: UZGIZ Publishers. 448 pp. (In Russian.)

Keibel, F. (1899) Zur Entwicklungsgeschichte des Rehes. *Anat. Anz.* **16** 64–65.

Kerechun, S.F. (1979) The effect of predators on ungulate numbers in the Karpati. *Ecological Principles of Predatory Mammal Management*. Moscow: Nauka Publishers. pp. 43–44. (In Russian.)

Kerschagl, W. (1952) *Rehwildkunde*. Vienna: Hubertusverlag. 272 pp.

Khaveson, Ya.I. (1960) On the importance of form hybridizability and hybrid fertility for systematic category rank determination. *Scientific Conference on the results of 1958–1959 research work*. Vologda. pp. 73–75. (In Russian.)

Kholodova, M.V. (1986) Seasonal variations of food requirements in some ungulates. *IV Congress of the All-Union Theriological Society: Theses of reports*. Moscow. Vol. 1. pp. 367–368. (In Russian.)

Kiili, Ya. (1965) Analysis of factors determining roe deer population dynamics. *Transactions of the Estonian Academy of Sciences. Biology*. **34** (3) 216–225. (In Russian.)

Kioroglanidis, J. (1981) Nahrungswahl und Uberlegungen zur Nahrungsnische bei Reh (*Capreolus capreolus*), Rothirsch (*Cervus elaphus*) und Gemse (*Rupricapra rupricapra*) im Nationalpark Berchtesgaben. Diss., Ludwig-Maxmilian-Univ. Munich. 175 pp.

Kirikov, S.V. (1952) *Birds and Mammals in Landscapes of the Southern Urals*. Moscow: USSR Academy of Sciences' Publishing House. 412 pp. (In Russian.)

Kirikov, S.V. (1959) *Changes in the Fauna of Geographical Zones of the USSR (the 13th to 19th Centuries): the Steppe Zone and Forest Steppe*. Moscow: USSR Academy of Sciences' Publishing House. 175 pp. (In Russian.)

Kirikov, S.V. (1966) *Game Animals, Environment and Man*. Moscow: Nauka Publishers. 346 pp. (In Russian.)

Kirikov, S.V. (1983) *Man and Nature of the Steppe Zone*. Moscow: Nauka Publishers. pp. 52–63. (In Russian.)

Kirilov, N. (1902) Wildlife management in the Trans-Baikal Region. *Priroda i okhota*. January. pp. 32–40. (In Russian.)

Kiselev, A.A. (1974) The roe deer and lynx in the Sverdlovsk Region. *Okhota i Okhotnich'e Khozyaistvo* **12** 16–17. (In Russian.)

Kiselev, A.A. (1975) More on the morphology of roe deer of two Ural populations. *Proceedings of the Institute of Plant and Animal Ecology of the USSR Academy of Sciences' Ural Research Centre*. Iss. 96. pp. 112–115. (In Russian.)

Kiselev, A.A. (1976) Distribution and numbers of the roe deer in the Middle Urals. *Hunting Science (Management of Forest Ungulates)*. Moscow: Lesnaya Promishlennost Publishers. pp. 63–70. (In Russian.)

Kiselev, A.A. (1979) The roe deer in the Urals. Author's abstract of the dissert. for the degree of Cand. Sc. (Biology). Sverdlovsk. 23 pp. (In Russian.)

Klotzli, F. (1965) *Qualitat und Quantitat der Rehasung*. Bern: Hans Huber. 186 pp.

Knorr, H. and Briedermann, L. (1976) Die Ernahrung des Feldrehes. *Jagd-Inform. Inst. Forstwissensch. Eberswalde*. pp. 40–49.

Kolb, A. (1979) Das Scheinasen des Rehes. *Zeitschrift fur Jagdwissenschaft* **25** (4) 201–207.

Komarek, J. (1954) *Lesnicka Zoologie*. III. Praha. 287 pp.

Korelov, M.N. (1956) Vertebrate fauna of the Bostandik District. *Nature and Economic Conditions of the Mountainous Bostandik*. Alma-Ata. pp. 259–325. (In Russian.)

Korneyev, O.P. (1952) *A Key to Animals of the Ukrainian Workers' Socialist Republic*. Kiev: Radyanska Shkola Publishers. 216 pp. (In Ukrainian.)

Kornilov, I.P. (1856) Deer of Eastern Siberia. Roe deer. *Vestnik Estestv. Nauk*. **3** (16) 494–504. (In Russian.)

Korotkevich, Ye.L. (1970) *Late Neogenic Deer of Steppes Lying to the North of the Black Sea*. Kiev: Naukova Dumka Publishers. 196 pp. (In Russian.)

Korotkevich, Ye.L. (1974) A new member of the genus *Procapreolus* of steppes lying to the north of the Black Sea. *Vest. Zool.* **6** 68–77. (In Russian.)

Korotkevich, Ye.L. (1984) Background groups of Late Miocenic ruminants and the history of the Eastern Europe hipparionic fauna. Author's abstract of the dissert. for the degree of D.Sc. (Biology). Kiev. 47 pp. (In Russian.)

Korsh, P. (1941) Wild goats in the Omsk Region. *Sov. okhotnik.* **4** 33 (In Russian.)

Kosintsev, P.A. (1981) The roe deer of archeological sites of the Urals and Western Siberia (the Neolithic Epoch to the Middle Ages). *Theriology in the Urals.* Sverdlovsk. pp. 39–43. (In Russian.)

Kossak, S. (1976) The complex character of the food preferences of the *Cervidae* and phytocenosis structure. *Acta Theriologica* **21** (24–31) 359–375.

Kossak, S. (1981a) Hand-rearing and care of a group of roe deer. *Acta Theriologica* **26** (11) 207–219.

Kossak, S. (1981b) Development of food habits in roe-deer. *Acta Theriologica* **26** (33) 483–494.

Kossak, S. (1983) Trophic relations of roe deer in a fresh deciduous forest. *Acta Theriologica* **28** (6) 83–128.

Kozlov, P.K. (1924) The Mongolian Bogda-Ula Preserve. *Transactions of the Russian Geographical Society* **56** (1) 169–173. (In Russian.)

Krushinski, L.V. (1960) *The Formation of Behaviour in Health and Disease*. Moscow: Moscow State University Press. 263 pp. (In Russian.)

Kucherenko, S.P. (1976) Hoofed mammals of the Amur-Ussuri Region. *The Fauna and Wildlife Management of the Far East*. Vladivostok. pp. 97–125. (In Russian.)

Kucherenko, S.P. (1979) The wolf in the south of the Far East. *Ecological Principles of Predatory Mammal Management*. Moscow: Nauka Publishers. pp. 117–118. (In Russian.)

Kucherenko, S. and Shvets, V. (1977) The roe deer of the Amur-Ussuri Region. *Okhota i Okhotnich'e Khozyaistvo* **3** 22–23. (In Russian.)

Kucherenko, S. and Zubkov, Yu. (1980) The wolf in the south of the Far East. *Okhota i Okhotnich'e Khozyaistvo* **1** 20–23. (In Russian.)

Kuen, H. and Bubenik, A.B. (1980) Botanische Pansenanalysen bei Rotwild (*Cervus elaphus hippelaphus*), Rehwild (*Capreolus capreolus*) und Gamswild (*Rupicapra rupicapra*). Alpine Umwelt Probleme. *Ergebnisse des Forschungsprojekts Achenkirch*. V–IX. 37. Beitr. Imweltgestaltung. pp. 41–60.

Kumerloeve, H. (1975) Die Saugetiere (Mammalia) der Turkei. Die Saugetiere (Mammalia) Syriens and des Libanon. *Veroff. Zool. Staatssamml. Munich* **18** 69–225.

Kurazhkovski, Yu.N. (1952) On the study of the ecology of roe deer inhabiting timber stands of arid areas. *Okhrana Prirodi* **15** 122–126. (In Russian.)

Kurt, F. (1965) Zur Rolle des Eieruchs inn Verhalten des Renwiles. verh. (1964) Schweiz. Nat. Eies. Zwich. pp. 140–142.

Kurt, F. (1966) Feldbeobachtungen und Versuche uber das Revierverhalten der Rehbocke (*Capreolus capreolus* L.). *Revue Suisse de Zoologie* **73** (3) 408–421.

Kurt, F. (1968a) *Das Sozialverhalten des Rehes (*Capreolus capreolus *L.)*. Hamburg-Berlin: Paul Parey. 102 pp.

Kurt, F. (1968b) Zusammenhange zwichen Verhalten und Fortpflahzungsleistung beim Reh (*Capreolus capreolus*). *Zeitschrift fur Jagdwissenschaft* **14** (3) 97–106.

Kurt, F. (1970) *Rehwild*. Munich: BLV Verlag. 174 pp.

Kurt, F., Hartl, G.B. and Volk, F. (1993) Breeding strategies and genetic variation in European roe deer *Capreolus capreolus* populations. *Acta Theriologica* **38**, Suppl. 2, 187–194.

Kurten, B. (1968) Pleistocene mammals in Europe. *Striae* **24** 47–49.

Kuznetsov, B.A. (1948) *Mammals of Kazakhstan*. Moscow: MNS Press. 226 pp. (In Russian.)

Kuzmina, Ye.I. (1971) Formation of the Northern Urals mammalian fauna in the Late Anthropogene. *Proceedings of the USSR Academy of Sciences' Zoological Institute* **49** 44–122. (In Russian.)

Lapsin, G.M. and Sokolov, G.A. (1971) The ecology and economic value of the roe deer of spurs of the Kuznetsk Alatau. *Problems of Wildlife Management in the Krasnoyarsk Territory*. Krasnoyarsk. pp. 65–70.

Laptev, I.P. (1955) On distributions of some mammals in Western Siberia. *Notes on the Siberian flora and fauna*. Tomsk. Iss. 18. pp. 3–6. (In Russian.)

Laptev, I.P. (1958) *Mammals of the Taiga Zone of Western Siberia*. Tomsk. Tomsk University Press. 285 pp. (In Russian.)

Lavov, M.A. (1966). The roe deer population in the Vitim Plateau. *Voprosi zoologii*. Tomsk: Tomsk University Press. pp. 206–207. (In Russian.)

Lavov, M.A. (1970) Roe deer. *Okhota i Okhotnich'e Khozyaistvo* **8** 16–19. (In Russian.)

Lavov, M.A. (1971) Structure of the Vitim Plateau roe deer population. *MNS Bull., Biol. Dept* **76** (2) 56–59. (In Russian.)

Lavov, M.A. (1974) Structure and reproduction of the Vitim Plateau roe deer population. *AURIHP collected scientific & technical information of 'Hunting, furs, game'*. Kirov. Iss. 43. pp. 8–12. (In Russian.)

Lavov, M.A. (1975) Distributions and approximate numbers of ungulate mammals in the Krasnoyarsk Territory. *Voprosi Ekologii. The Proceedings of the Stolbi State Preserve*. Krasnoyarsk. Iss. 10. pp. 339–347. (In Russian.)

Lavov, M.A. (1978) Roe deer. *Large Predators amd Ungulates*. Moscow: pp. 190–220. (In Russian.)

Lavov, M.A. (1982) Wolves – myths and reality. *Okhota i Okhotnich'e Khozyaistvo* **7** 20–21. (In Russian.)

Lavrov, N.P. (1927) Distribution of the roe deer in Siberia. *Okhotnik*. **5** 14–16. (In Russian.)

Lavrov, N.P. (1929) Geographical distribution, biology and economic value of the roe deer in the USSR. *Proceedings on Forestry Experiments of the Central Experimental Forestry Station* **6** 49–82. (In Russian.)

Lehmann, E.R. (1958) Chevreuil d'Asia et d'Europe. Contribution a l'étude du genre *Capreolus*. *Mammalia* **22** (2) 262–270.

Lehmann, E. (1960a) Das Problem der Grossenabnahme (Dominutions Tedenz) beim Reh. *Zeitschrift fur Jagdwissenschaft* **6** (2) 41–51.

Lehmann, E. (1960b) Warum brunftet das Reh im Hochsommer? *Der Deutsche Jager* **10** 87.

Lehmann, E. (1971) Das Sommerkleid des Rehe im Klimagefalle. *Umschau* **12** 428–429.

Lehmann, E. (1976) Einige Bemerkungen zum Sibirischen Reh (*Capreolus capreolus /pygargus/ caucasicus* Dinnik, 1910) in MittelEuropa. *Zeitschrift fur Jagdwissenschaft* **22** (2) 75–84.

Lehmann, E. and Sagesser, H. (1986) *Capreolus capreolus* Linnaeus, 1758 – Reh. *Handbuch der Saugetiere Europas*. Wiesbaden. 2. pp. 233–264.

Lent, P. (1971) Mother–infant relationships in ungulates. *JVCN Publishers N.S.* **1** (24) 14–53.

Liepmann, H. (1972) Rohe gegen Solanin unempfindlich? *Dtsch. Jager-Ztg.* **90** (12) 488.

Lincoln, G.A. (1972) The role of antlers in the behaviour of red deer. *Experimental Zoology* **182** (2) 233–250.

Lincoln, G.A. and Guinness, F.E. (1972) Effect of altered photoperiod on delayed implantation and moulting in roe deer. *Journal of Reproduction and Fertility* **31** (3) 455–457.

Linnaeus, C. (1758) *Systema naturae per regna tria naturae, secundum classes, ordines, genera, species, cum characteribus, differentiis, synonimis, locis*. 10th edn. Laurentii Salvii, Stockholm, 824 pp.

Lister, A.M. (1984) Evolutionary and ecological origins of British deer. *Proceedings of the Royal Society of Edinburgh* **82B** 205–229.

Litvinov, V.F., Iliasevich, V.A. and Dunin, V.F. (1979) The role of predators in ungulate mortality. *Ecological Principles of Predatory Mammals Management*. Moscow: Nauka Publishers. pp. 44–46. (In Russian.)

Lobachev, V.S. (1972) The roe deer in deserts near the Aral Sea. *Okhota i Okhotnich'e Khozyaistvo* **9** 11. (In Russian.)

Lochman, J. (1965) Pastevni rytmus s denni rezim jeleni a srnci zvere. *Prace vyzk. ustavu lesnickych CSSR*. Praha. Vol. 30. pp. 103–140.

Lochman, J. and Barth, A. (1965) Otazka zkod zpusobovanych zveri s hlediska fysiologie vyzivy sparkate zvere. *Zpravy les. vyzkumu. Zbraslav-Strnady* **11** (5–6) 16–17.

Lochman, J., Tvrdikova, A. and Melicharova, A. (1961) Spotreba zivin u srnci zvere (*Capreolus capreolus* L.). *Sbornik Ceskoslovenske Akademie Zemedelskych Ved. Lesnictvi* **6** 551–570. (In Slovakian).

Lochman, J., Sesulkova, E. and Melicharova, A. (1974) Vyuzivani krmiv sparkatou zveri. III. Vysledky zjistovani koeficientu stravitelnosti a delky doby pruchodu potravy zazivacim traktem. *Prace VULHM* **45** 39–60.

Long, M. (1980) Fair play for roe deer. *Scottish Forestry* **34** (2) 93–97.

Lorenz, K. (1967) *On Aggression*. London: Methuen. 306 pp.

Lott, D.L. (1991) *Intraspecific Variation in the Social Systems of Wild Vertebrates*. Cambridge University Press.

Loudon, A.S.I. (1978) The control of roe deer populations: a problem in forest management. *Forestry* **51** (1) 73–83.

Loudon, A.S.I. (1982) Too many deer for the trees? *New Scientist* **93** (1297) 708–711.

Loudon, A.S.I. (1987) The influence of forest habitat structure on growth, body size and reproduction in roe deer (*Capreolus capreolus* L.). In: *Biology and Management of the Cervidae*, (ed. C.M. Wemmer), pp. 559–567. Smithsonian, Washington.

Lund, J.J. (1989) Activity patterns of free-ranging roe deer (*Capreolus capreolus*) at Kalo. *Danish Review of Game Biology* **13** (8) 1–32.

Lund, J.J. (1990) Home range and movements of free-ranging roe deer (*Capreolus capreolus*) at Kalo. *Danish Review of Game Biology* **14** (1) 1–14.

Maizeret, C. and Sung, D.T.M. (1984) Etude du régime alimentaire recherche du déterminisme fonctionnel de la sélectivité chez le chevreuil (*Capreolus capreolus*) des landes de Gascogne. *Giber Faune Sauvage* **3** 63–103.

Malafeyev, Yu.M. and Kryazhimski, F.V. (1984) Lynx feeding and its relations with ungulates in the Middle Urals. *MNS Bull., Biol. Dept.* **89** (4) pp. 70–81. (In Russian.)

Manolache, L. and Babulia, T. (1969) *Cercetari asupra hranei suplimentare la caprior si lepure*. Buchurest. 80 pp.

Markov, G.G. (1985) A comparative population – morphological and genetical analysis of the genus *Capreolus* Gray, 1821. Author's abstract of the dissert. for the degree of Cand. Sc. (Biology). Sofia. 30 pp. (In Bulgarian.)

Markov, G.G. and Dobrianov, D. (1985) A karyotaxonomic study of the roe deer (*C. capreolus* L., 1758) in Bulgaria. *Acta Zool. Bulg.* **28** 241–244. (In Bulgarian.)

Markov, K.K. and Velichko, A.A. (1967) *The Quaternary Period.* Moscow: Nedra Publishers. Vol. 3. 440 pp. (In Russian.)

Markov, K.K., Lizukov, G.I. and Nikolayev, V.V. (1965) *The Quaternary Period.* Moscow: Moscow State University Press. Vol. 2. 435 pp. (In Russian.)

Markov, G.G., Kekhaiov, I., Danilkin, A.A. *et al.* (1985a) A comparative immunological study of the European (*Capreolus capreolus* L.) and Siberian (*C. pygargus* Pall.) roe deer. *Reports of the USSR Academy of Sciences* **281** (5) 1277–1281. (In Russian.)

Markov, G.G., Danilkin, A.A., Gerasimov, S. and Nikolov, C.M. (1985b) A comparative craniometric analysis of the European roe deer (*Capreolus capreolus* L.). *Reports of the USSR Academy of Sciences* **282** (2) 489–493. (In Russian.)

Markowski, J. (1993) Fluctuating asymmetry as an indicator for differentiation among roe deer *Capreolus capreolus* populations. *Acta Theriologica* **38**, Suppl. 2, 19–31.

Markowski, J. and Markowska, M. (1988) Non-metrical variation in roe deer from three populations. *Acta Theriologica* **33** 519–536.

Matschie, P. (1907) *Weidwerk in Wort und Bild,* **19** 263.

Matyushkin, Ye.N. (1985) Relations with other predatory mammals. *The Wolf.* Moscow: Nauka Publishers. pp. 355–370. (In Russian.)

Maublanc, M.L., Bideau, E. and Vincent, J.P. (1987) Flexibilité de l'organisation sociale du chevreuil en fonction des caracteristiques de l'environnement. *Revue d'Ecologie (La Terre et la Vie)* **42** (2) 109–133.

Mayer, H. (1977) Waldstruktur-Wildstand-Wildschaden-Jagdtechnik. *Wald und Wild Seminar*. pp. 28–34.

Mayr, E. (1968) *Animal Species and Evolution.* Moscow: Mir Publishers. 513 pp. (In Russian.)

Mayr, E. (1974) *Population, Species and Evolution.* Moscow: Mir Publishers. 460 pp. (In Russian.)

Mayr, E., Linsley, E. and Usinger, R. (1956) *Methods and Principles of Zoological Systematics.* Moscow: Foreign Literature Publishers. 352 pp. (In Russian.)

McDiarmid, A. (1974) Mortality in deer. *Mammal Review* **4** (3) 75–78.

Meikle, I. (1975) Return of the roe. *Wildlife* **17** (1) 4–7.

Melichar, J. (1960) Das Abasen der Waldvegetation durch das Rotund Rehwild. *Zeitschrift fur Jagdwissenschaft* **6** 78.

Melichar, J. (1961) Vyzivne hodnoty v potrave drevin pro parohatou zver. *Myslivost.* **10** 151–152.

Melnikov, V.K. (1978) Periods of the autumnal migration of the roe deer, Siberian red deer and bear in the Maly Abakan basin (Western Sayani). *Siberian Fauna and its Exploitation.* Irkutsk. pp. 3–5. (In Russian.)

Mertens, S. (1984) Zur Verhaltens-Synchronisation bei gruppenweise gehaltenen Rehen (*Capreolus capreolus*). *Behaviour* **89** (3–4) 220–240.

Metelski, A.P. (1975) Factors determining roe deer distribution in mountainous landscapes of Siberia and the Far East. *Ungulates of the USSR.* Moscow: Nauka Publishers. pp. 114–115. (In Russian.)

Metelski, A.P. (1976) The distribution and numbers of the roe deer in Siberia and the Far East. *Hunting Science (Management of Forest Ungulates).* Moscow: Lesnaya Promishlennost Publishers. pp. 23–63. (In Russian.)

Meyer, P. (1968) Territoriumsmarkierung beim Reh und Morphologie des sogenannten Stirnorgans. Inang. Dissert. Hannover. 54 pp.

Meyer-Brenken, H. (1986) Schwarzes Rehwild. *Wild und Hund* **89** (9) 40–45.

Migulin, O. (1927) *Herbivorous and Carnivorous Mammals of the Ukraine*. Kharkov: Radyanski Selyanin Publishers. 166 pp. (In Ukrainian.)

Migulin, O. (1929) *A Key to Animals of the Ukraine*. Kharkov: The Ukranian State Publishing House. 95 pp. (In Ukrainian.)

Mohring, G. (1963) Zur Beerenahrung des Rehes (*Capreolus capreolus* L.). *Waldhygiene* **5** 68–69.

Morozov, K.A. (1975) The size of the roe deer population of the Upper Amur Area. *Ungulates of the USSR*. Moscow: Nauka Publishers. pp. 56–57. (In Russian.)

Morozov, K.A. (1982) An experiment of artificial mineral feeding of wild ungulates in the Amursk Region. *Increasing Productivity of Hunting Grounds*. Moscow. pp. 43–49. (In Russian.)

Mottl, S. (1954) Potrava srnci zvere (*Capreolus capreolus capreolus* L.). *Biologia* **12** (1) 29–43.

Mottl, S. (1957) Die Jagdwirtschaftlich erforderliche Mindestrosse von Rehwildrevieren im Wald. *Zeitschrift fur Jagdwissenschaft* **3** 64–69.

Mottl, S. (1962) Zur Frage der Wilddichte und der Qualitat des Rehwildes. *Beitr. Jagd- und Wildforsch.* **11** 35–40.

Mottl, S. and Janda, M. (1955) Nektere vysledky studia o vlivu prostredi na srnci zver. *Sbornik Cs. akad. zemedels. ved., r. les.* **28** (4) 613–628.

Muller, K. (1973) Okologische Untersuchungen der Liegezonen beim Reh (*Capreolus capreolus*). *Revue Suisse de Zoologie* **80** (3) 667–670.

Myrberget, S. (1965) Merkingsresultaten XVI. Returns of Game Species Marked by the Norwegian Hunting and Fishing Association 1947–1962. *Medd. fra States Viltundersokelser* Ser. 2 **21** 24 pp.

Myrberget, S. (1973) Merking av radyr i Norge. *Fauna (Norge)* **26** (2) 97–101.

Myrberget, S. and Milton, R. (1971) *Capreolus capreolus* with 5 embryos. *Fauna Norsk Zoologisk Forening Tidsk.* **1** 67.

Nagy, J.G. and Regelin, W.L. (1975) Comparison of digestive organ size of three deer species. *Journal of Wildlife Management* **39** (3) 621–624.

Nasimovich, A.A. (1938) On mineral nutrition of wild animals of the Caucasian Preserve. *Proceedings of the Caucasian State Preserve* **1** 103–150. (In Russian.)

Nasimovich, A.A. (1939) Winter in the life of ungulates of western Caucasus. *Problems of Ecology and Biocenology*. Leningrad. Iss. 7. pp. 3–91. (In Russian.)

Nasimovich, A.A. (1955) *The Snow Cover Role in the Life of Ungulate Animals of the USSR*. Moscow: USSR Academy of Sciences Publishing House. 401 pp. (In Russian.)

Nasimovich, A.A. and Shubnikova, O.N. (1976) Game animal management. *Natural Resources of the Russian Plain in the Past, at Present and in the Future*. Moscow: Nauka Publishers. pp. 202–217. (In Russian.)

Naumov, N.P. (1963) *Ecology of Animals*. Moscow: Vysshaja Shkola Publishers. 618 pp. (In Russian.)

Naumov, N.P. (1967) Population structure and dynamics of terrestrial vertebrates. *Zoologicheskii Zhurnal* **46** (10) 1470–1476. (In Russian.)

Naumov, N.P. (1971) Spatial structures of mammalian species. *Zoologicheskii Zhurnal* **50** (7) 965–980. (In Russian.)

Naumov, N.P. (1972) Ethological structure of terrestrial vertebrate populations. *Behaviour of Animals*. Moscow: Nauka Publishers. pp. 37–39. (In Russian.)

Naumov, N.P. (1977) Biological (signal) fields and their role in the life of mammals. *Advances of Modern Theriology*. Moscow: Nauka Publishers. pp. 93–108. (In Russian.)

Necas, J. (1960) Prispevek k paznani pocetnosti, skladby a organizace tlup srnci zvere. *Prace vyzk, Ust.lesn.* **20** 111–142.

Necas, J. (1963) *Srnci zver*. Praha: SZN. 283 pp.

Neubert, E. (1970) Beschlagenes Rehkitz. *Unsere Jagd.* **20** (7) 212.

Nikiforov, M.N. (1969) Wild ungulate stocks in the Tuva ASSR and their exploitation. *Capability and Productivity of Hunting Grounds of the USSR*. Kirov. Pt. 2. pp. 28–31. (In Russian.)

Nikolandic, D. (1968) Ekoloske karakteristike populacije srna na Belju. *Jelen.* **7** 73–95.

Nikolayev, V. (1890) The 1889 Review. *Priroda i okhota*. January. pp. 105–110. (In Russian.)

Nikolayev, V.V. (1980) On the Siberian roe deer (*Capreolus capreolus* L.) of the middle of the Vitim Plateau. *Fauna and Ecology of Siberian Vertebrates*. Novosibirsk. pp. 158–160. (In Russian.)

Nikolayev, V.V. (1982) Distribution and numbers of the roe deer in the forest steppe of Western Siberia. *Distribution and Numbers of Siberian Vertebrates*. Novosibirsk: Nauka Publishers (Siberian Branch). pp. 254–258. (In Russian.)

Nikolski, A.A. (1976) Vocal danger signals of marmots (Marmota) as a specific character. *Zoologicheskii Zhurnal* **65** (8) 1214–1224. (In Russian.)

Nikolski, A.M. (1883) A 1882 summer travel to the Altai Mountains. *Proceedings of Saint-Petersbourg Naturalists' Society* **14** (1) 150–218. (In Russian.)

Nikolski, S. and Pozov, S. (1973) Tick-borne toxicosis of roe deer. *Okhota i Okhotnich'e Khozyaistvo* **10** 19. (In Russian.)

Novakova, E. and Hanzl, R. (1974) Prispevek k urceni potencialu krajiny pro chov nekterych druhu zvere. *Quaest. geobiol.* **13** 5–82.

Novikov, G.A. and Timofeyeva, Ye.K. (1964) Feeding and forest role of the roe deer in forest–steppe oak groves. *MNS Bull., Biol. Dept.* **69** (2) 39–53. (In Russian.)

Novikov, G.A. and Timofeyeva, Ye.K. (1965) On the ecology of the roe deer of forest–steppe oak groves. *Zoologicheskii Zhurnal* **44** (3) 442–451. (In Russian.)

Nudds, T.D. (1980) Forage 'preference': theoretical consideration of diet selection by deer. *Journal of Wildlife Management* **44** (3) 735–739.

Obrtel, R., Holisova, V. and Kozena, I. (1985) The winter diet of sika deer (*Cervus nippon*) in the Bouzovsko area. *Folia Zoologica* **34** (1) 1–22.

Orlov, V.N., Chudinovskaya, G.A. and Kryukova, Ye.P. (1976) *Studies of Mammalian Chromosome Sets*. Moscow: Nauka Publishers. 36 pp. (In Russian.)

Oszlanyi, J. (1982) Wood, bark, needle, leaf and root energy values of *Pinus silvestris* L., *Picea excelsa* Link. and *Fagus silvatica* L. *Ekologia (CSSR)* **1** (3) 289–296.

Otskheli, T.A. (1964) More on roe deer breeding. *Reports of the Georgian Academy of Sciences* **34** (3) 677–682. (In Russian.)

Paaver, K.L. (1965) *Formation of mammalian fauna and mammal variability in the Baltic Region in the Recent Period.* Tartu. 494 pp. (In Russian.)

Padaiga, V.I. (1965) The role of the roe deer in forestry and protection of new timber stands in the Lithuanian SSR. Author's thesis of the dissert. for the degree of Cand. Sc. (Biology). 25 pp. (In Russian.)

Padaiga, V.I. (1971) Factors determining winter mortality of the roe deer (*Capreolus capreolus*). *Zoologicheskii Zhurnal* **50** (10) 1546–1552. (In Russian.)

Padaiga, V.I. (1975a) Main environmental factors determining population density of wild ungulate animals in hunting forests. *Wildlife Management in Intensive Forestry*. Kaunas: Girionis Publishers. pp. 37–39.

Padaiga, V.I. (1975b) Okologie und Prinzipien der Wirtschaftlichen Nutzung des Rehwildes (*Capreolus capreolus*) in den Waldern des sudlichen Teils der Baltischen Sowjetrepubliken. Beitr. *Jagd-und Wildforschg*. **9** 276–287.

Padaiga, V.I. (1984) Ecological principles of deer number control in the Lithuanian SSR. Author's abstract of the dissert. for the degree of D.Sc.(Biology). Tartu. 38 pp. (In Russian.)

Padaiga, V.I. and Marma, B.B. (1970) The relationship of the degree of roe deer (*Capreolus capreolus* L.) infestation with some parasites and the population density and living conditions. *Proceedings of the IXth International Congress of Game Biologists*. Moscow. pp. 667-672. (In Russian.)

Pallas, P.S. (1771) *Reise durch verschiedene Provinzen des Russischen Reiches*. St. Petersburg, Pt. 1. 504 pp.

Pallas, P.S. (1809) *A travel to various provinces of the Russian Empire*. Pt. 1. Saint-Petersburg. 657 pp. (In Russian.)

Panichev, A.M. (1987) Natural mineral ion exchangers as regulators of ion balance in lithofagous animals. *Reports of the USSR Academy of Sciences* **292** (4) 1016–1019.

Panov, Ye.N. (1969) Demonstrative behaviour of animals. *Priroda* **1** 51–59. (In Russian.)

Panov, Ye.N. (1983) *Behaviour of Animals and the Ethological Structure of Populations*. Moscow: Nauka Publishers. 423 pp. (In Russian.)

Papageorgiou, N., Neophytou, C., Spais, A. and Vavalekas, C. (1980) Food preferences and protein and energy requirements for maintenance of roe deer. *Journal of Wildlife Management* **45** (3) 728–732.

Pasternak, F.A. (1952) A monograph on the roe deer. Author's abstract of the dissert. for the degree of Cand. Sc. (Biology). Moscow. 9 pp. (In Russian.)

Pasternak, F.A. (1955) Materials on the systematics and biology of the roe deer. *Learned Transactions of the Potemkin Moscow State Pedagogical Institute* **38** (3) 29–140. (In Russian.)

Pavlov, I.P. (1938) *Twenty Years of Objective Studies of Higher Nervous Activity (Behaviour) of Animals*. Moscow–Leningrad: Biomedgiz Publishers. 771 pp. (In Russian.)

Pavlov, M.P., Korsakova, I.B. and Lavrov, N.P. (1974) *Acclimatization of Game Animals and Birds in the USSR*. Kirov. 458 pp. (In Russian.)

Pedroli, J.C., Blant, M., Chappuis, F. and Maire, J.A. (1981) La biologie du chevreuil *Capreolus capreolus* (L.) dans le canton de Neuchatel (Suisse). I. Recensement de la population: comparaison de deux methodes. *Mammalia* **45** (4) 467–472.

Pelosse, J. (1976) Etude sur le comportement d'attention de chevreuils (*Capreolus capreolus*). *Zool. Gart.* **46** (6) 441–451.

Pereira, M.S.R. (1985) Effect of human and wolf (*Canis lupus*) presence on a roe deer (*Capreolus capreolus*) population in Northeastern Portugal (Sierra da Nogueira). *XVII Congress of the International Union of Game Biologists*. Brussels. Pt. 2. pp. 671–678.

Perzanowski, K. (1978) The effect of winter food composition on roe deer energy budget. *Acta Theriologica* **23** (31) 451–468.

Petrak, Von M. (1993) Nischenbreite und Nischenuberlappung bei der Nahrungswahl von Rothirsch (*Cervus elaphus* L.) und Reh (*Capreolus capreolus* L.) in der Nordwesteifel. *Zeitschrift fur Jagdwissenschaft* **39** 161–170.

Petrov, P., Dragoyev, P. and Kolev, I. (1968) *The Roe Deer in Bulgaria*. Sofia: Zemizdat Publishers. 236 pp. (In Bulgarian.)

Petrusewicz, K. (1966) Dynamics, organization and ecological structure of population. *Ecol. Pol.* Ser. A. **14** (25) 412–436.

Picard, J.F., Caburet, A. and Oleffe, P. (1985) Etude du regime alimentaire automnal et hivernal du Cerf (*Cervus elaphus* L.) et du Chevreuil (*Capreolus capreolus* L.) par l'analyse des contenus stomacaus. *XVII Congress of the International Union of Game Biologists*. Brussels. Pt. 1. pp. 439–446.

Pidoplichko, I.G. (1951a) *On the Glacial Period*. Kiev: Ukrainian Academy of Sciences Publishing House. Iss. 2. 264 pp. (In Russian.)

Pidoplichko, I.G. (1951b) A new barking deer (*Muntiacus*) form from Pleistocenic deposits of the south of the Ukrainian SSR. *Reports of the USSR Academy of Sciences* **34** (6) 647–679. (In Russian.)

Pielowski, Z. (1970) *Sarna*. Warsaw: PWRIL. 220 pp.

Pielowski, Z. (1984) Some aspects of population structure and longevity of field roe deer. *Acta Theriologica* **29** (2) 17–33.

Pielowski, Z. and Bresinski, W. (1982) Population characteristics of roe deer inhabiting a small forest. *Acta Theriologica* **27** (28) 409–425.

Pikula, J., Koubek, P., Kratovchil, Z. and Kux, Z. (1985a) Age composition of roe deer population in Czechoslovakia. *Acta Sc. Nat. Brno.* **19** (5) 1–46.

Pikula, J., Koubek, P., Kratochvil, Z. and Kux, Z. (1985b) Reproduction of roe deer population in Czechoslovakia. *Acta Sc. Nat. Brno.* **19** (6) 1–47.

Pleshak, T.V. (1980) The roe deer in the Trans-Ili Alatau. *Ungulates of the USSR*. Moscow: Nauka Publishers. pp. 195–196. (In Russian.)

Podarevski, V.B. (1936) *Problems of Game Animals Acclimatization in Eastern Siberia*. Irkutsk: The East Siberian Territorial Publishing House. 118 pp. (In Russian.)

Pole, V.B. (1973) Breeding of the roe deer in Kazakhstan. *Proceedings of the Kazakhstan Academy of Sciences' Institute of Zoology* **34** 135–144. (In Russian.)

Pole, V.B. (1974) The systematical status of the roe deer (*Capreolus capreolus*) in Kazakhstan. *Zoologicheskii Zhurnal* **53** (12) 1833–1839. (In Russian.)

Popov, M.V., Solomonov, N.G., Mordosov, I.I. and Labutin, Yu.V. (1980) *Biology of Game Animals of Yakutia*. Novosibirsk: Nauka Publishers (Siberian Branch). 158 pp. (In Russian.)

Popovici, N., Crisan, A. and Gotea, J. (1978) Modificari sezoniere ale structurii ovarului la caprioara (*Capreolus capreolus* L.). *Stud. Univ. Babes-Bolyai Biol.* **23** (2) 22–32.

Prieditis, A.A. (1970) The effect of helminth infestation on winter mortality of the roe deer (*Capreolus capreolus*). *Proceedings of the IXth International Congress of Game Biologists*. Moscow. pp. 709–713. (In Russian.)

Prieditis, A.A. (1975) Factors determining the winter mortality of roe deer. *Wildlife Management in Intensive Integrated Forestry*. Kaunas: Girionis Publishers. pp. 62–64. (In Russian.)

Prieditis, A. (1984) Influence of dry food and needles on body weight and consumption of food substances in roe deer, *Capreolus capreolus* L. *Acta Zoologica Fennica* **171** 213–216.

Prieditis, A.A. (1985) An evaluation of the state of the Latvian roe deer population. *Control of Wild Ungulate Populations*. Moscow. pp. 79–94. (In Russian.)

Prior, R. (1968) *The Roe Deer of Cranborne Chase. An Ecological Survey*. London: Oxford University Press. 119 pp.

Prior, R. (1987) *Roe Stalking*. Dorset: Bourne Press Limited. 130 pp.

Prostakov, N.I. and Obtemperanski, S.I. (1986) Living conditions and spatial distribution of roe deer in island forests in the north of the Chermozem Centre.

Study and Conservation of Nature of Small Reserves. Voronezh University. pp. 160–164. (In Russian.)

Prusaite, J.A. and Baleishis, R.M. (1975) Fertility of the roe deer in Northern Lithuania. *Wildlife Management in Intensive Integrated Forestry.* Kaunas: Girionis Publishers. pp. 129–130. (In Russian.)

Prusaite, J.A., Blajis, A.S., Mitskus, A.V. and Bluzma, P.P. (1973) Dynamics and structure of a non-exploited roe deer population. *Proceedings of the Lithuanian Academy of Sciences* Ser. B(V). **2** (62) 115–125. (In Russian.)

Prusaite, J.A., Blajis, A.S. and Baleishis, R.M. (1974. Breeding intensity and growth of the Lithuanian *Capreolus capreolus* L. (*Cervidae*) population. *I International Congress on Mammals.* Moscow: The All-Union Institute for Information on Science and Technology. Vol. 2. pp. 144. (In Russian.)

Prusaite, J.A., Blajis, A.S. and Baleishis, R.M. (1977) Breeding intensity and fertility of the European roe deer in Northern Lithuania. *Proceedings of the Lithuanian Academy of Sciences* Ser. B(V). **3** (79) 105–110. (In Russian.)

Prusaite, J.A., Baleishis, R.M. and Bluzma, P.P. (1983) The relationship between the set of foods of the European roe deer and the degree of woodedness of its habitat. *Proceedings of the Lithuanian Academy of Sciences* Ser. B(V). **4** (84) 84–98.

Pryadko, E.I. (1972) On the systematic status of the genus *Capreolus* in the light of helminthological data. *Zoologicheskii Zhurnal* **51** (7) 1092–1093. (In Russian.)

Przewalski, N.M. (1870) An 1868–1869 travel in the Ussuri Territory. Saint-Petersburg. 297 pp. (In Russian.)

Pujauskas, R. (1970) The relationship between Lithuanian roe deer (*Capreolus capreolus*) infestation with helminths and ecological factors. *Proceedings of the IXth International Congress of Game Biologists.* Moscow. pp. 685-687. (In Russian.)

Pulliainen, E. (1980) Occurrence and spread of the roe deer (*Capreolus capreolus* L.) in eastern Fennoscandia since 1970. *Mem. Soc. Fauna Flora Fenn.* **56** (1) 28–32.

Putman, R.J. (1986) Foraging by roe deer in agricultural areas and impact on arable crops. *Journal of Applied Ecology* **23** 91–99.

Putman, R. (1988) *The Natural History of Deer.* Christopher Helm. 191 pp.

Putman, R.J., Langbein, J., Hewison, A.J.M. and Sharma, S. (in press) Population dynamics of three British species of deer. *Mammal Review.*

Raesfeld, F. (1956) *Das Rehwild.* Hamburg–Berlin: Paul Parey. 328 pp.

Rakov, N.V. (1965) Distribution and numbers of the roe deer in the Amur Area. *Game Animals.* Moscow: Rosselkhozizdat Publishers. pp. 93–107. (In Russian.)

Rakov, N.V. (1979) On the role of predators in limiting ungulate numbers in the Amur Area and Primorski Territory. *Ecological Principles of Predatory Mammals Exploitation.* Moscow: Nauka Publishers. pp. 58–59. (In Russian.)

Randweer, T.E. (1985) The roe deer in Estonia. *Control of Wild Ungulate Populations.* Moscow. pp. 94–111. (In Russian.)

Randweer, T.E. (1989) Ecological peculiarities and exploitation of the Estonian roe deer (*Capreolus capreolus* L.) population. Author's thesis of the dissert. for the degree of Cand. Sc. (Biology). Moscow. 16 pp. (In Russian.)

Rasewig, V.A. (1909) The Fergan roe deer, *Capreolus pygargus,* var.? *Semia okhotnikov* **8** 160. (In Russian.)

Ratcliffe, P.R. and Rowe, J.J. (1985) A biological basis for managing red and roe deer in British commercial forests. *XVIIth Congress of the International Union of Game Biologists.* Brussels. pp. 917–925.

Ravski, E.I., Aleksandrova, L.I., Wangenheim, E.A. *et al.* (1964) Anthropogenic deposits in the south of Eastern Siberia. *Proceedings of the Geological Institute.* Iss. 105 280 pp. (In Russian.)

Reichholf, J. (1980) Jahreszeit- und Biotopabhangigkeit der Rudelbildung beim Rehwild (*Capreolus capreolus* L.). *Spixiana* Suppl. 3, No. 2. pp. 193–208.

Reimoser, F. (1982) Rehwildbejagung in einem deckungsreichen Gebirgsrevier bei waldbaulicher Betriebsumstellung. *Centralblatt. Gesamte Forstw.* **99** (3) 167–170.

Reimoser, F. and Zandl, J. (1986) Zwichenergebnisse der n-o. Rehwildmarkierung. *Wiedwerk.* **5** 6–10.

Reimoser, F. and Zandl, J. (1987) Rehwildmarkierung. *Wiedwerk.* **5** 4–5.

Rieck, W. (1955) Die Setzzeit dei Reh-, Rot- und Damwild in Mitteleuropa. *Zeitschrift fur Jagdwissenschaft* **2** (1) 69–75.

Rikovski, A.S. (1970) On the place and role of helminths in the dynamics of biocenoses. *Proceedings of the IXth International Congress of Game Biologists.* Moscow. pp. 592–597. (In Russian.)

Robin, K. (1975) Raumliche Verschiebungen von markierten Rehen (*Capreolus capreolus* L.) in einem voralpinen Gebiet der Ostschweiz. *Zeitschrift fur Jagdwissenschaft* **21** (3) 145–163.

Romanov, Yu.M. and Romashin, A.V. (1982) Materials on the ecology of the European roe deer of the Kaliningrad Region. *Vestn. Zool.* **3** 48–52. (In Russian.)

Roucher, F., Turchheim, B. and Turchheim, E. (1985) Roe deer and trees joint management on an overcrowded woodland estate in Northern Vosges, France. *XVIIth Congress of the International Union of Game Biologists.* Brussels. Pt. 1. pp. 415–420.

Rubis, L.V. (1982) Feeding of the roe deer (*Capreolus capreolus* L.) in autumn and winter. *Reports of the Byelorussian Academy of Sciences. Biol. Ser.* Iss. 3. pp. 109–111. (In Russian.)

Rukhlyadev, D.P. (1941) Parasitic fauna, diseases and causes of mortality of wild mammals of the Crimean Preserve. *Transactions on Scientific Methods.* Iss. VIII. pp. 78–79. (In Russian.)

Rukhlyadev, D.P. (1948) Parasites and parasitoses of wild ungulates and predators of mountain forests of the Crimea. *Parasitic Fauna and Diseases of Wild Animals.* pp. 3–102.

Rusakov, O.S. (1978) Range dynamics and roe deer numbers in the north-west of the USSR. *Fauna and Ecology of Birds and Mammals of Taiga in the North-West of the USSR.* Petrozavodsk. pp. 175–181. (In Russian.)

Sabaneev, L.P. (1872) *Mammal Hunting in the Ural Mountains.* Beseda. Book 6. (In Russian.)

Sabaneev, L.P. (1875) *Roe Deer and its Take in the Ural Mountains.* Priroda. Book 4. pp. 1–21. (In Russian.)

Sablina, T.B. (1955) Ungulates of the Bialowieza Primeval Forest. *Proceedings of the Severtsov Institute of Animal Morphology* **15**. 191 pp. (In Russian.)

Sablina, T.B. (1959) Adaptive peculiarities of feeding of some ungulate species and the influence of these species on plant succession. *Reports of the Forest Institute.* Iss. 13. pp. 32–43. (In Russian.)

Sagesser, H. (1966a) Uber den Einfluss des Standortes auf das Gewicht des Rehwildes (*Capreolus c. capreolus* Linn. 1758). *Zeitschrift fur Jagdwissenschaft* **12** (2) 54–62.

Sagesser, H. (1966b) Uber den Einfluss der Hohe auf einige biologische Erscheinungen beim Reh (*Capreolus c. capreolus*) und bei der Gemse (*Rupicapra r. rupicapra*). *Revue Suisse de Zoologie* **73** (2) 422–433.

Sagesser, H. (1968) Analyse der Setzzeiten (1965–1967) beim Reh im Schweizerischen Mittelland. *Beitr. Jagd- und Wildforsch.* **6** 35–46.

Sagesser, H. and Kurt, F. (1966) Uber die Setzzeit 1965 beim Reh (*Capreolus c. capreolus* L.). *Mitt. Nat. Ges. Bern.* **23** 21–38.

Sandell, M. (1990) The evolution of seasonal delayed implantation. *Quarterly Review of Biology* **65** 23–42.

Sapozhenkov, Yu.F. (1975) Distribution of ungulates in the Kostroma Region. *Natural and agricultural zones of the USSR: Materials of the 7th All-Union Scientific Conference on Natural and Economic Geographical Zoning of the USSR for Agricultural Purposes*. Moscow. p. 115. (In Russian.)

Satunin, K.A. (1906) Ein neues Reh vom Tjan-Shan. *Zoologischer Anzeiger*, **30** (17–18) 527–528.

Savinov, V. (1961) *Hunting Grounds of the Altai Territory*. Barnaul: The Altai Publishing House. 25 pp. (In Russian.)

Savinov, V. (1967) *Mammals and Birds of the Altai*. Barnaul: The Altai Publishing House. 187 pp. (In Russian.)

Savinov, Ye.F. (1962) Breeding and growth of the Siberian ibex (*Capra sibirica*) in the Djungar Alatau. *Proceedings of the Kazakhstan Academy of Sciences' Institute of Zoology* **17** 167–182. (In Russian.)

Schafer, E. (1974) *Hegen und Ansprechen von Rehwild*. Munich: BLV. 189 pp.

Schaffer, J. (1940) *Die Hautdrusenorgane der Saugetiere mit besonderer Berucksichtigung ihres histologischen Aufbaues und Bemerkingen uber die Proktodaldrusen*. Berlin–Vienna: Urban und Schwarzenberg. 464 pp.

Schams, D. and Barth, D. (1982) Annual profiles of reproductive hormones in peripheral plasma of the male roe deer (*Capreolus capreolus*). *Journal of Reproduction and Fertility* **66** (2) 463–468.

Scharff, R.F. (1918) *Animals of Europe*. Moscow: Priroda Publishers. 233 pp. (In Russian.)

Schlosser, M. (1924) Tertiary vertebrates from Mongolia. *Palaeontol. Sinica*, Ser. C. Peking. **1** (1). 119 pp.

Schroder, J. and Schroder, W. (1982) Niche breadth and overlap in red deer (*Cervus elaphus* L.), roe deer (*Capreolus capreolus*) and chamois (*Rupicapra rupicapra*). *III International Theriological Congress Abstracts of Papers*. Helsinki. pp. 222.

Schumacher, S. (1939) *Jagd- und Biologie*. Berlin. 136 pp.

Schwarz, S.S. (1967) Population structure of the species. *Zoologicheskii Zhurnal* **46** (10) 1456–1469. (In Russian.)

Schwarz, S.S. (1980) *Ecological Laws of Evolution*. Moscow: Nauka Publishers. 278 pp. (In Russian.)

Sedlaczek, E. (1964) *Zehn Jahre Here im Rehwildrevier*. Pirsch. No. 3. pp. 148–155.

Sempere, A. (1979) Utilisation et evolution du domaine vital chez le chevreuil male european determinées par radiotracking. *Biological Behaviour* **4** (1) 75–87.

Sempere, A. (1982) *Fonction de reproduction et caracteres sexuels secondaires chez le chevreuil (*Capreolus capreolus *L.): variations saisonnieres et incidences sur l'utilisation du budget temps–espace*. Univ. Frangois-Rabelais (Tours). 306 pp.

Serzhanin, I.N. (1955) *Mammals of the Byelorussian SSR*. Minsk: Byelorussian Academy of Sciences' Publishing House. 311 pp. (In Russian.)

Severtsov, S.A. (1941) *Dynamics of Populations and Adaptive Evolution of Animals*. Moscow-Leningrad. 316 pp. (In Russian.)

Severtsov, S.A. (1951a) *On Congruences and the Concept of Species Integrity*. Moscow: USSR Academy of Sciences' Publishing House. Vol. 1. pp. 30–57. (In Russian.)

Severtsov, S.A. (1951b) Evolution of antlers of some artiodactyls as a weapon used in fights for doe. *Problems of Animal Ecology*. Moscow: USSR Academy of Sciences' Publishing House. Vol. 1. pp. 58–59. (In Russian.)

Shakhmardanov, Z.A. and Ramazanov, K.M. (1980) Areas of Dagestanian ungulates. *Ungulates of the USSR*. Moscow: Nauka Publishers. pp. 113. (In Russian.)

Shargayev, M.A. (1966) On the distribution of the Siberian roe deer in the north of the Ob-Irtish Area. *IV Zoogeographic Conference of Higher Educational Institutions*. Odessa. pp. 310–311. (In Russian.)

Shargayev, M.A. (1972) Ungulates of the Vitim Plateau: population and distribution dynamics versus the anthropogenic factor. *Zoological Problems of Siberia*. Novosibirsk: Nauka Publishers. pp. 496–497. (In Russian.)

Shchetinin, V.I. (1975) A local roe deer grouping of the Zeya Preserve. *Ungulates of the USSR*. Moscow: Nauka Publishers. pp. 147. (In Russian.)

Shen, S., Ables, E.D. and Qian-zhu, X. (1982) The Chinese view of wildlife. *Oryx* **16** (4) 340–347.

Shilov, I.A. (1972) The relationship of the spatial and ethological structures in vertebrate populations. *Behaviour of Animals*. Moscow: Nauka Publishers. pp. 12–14. (In Russian.)

Shilov, I.A. (1977) *Ecological and Physiological Basis of Population Relations of Animals*. Moscow: MSU Press. 261 pp. (In Russian.)

Short, R.V. and Hay, M.F. (1966) Delayed implantation in the roe deer *Capreolus capreolus*. *Symposium of the Zoological Society of London* **15** 173–194.

Short, R.V. and Mann, T. (1965) Androgenic activity in a seasonally breeding animal, the roe buck (*Capreolus capreolus*). *Journal of Endocrinology* **31** 1–2.

Short, R.V. and Mann, T. (1966) The sexual cycle of a seasonally breeding animal, the roebuck (*Capreolus capreolus*). *Journal of Reproduction and Fertility* **12** (2) 337–351.

Shostak, S.V. (1978) Numerical ratios of the European red deer to other ungulates. *Preserves of Byelorussia*. Iss. 2. pp. 130–138. (In Russian.)

Showen, R. (1972) *Behaviour of Animals*. Moscow: Mir Publishers. 487 pp. (In Russian.)

Shukhov, N. (1929) Introduction of roe deer in the forest–steppe zone of Western Siberia. *Okhotnik i ribak Sibiri* **1** 20–21. (In Russian.)

Shurkhal, A.V., Danilkin, A.A., Markov, G.G. *et al.* (1985) Genetic variability of the roe deer, as determined by a set of biochemical gene markers. *Phenetics of populations (Materials of the 3rd All-Union Conference, Saratov)*. Moscow. pp. 273–274. (In Russian.)

Shurupov, I.I. (1982) The Zoological Association role in introduction of game animals. *Fauna Enrichment and Game Animal Breeding*. Kirov. pp. 116. (In Russian.)

Shvets, V.G. (1975) Decrease of roe deer numbers in the Khabarovsk-Region part of the Amur Area. *Ungulates of the USSR*. Moscow: Nauka Publishers. pp. 142–143. (In Russian.)

Shvets, V.G. (1980) The effect of man's economic activity on population numbers in the Amur Area. *Ungulates of the USSR*. Moscow: Nauka Publishers. pp. 54–57. (In Russian.)

Siefke, A. (1966) Die jagdwirtschaftliche Bedeutung parasiticher Wurmer beim Reh- und Damwild. *Beitr. Jagd- und Wildforsch.* **4** 135–145.

Silantiev, A.A. (1898) *A Survey of Russian Hunting*. Saint-Petersbourg. 619 pp. (In Russian.)

Simpson, G. (1945) The principles of classification and a classification of mammals. *Bulletin of the American Museum of Natural History*. New York. Vol. 85. 350 pp.

Siuda, A., Zurowski, W. and Siuda, H. (1969) The food of the roe deer. *Acta Theriologica* **14** (18) 247–262.

Skalon, V.N. (1946) On the expansion of the roe deer's area in Siberia. *Priroda*. **6** 73. (In Russian.)

Skalon, V.N. (1960) Distributions and a survey of life of wild ungulates in the Olyokma Basin. *Collected Biol. papers*. Irkutsk. pp. 150–158. (In Russian.)

Skriba, G.B. (1975) Relations of European deer with moose and roe deer. *Wildlife Management in Intensive Integrated Forestry*. Kaunas: Girionis Publishers. pp. 101–102. (In Russian.)

Sludski, A.A. (1953) Emigration of 'taiga' mammals to forest steppes and steppes of Western Siberia and Kazakhstan. *MNS Bull., Biol. Dept*. **58** (2) 14–32. (In Russian.)

Sludski, A.A. (1956) Breeding of Persian gazelle (*Gazella subgutturosa*). *Proceedings of the Kazakhstan Academy of Sciences' Institute of Zoology* **6** 78–108. (In Russian.)

Sludski, A.A. and Baidavletov, R.Z. (1984) *Mammals of Khazakhstan*. Alma-Ata: Nauka Kazakhstan SSR Publishers. Vol. 3, Pt. 4. 231 pp. (In Russian.)

Smirnov, M.N. (1970) More on territorial differentiation of roe deer population versus living conditions in the west of the Trans-Baikal Area and in the Eastern Sayan. *Transactions of the Irkutsk Agricultural Institute* **26** (3) 91–104. (In Russian.)

Smirnov, M.N. (1974) The roe deer in the west of the Trans-Baikal Area. Author's abstract of the dissert. for the degree of Cand. Sc. (Biology). Irkutsk. 29 pp. (In Russian.)

Smirnov, M.N. (1975a) Roe of Buryatia: distribution in the past and at present. *Zoological Studies in the Trans-Baikal Area: Proceedings of the Buryatian Institute of Natural Sciences*. Ulan-Ude. Iss. 13. pp. 218–230. (In Russian.)

Smirnov, M.N. (1975b) On vocal activities of roe deer inhabiting the west of the Trans-Baikal Area. *Proceedings of the Buryatian Institute of Natural Sciences of the Buryat Branch of the USSR Academy of Sciences' Siberian Section*. Iss. 13, Zool. Ser., Iss. 1. pp. 218–230. (In Russian.)

Smirnov, M.N. (1977) An outline of behaviour of roe deer (*Capreolus capreolus pygargus* Pall.) inhabiting the west of the Trans-Baikal Area. *Behaviour of Mammals*. Moscow: Nauka Publishers. pp. 221–243. (In Russian.)

Smirnov, M.N. (1978) *The Roe Deer in the West of the Trans-Baikal Area*. Novosibirsk: Nauka Publishers. 189 pp. (In Russian.)

Smirnov, M.N. (1980) Wild ungulates of the Tuva ASSR. *Ungulates of the USSR*. Moscow: Nauka Publishers. pp. 104–105. (In Russian.)

Smirnov, M.N. (1985) Distribution, resources and prospects of exploitation of wild ungulates of Tuva. *Game Resources and their Exploitation*. Moscow. pp. 141–152. (In Russian.)

Smith, F., Beeson, K. and Price, W. (1956) Chemical composition of herbage browsed by deer in two wildlife mamagement areas. *Journal of Wildlife Management* **20** (4) 359–367.

Smoktunovich, Ye.A. (1980) Some factors determining roe population dynamics in the Bialowieza Primeval Forest. *Preserves of the Byelorussia*. Iss. 4. pp. 132–139. (In Russian.)

Smoktunovich, Ye.A. (1983) The deer and roe in co-inhabited areas. *Okhota i Okhotnich'e Khozyaistvo* **10** 19. (In Russian.)

Sobanski, G.G. (1986) Ungulate area expansion and population growth in the Altai. *MNS Bull., Biol. Dept*. **91** (1) 25–27. (In Russian.)

Sobanski, G.G. (1987) Migrations of the roe deer (*Capreolus capreolus pygargus* Pall.) in the Altai. *Fauna, Taxonomy, Ecology of Mammals and Birds (Fauna of Siberia)*. Novosibirsk: Nauka Publishers (Siberian Branch). pp. 121–128. (In Russian.)

Sokolov, E.N. (1966) Neuronal mechanism of the orientative reflex. *Symposium of the XVIIIth International Congress of Psychology*. Moscow: Nauka Publishers. pp. 31–33. (In Russian.)

Sokolov, I.I. (1959) Ungulates (the orders Perissodactyla and Artiodactyla). *Fauna of the USSR. Mammals*. Moscow–Leningrad: Nauka Publishers. Vol. 1, Iss. 3. 640 pp. (In Russian.)

Sokolov, V.E. (1979) *Systematics of Mammals*. Moscow: Vysshaja Shkola Publishers. 528 pp. (In Russian.)

Sokolov, V.E. and Danilkin, A.A. (1977) Territory marking by Siberian roebucks (*Capreolus capreolus pygargus*) in the Southern Urals. *Zoologicheskii Zhurnal* **56** (10) 1548–1556. (In Russian.)

Sokolov, V.E. and Danilkin, A.A. (1980) Einige Aspekte der Systematik, Okologie und des Verhaltens des Sibirischen Rehs (*Capreolus capreolus pygargus*, Pallas, 1771). *Beitrage Jagd-und Wildforsch.* **11** 276–288.

Sokolov, V.E. and Danilkin, A.A. (1981) *The Siberian Roe Deer*. Moscow: Nauka Publishers. 144 pp. (In Russian.)

Sokolov, V.E. and Gromov, V.S. (1985a) Morphometric studies of skulls belonging to various subspecies of the roe deer (*Capreolus capreolus*). *Zoologicheskii Zhurnal* **64** (5) 738–749. (In Russian.)

Sokolov, V.E. and Gromov, V.S. (1985b) Experimental hybridization of European and Asiatic roe deer. *Reports of the USSR Academy of Sciences* **285** (4) 1022–1024. (In Russian.)

Sokolov, V.E., Chudinovskaya, G.A., Danilkin, A.A. and Orlov, V.N. (1975) The chromosomal set of the roe deer *Capreolus capreolus* L. *Systematics and cytogenetics of mammals: Proceedings of the All-Union Symposium*. Moscow: Nauka Publishers. pp. 45–46. (In Russian.)

Sokolov, V.E., Orlov, V.N., Chudinovskaya, G.A. and Danilkin, A.A. (1978) Chromosomal differences between two roe subspecies (*Capreolus capreolus* L. and *Capreolus capreolus pygargus* Pall.). *Zoologicheskii Zhurnal* **57** (7) 1109–1112. (In Russian.)

Sokolov, V.E., Orlov, V.N., Chudinovskaya, G.A. and Danilkin, A.A. (1980) On the systematical status of the European and Siberian roe deer. *Ungulates of the USSR*. Moscow: Nauka Publishers. pp. 105–106. (In Russian.)

Sokolov, V.E., Danilkin, A.A. and Dulamtseren, S. (1982a) The present day distribution and numbers of forest ungulates in Mongolia. *Zoological Studies in the Mongolian People's Republic*. Moscow: Nauka Publishers. pp. 37–56. (In Russian.)

Sokolov, V.E., Danilkin, A.A. and Romanenko, Ye.V. (1982b) Radio tracking of European roebucks in the Central Chermozem Preserve. Mammals. *The III Congress of the All-Union Theriological Society: Theses of Reports* **1** 300–301. (In Russian.)

Sokolov, V.E., Markov, G.G., Danilkin, A.A. *et al.* (1985a) On the specific status of the European (*Capreolus capreolus* L.) and Siberian (*C. pygargus* Pall.) roe deer: a craniometric study. *Reports of the USSR Academy of Sciences* **280** (6) 1505–1509. (In Russian.)

Sokolov, V.E., Markov, G.G. and Danilkin, A.A. *et al.* (1985b) A comparative craniometric study of the development of the European (*Capreolus capreolus* L.) and Siberian (*C. pygargus* Pall.) roe deer. *Reports of the USSR Academy of Sciences* **282** (1) 243–248. (In Russian.)

Sokolov, V.E., Gromov, V.S. and Danilkin, A.A. (1985c) The ontogeny of Siberian roe deer (*Capreolus capreolus pygargus*) behaviour. *Zoologicheskii Zhurnal* **64** (6) 915–926. (In Russian.)

Sokolov, V.E., Markov, G.G. and Danilkin, A.A. (1986a) Morphometries of karyotypes of the European (*Capreolus capreolus* L.) and Siberian (*C. pygargus* Pall.) roe deer. *Reports of the USSR Academy of Sciences. Biol. Ser.* Iss. 5. pp. 780–784. (In Russian.)

Sokolov, V.E., Shurkhal, A.V., Danilkin, A.A. *et al.* (1986b) A comparative analysis of electrophoretic spectra of blood and muscle proteins of the European (*Capreolus capreolus* L.) and Siberian (*C.pygargus* Pall.) roe deer. *Reports of the USSR Academy of Sciences* **288** (5) 1274–1276. (In Russian.)

Sokolov, V.E., Danilkin, A.A. and Markov, G.G. (1986c) A craniometric study of the Siberian subspecies of the Siberian roe deer (*Capreolus pygargus pygargus* Pall.). *Transactions of the USSR Academy of Sciences. Biol. Ser.* Iss. 3. pp. 445–448. (In Russian.)

Sokolov, V.E., Danilkin, A.A. and Minayev, A.N. (1986d) Home ranges of the European roe deer (*Capreolus capreolus* L.) in the east of its area. *Reports of the USSR Academy of Sciences* **291** (5) 1267–1271. (In Russian.)

Sokolov, V.E., Gromov, V.S. and Rutovskaya, M.V. (1987) Vocal communication in the European (*Capreolus capreolus*) and Siberian (*C. pygargus* Pall.) roe deer. *Zoologicheskii Zhurnal* **66** (3) 430–443. (In Russian.)

Sokolov, V.E., Danilkin, A.A., Darman, Yu.A. and Minayev, A.N. (1991) Radiotracking of a migratory population of the Siberian roe deer. *Reports of the USSR Academy of Sciences* **320** (4) 1018–1024. (In Russian.)

Soloviev, D.K. (1922) Ungulate animal and forest game hunting in Russia. *Yearbook of the All-Russian Union of Hunters.* pp. 53–73. (In Russian.)

Southern, H.N. (1964) *The Handbook of British Mammals.* Oxford: Blackwell Scientific. 465 pp.

Sperk, F. (1885) The Russian Far East. *General Geographical Transactions of the Russian Geographical Society.* Saint Petersburg. Vol. 14. p. 503. (In Russian.)

Staines, B.W. and Welch, D. (1984) Habitat selection and impact of red (*Cervus elaphus* L.) and roe (*Capreolus capreolus* L.) deer in a Sitka spruce plantation. *Proceedings of the Royal Society of Edingburgh* **82B** 303–319.

Stangl, W.G. and Margl, H. (1977) Rehwildstandserhebungen in einem Mittelwaldbetrieb in Weinviertel. *Mitt. Forstl. Bundes-Versuch. Vienna* **122**, 97–113.

Stangl, W.G. and Wenter, W. (1977) Versuch einer Darstellung der Wildstandsentwicklung auf Grund von Abschussmeldungen bei Rehwild. *Mitt. Forstl. Bundes-Versuch. Vienna* **122** 63–83.

Stark, D. (1966) Territoriumsmarkierung bei Rehboken (*Capreolus capreolus*). *Saugetierkundl. Mitt.* **14** (4) 325–326.

Steklenev, Ye.P. (1975) Morphogenetics of sperms of cervids and the latter's hybridization. *Proceedings of the II All-Union Conference on Mammals.* Moscow: Moscow University Press. pp. 55–62. (In Russian.)

Steklenev, Ye.P. (1980) Breeding peculiarities of the European roe deer in the south of the Ukraine. *Ungulates of the USSR.* Moscow: Nauka Publishers. pp. 202–203.

Steklenev, Ye.P. (1986) Interspecific hybridization of the red (*Cervus elaphus* L.) and sika (*Cervus nippon hortulorum* Temm.) deer. *Zitologiya i genetika* **20** (2) 138–142. (In Russian.)

Sterba, F. and Zamek, L. (1985) Parazitarni priciny ztrat srnci zvere v letech 1972–1981. *Veterinarstvi.* **35** (3) 131–135.

Stieve, H. (1950) Anatomisch biologische Untersuchungen uber die Fortpflanzungsfahigkeit des europaischen Rehes. *Z. Mikr. Anat. Forschung.* **55** 427–530.

Strandgaard, H. (1972a) The roe deer (*Capreolus capreolus*) population at Kalo and the factors regulating its size. *Danish Review of Game Biology* **7** (1). 205 pp.

Strandgaard, H. (1972b) An investigation of corpora lutea, embryonic development, and time of birth of roe deer (*Capreolus capreolus*) in Denmark. *Danish Review of Game Biology* **6** (7) 1–22.

Strandgaard, H. (1977) Untersuchingen uber danisches Rehwild in seiner Beziehung zur Umwelt. *Beitr. Jagd- und Wildforsch.* **10** 234–243.

Strandgaard, H. (1978) Densite population et croissance des animaux chez le chevreuil: perspectives nouvelles. *Bulletin Mensuel, Office National de la Chasse, No. Sp. Scien. Tech.* pp. 149–159.

Strothman, A. and Oslage, H.-J. (1986) Untersuchungen zur Jahresrhythmik der Nahrstoff- und Energieaufnahme von Rehwild. *Kleintierpraxis* **31** (3) 153–156.

Stubbe, C. (1987) Das Rehwild in unserer Kulturlandschaft. *Unsere Jagd.* **37** (5) 136–137.

Stubbe, C. and Danilkin, A. (1992) Breeding. *European and Siberian Roe Deer.* Moscow: Nauka Publishers, pp. 140–159 (In Russian.)

Stubbe, C. and Passarge, H. (1979) *Rehwild.* Berlin: VEB Deutsch Landwirtschaftsverlag. 432 pp.

Stubbe, C. and Stubbe, W. (1985) Ergebnisse der Rehwildmarkierung im Wildforschungsgebiet Hakel. *Unsere Jagd.* **35** (6) 172–173; (12) 355.

Stubbe, C., Stubbe, M. and Stubbe, I. (1982) Zur Reproduktion der Rehwildpopulation – *Capreolus c. capreolus* (L.,1758) – des Wildforschungsgebietes Hakel. *Hercynia* **19** (1) 97–109.

Stubbe, H. and Bruchholz, S. (1979) Experiments in the hybridization of the European and Siberian roe deer (*Capreolus capreolus capreolus* L., 1758 x *Capreolus capreolus pygargus* Pall., 1771). *Zoologicheskii Zhurnal* **58** (9) 1398-1403. (In Russian.)

Stubbe, H. and Bruchholz, S. (1980) Uber Bastardierungsversuche zwischen Europaischen und Sibirischen Rehen (*Capreolus capreolus* L., 1758 x *Capreolus capreolus pygargus* Pallas, 1771). *Beitr. Jagd- und Wildforsch.* **11** 289-303.

Subbotin, A.M. (1973) On roe deer migrations in central parts of the Krasnoyarsk Territory. *Proceedings of the Kirov Agricultural Institute (1971).* Iss. 28. pp. 72–77. (In Russian.)

Sysa, P.S. and Kaluzinski, J. (1984) Possibility of freemartinism in roe deer. *Acta Theriologica* **29** (1–10) 133–137.

Szaniawski, A. (1973) Proportions of the alimentary tract in deer. *Acta Theriologica* **18** (18) 343–347.

Szederjei, A. and Szederjei, M. (1971) *Geheimnis der Weltrekordes – das Reh.* Budapest: Terra. 403 pp.

Tatarinov, K.A. (1970) Neogenic and Anthropogenic vertebrate fauna of Podolia and Ante-Karpati Area, its history and present day state. Author's abstract of the dissert. for the degree of D. Sc.(Biology). Kiev. 56 pp. (In Russian.)

Tavrovski, V.A., Yegorov, O.V., Krivosheyev, V.G. *et al.* (1971) *Mammals of Yakutia.* Moscow: Nauka Publishers. 650 pp. (In Russian.)

Timchenko, N.G. (1972) *On the History of Hunting and Animal Breeding in Kievan Rus (the Middle Dnieper Area).* Kiev: Naukova Dumka Publishers. 204 pp. (In Russian.)

Timofeyeva, Ye.K. (1965) *The Roe Deer.* Leningrad: LSU Press. 224 pp. (In Russian.)

Timofeyev-Resovski, N.V., Vorontsov, N.N. and Yablokov, A.V. (1977) *An Outline of Evolution Theory.* Moscow: Nauka Publishers. 301 pp. (In Russian.)

Tinbergen, N. (1969) *Behaviour of Animals*. Moscow: Mir Publishers. 93 pp. (In Russian.)

Tinbergen, N. (1970) *Wasps, Birds, Men*. Moscow: Mir Publishers. 334 pp. (In Russian.)

Tinbergen, N. (1974) *The World of the Herring Gull*. Moscow: Mir Publishers. 272 pp. (In Russian.)

Tolstov, S.P. (1948) *A Treatise on the History and Archeology of Ancient Khorezm*. Moscow: MSU Press. 352 pp. (In Russian.)

Troitski, V.N. (1930) Hunting in the Chuna-Angara District of the former Kansk Area. *Transactions of the Siberian Territorial Hunting-Science Station*. Novosibirsk: Siberian Okhotpromstantsiya Press. Iss. 1. pp. 3–35. (In Russian.)

Tsalkin, V.I. (1963) On the history of mammals of the Eastern European steppe. *MNS Bull., Biol. Dept*. **68** (2) 43–62. (In Russian.)

Tsaplyuk, O.E. (1977) Age-related and seasonal peculiarities of the reproductive biology of the roe deer (*Capreolus capreolus* L.) of Kazakhstan. *Zoologicheskii Zhurnal* **56** (4) 611–618. (In Russian.)

Tsarev, Yu.S. (1969) Distributions and numbers of the moose, roe deer and reindeer. *Biological Zones of the Novosibirsk Region*. Novosibirsk. pp. 182–190. (In Russian.)

Turcek, F.J. (1951) Effect of introductions on two game populations in Czechoslovakia. *Wildlife Management* **15** (1) 113–114.

Turcek, F.J. (1962) Untersuchingen uber das Fegen der Rehbocke an Geholzen. *Z. Angew. Zool*. **49** (4) 429–438.

Turkin, N.V. and Satunin, K.A. (1902) *Mammals of Russia*. Moscow: N.V. Turkin Publishers. 506 pp. (In Russian.)

Turner, D.C. (1978) Aktivitatsmuster freilebender Rehe im Verlauf des Fruhjahrs: optimale Ausnutzung der Tageszeit. *Revue Suisse de Zoologie* **85** (4) 710–712.

Turner, D.C. (1979) An analysis of time-budgeting by roe deer (*Capreolus capreolus*) in an agricultural area. *Behavior* **71** (3–4) 246–279.

Turner, D.C. (1980) A multi-variate analysis of roe deer (*Capreolus capreolus*) population activity. *Revue Suisse de Zoologie* **87** (4) 991–1002.

Ueckermann, E. (1952) Rehwild und Standort. *Anblick*. **7** 32.

Ueckermann, E. (1957) *Wildstandsbewirtschaftung und Wildschadensverhutung beim Rehwild*. Neuwied. 111 pp.

Ueckermann, E. (1964) *Der Rehwildabschus*. Hamburg–Berlin. Paul Parey 109 pp.

Ueckermann, E. (1985) Forstwirtschaft und Wild. *XVIIth Congress of the International Union of Game Biologists*. Brussels. Pt. 1. pp. 371–383.

Ullrey, D., Jouatt, W., Johnson,. and Fay, L. (1967) Protein requirement of white-tailed deer fawns. *Journal of Wildlife Management* **31** (4) 679–685.

Ushkov, S.L. (1947) The game fauna of the Ilmen State Preserve. *Proceedings of the Ilmen Preserve*. Iss. 3. Pt. l. pp. 87–159. (In Russian.)

Ushkov, S.L. (1954) Roe deer migrations in the Southern Urals. *MNS Bull., Biol. Dept*. **59** (5) 9–12. (In Russian.)

Uspenski, G.A. and Lozan, M.N. (1965) Roe deer in Moldavia. In: *Moldavian Nature Conservation*. Kishinev. Iss. 3. pp. 150–166. (In Russian.)

Ustinov, S.K. (1967) The musk of *Moschus moschus*. *Okhota i Okhotnich'e Khozyaistvo* **3** 9. (In Russian.)

Ustinov, S.K. and Lobanov, P.N. (1983) More on the ecology and morphology of the roe deer of the Kulinga River basin. *Ecologiya* **2** 81–82. (In Russian.)

Vatolin, B.A. (1975) The Bryansk Region roe deer population: distribution, numbers and the effect of ecological factors. *Ungulates of the USSR*. Moscow: Nauka Publishers. pp. 69–70. (In Russian.)

Vereshchagin, N.K. (1959) *Mammals of the Caucasus*. Moscow–Leningrad: USSR Academy of Sciences Publishing House. 703 pp. (In Russian.)

Vereshchagin, N.K. and Rusakov, O.S. (1979) *Ungulates of the North-West of the USSR*. Leningrad: Nauka Publishers (Leningrad Branch). 308 pp. (In Russian.)

Vershinin, A.A. (1969) Dispersal of game animals in the Moscow Region. *Increasing Productivity of Hunting Grounds*. Moscow. pp. 50–99. (In Russian.)

Vetlitsin, P.I. (1902) On migratory roe deer in the Amursk Region. *Priroda i okhota* **11** 1–8. (In Russian.)

Vincent, J.P. (1982) Schatzung der Rehwildpopulation im Walde. Vorstellung einer einfachen Methode. *Zeitschrift fur Jagdwissenschaft* **28** (1) 58–63.

Vincent, J.P. and Bideau, E. (1987) Influence de la densité sur l'occupation de l'e-space chez le chevreuil forestier. *Bull. SFECA* **2** (1) 165–168.

Vincent, J.P., Janeau, G., Quere, J.P. and Spitz, F. (1979) Note sur la repartition et le rythme d'activite du chevreuil (*Capreolus capreolus*) en foret ouverte. *Ann. Zool. Ecol. Anim.* **11** (2) 145–157.

Vincent, J.P., Bideau, E., Quere, J.P. and Angibault, J.M. (1983) Occupation de l'e-space chez le chevreuil (*Capreolus capreolus* L.). II. Cas des femelles. *Acta Oecologica-Oecologica Applicata* **4** (4) 379–389.

Vincent, J.P., Bideau, E., Hewison, A.J.M. and Angibault, J.M. (1995) The influence of increasing density on body weight, kid production, home range and winter grouping in roe deer. *Journal of Zoology* **236** 371–382.

Vislobokova, I.A. (1973) *On Pavlodar-Region Eopleistocenic mammal sites near the Irtish. Geology and geophysics*. Novosibirsk. No. 5. pp. 123–125. (In Russian.)

Vislobokova, I.A. (1981) More on the early evolution of megacerines. *Paleontol. Zhurnal* **4** 105–117. (In Russian.)

Vladishevski, D.V. (1968) On factors determining numbers of the European roe deer. *Zoologicheskii Zhurnal* **47** (3) 438–443. (In Russian.)

Vladishevski, D.V. and Yelski, G.M. (1974) Some regularities in wintertime feed-ing of the deer and roe. *Population Ecology of Forest Animals of Siberia*. Novosibirsk. pp. 87–102. (In Russian.)

Wagner, H.A. (1961) Waldbau und Wild. *Der Anblick*. **16** 237–240.

Walther, F. (1958) Sum Kampf- und Paarungsverhalten einiger Antilopen. *Zeitscrift fur Tierpsychologie* **15** (3) 340–380.

Walther, F. (1963) Einige Verhaltensbeobachtungen am Dibatag (*Ammodoreas clarkei*, Thomas, 1891). *Zool. Garten, N.F.* **27** 233.

Walther, F. (1968) *Verhalten der Gazellen. (Die Neue Brehm Bucherei, 373)*. Wittenberg-Zutherstadt. 144 pp.

Walther, F.R. (1972) Territorial behaviour in certain horned ungulates with special reference to the examples of Thomson's and Grant's gazelles. *Zool. Afr.* **7** (1) 303–307.

Wandeler, A.I. (1975) Die Fortpflanzungsleistung des Rehs (*Capreolus capreolus* L.) im Berner Mittelland. *Jahrb. Natur. Hist. Bern*, 1972–1974, **5** 245–301.

Wandeler, A. and Huber, W. (1969) Gewichtswachstum und jahreszeitliche Gewichtsschwankungen bei Reh und Gemse. *Revue Suisse de Zoologie* **76** (3) 686–694.

Wangenheim, E.A. (1977) *The Paleontological Basis of the Anthropogene Stratigraphy of Northern Asia*. Moscow: Nauka Publishers. 171 pp. (In Russian.)

Wangenheim, E.A. and Flerov, K.K. (1982) A taxonomical review of main mam-malian orders. The order Artiodactyla. *Stratigraphy of the USSR. The Quaternary Period*. Half-volume 1. Moscow. pp. 326–329. (In Russian.)

Weiner, J. (1975) Model of the energy budget of an adult roe deer. *Polish Ecological Studies* **1** (2) 103–119.

Weiner, J. (1977) Energy metabolism of roe deer. *Acta Theriologica* **22** 3–24.

Welch, D., Staines, B.W., Catt, D.C. and Scott, D. (1990) Habitat usage by red (*Cervus elaphus*) and roe (*Capreolus capreolus*) deer in a Sitka spruce plantation. *Journal of Zoology, London* **221** 453–476.

Whitehead, G.K. (1964) *The Deer of Great Britain and Ireland*. Routledge and Keegan Paul, London.

Wishart, W. (1980) Hybrids of white-tailed and mule deer in Alberta. *Journal of Mammalogy* **61** (4) 716–720.

Wurster, D. and Benirschke, R. (1967) The chromosomes of twenty three species of the Cervoidea and Bovoidea. *Mammal. Chromosomes Newsl* **8** 226–228.

Wynne-Edwards, V.C. (1962) *Animal Dispersion in Relation to Social Behaviour*. Edinburgh: Oliver and Boyd. 653 pp.

Yanushevich, A. and Blagoveshchenski, I. (1952) *Game Mammals and Birds of Western Siberia*. Novosibirsk. 237 pp. (In Russian.)

Yanushevich, A.I., Aizin, B.M. *et al.* (1972) *Mammals of Kirghizia*. Frunze: Ilim Publishers. 464 pp. (In Russian.)

Yegorov, O.V. (1965) *Wild Ungulates of Yakutia*. Moscow: Nauka Publishers. 259 pp. (In Russian.)

Yelpatievski, P.V. and Panichev, A.M. (1980) Geochemical peculiarities of salt licks of the Sikhote-Alin. *MNS Bull., Biol. Dept*. **85** (6) 12–23. (In Russian.)

Yelski, G.M. (1975) The roe deer's selection of feeding grounds in the deep-snow period. *Ungulates of the USSR*. Moscow: Nauka Publishers. pp. 169–170. (In Russian.)

Yelski, G.M. (1976) The behaviour of migrating roe deer groups in the subtaiga zone of the Krasnoyarsk-Territory Angara Area. *Feeding Ecology of Forest Animals*. Novosibirsk. pp. 146–154. (In Russian.)

Yelski, G.M. (1979) Types of winter feeding grounds of the red deer and roe deer in Khamar-Daban. *The Fauna of Forests in the Lake Baikal Basin*. Novosibirsk. pp. 146–154. (In Russian.)

Yelski, G. (1983) The roe deer of the Ktasnoyarsk-Territory Angara Area. *Okhota i Okhotnich'e Khozyaistvo* **3** 22–23. (In Russian.)

Yelski, G.M. and Shishikin, A.S. (1979) Relationships of predatory and herbivorous mammals in winter concentration parts of the Angara Area. *Ecological Principles of Predatory Mammal Management*. Moscow: Nauka Publishers. pp. 265–266. (In Russian.)

Yemelianov, P. and Ribalov, V. (1968) The roe deer in the Stavropol Territory. *Okhota i Okhotnich'e Khozyaistvo* **8** 13. (In Russian.)

Yermolova, N.M. (1963) On the Paleolithic and Neolithic mammalian fauna of the Baikal Area. *Ethnographic Materials*. Leningrad: The All-Union Geographical Society Press. Iss. 3. pp. 27–64. (In Russian.)

Zakirov, A. (1974) Predators and artiodactyles. *Vertebrate Animals of the Fergan Valley*. Tashkent: Uzbekestan SSR Fan Publishers. pp. 185–201. (In Russian.)

Zalesski, P.M. (1934) Ungulate distribution in Western Siberia in the past and at present. *Okhotnik Sibiri*. **7–8** 28. (In Russian.)

Zarudni, N.A. (1915) An outline of hunting in the Sir Darya Region. *Turkestan. selskoye khozyaistvo*. **8** 756–778. (In Russian.)

Zejda, J. (1978) Field groupings of roe deer (*Capreolus capreolus*) in a lowland region. *Folia Zoologica* **27** (2) 111–122.

Zejda, J. and Bauerova, Z. (1985) Home ranges of field roe deer. *Acta Sc. Nat. Brno*. **19** (1) 1–43.

Zejda, J. and Danilkin, A. (1992) Environment. *European and Siberian Roe Deer*. Moscow: Nauka Publishers, pp. 86–100. (In Russian.)

Zejda, J. and Homolka, M. (1980) Habitat selection and population density of field roe deer (*Capreolus capreolus*) outside the growing season. *Folia Zoologica* **29** (2) 107–115.

Zejda, J., Rebickova, M. and Homolka, M. (1985) Study of behaviour in field roe deer (*Capreolus capreolus*). *Acta Sc. Nat. Brno.* **19** (12) 1–37.

Zernahle, K. (1980) Zytogenetische Untersuchungen am Europaischen Rehwild (*Capreolus c. capreolus* L., 1758), Sibirischen Rehwild (*Capreolus c. pygargus* Pallas, 1771) und deren Bastarden. *Beitr. Jagd- und Wildforsch.* **11** 304–309.

Zhang, T. (1984) Characteristics of mammal fauna and ecogeography in Beijung area. *Acta Theriologica Sinica.* **4** (3) 187–195.

Zhirnov, L.V. (1958) More on the biology of the roe deer of the Central Tien Shan. *Learned Transactions of the Potemkin Moscow State Pedagogical Institute* **84** (7) 137–148. (In Russian.)

Zhiryakov, V.A. (1979) The effect of big predators on wild mammal populations in the Alma-Ata Preserve. *Ecological Principles of Predatory Mammal Management*. Moscow: Nauka Publishers. pp. 37–38. (In Russian.)

Zhiryakov, V.A. (1980) More on the ecology of the roe deer of the Trans-Ili Alatau. *Ungulates of the USSR*. pp. 155–156. (In Russian.)

Zhiryakov, V.A. and Djansinayev, A.D. (1986) The snow leopard in the Alma-Ata Preserve. *Rare animals of Kazakhstan. Materials for the 2nd edition of the Red Data Book of the Kazakhstan SSR*. Alma-Ata. pp. 51–54. (In Russian.)

Zhuge, Y. (1982) On the geographical distribution and the mammalian fauna of Zhejiang province. *Acta Theriologica Sinica.* **4** (2) 157–166.

Zima, J. (1989) Non-metrical variability in the skull of the roe deer (*Capreolus capreolus*). *Folia Zoologica* **38** (2) 119–137.

Zima, J. and Havrankova, J. (1987) Character prouzkovani chronozomu v karyotypu srnce obecneho. *Veter. Med. (Praha). Roc.* **32(LX)** (1) 53–58.

Ziryanov, A.N. (1975) Wild ungulates of the Stolbi Preserve and adjoining areas. *Problems of Ecology: Proceedings of the Stolbi State Preserve*. Krasnoyarsk. Iss. 10. pp. 224–338. (In Russian.)

Zivkovic, S. and Isakovic, J. (1972) Analiza kariotipa divljaci kao metod geneticko-go testiranija pri jihovom unosenju u lovista. *Simp.lovstivu*. Beograd. pp. 104–122.

Zlobin, B.D. (1975) Rare ungulate species of the Kirov Region. *Ungulates of the USSR*. Moscow: Nauka Publishers. pp. 264–265. (In Russian.)

Zverev, M.D. (1937) Mammals of the Novosibirsk District. *Proceedings of the Novosibirsk Zoo, Vol. 1. Animals and birds of Western Siberia*. Novosibirsk. pp. 7–30. (In Russian.)

Index

Page references in **bold** refer to figures and page references in *italic* refer to tables.